T0335814

Thermodynamics

Principles and Applications

Thermodynamics
Principles and Applications

İsmail Tosun

Middle East Technical University, Turkey

World Scientific

NEW JERSEY · LONDON · SINGAPORE · BEIJING · SHANGHAI · HONG KONG · TAIPEI · CHENNAI

Published by

World Scientific Publishing Co. Pte. Ltd.

5 Toh Tuck Link, Singapore 596224

USA office: 27 Warren Street, Suite 401-402, Hackensack, NJ 07601

UK office: 57 Shelton Street, Covent Garden, London WC2H 9HE

British Library Cataloguing-in-Publication Data
A catalogue record for this book is available from the British Library.

THERMODYNAMICS
Principles and Applications

ISBN 978-981-4696-93-7

Printed in Singapore

To Çiğdem and Burcu
who brought joy and happiness to my life

Preface

Our everyday life is full of examples of the applications of thermodynamic laws though we usually fail to notice them. As stated by Albert Einstein:

"Thermodynamics is the only physical theory of universal content which I am convinced, within the framework of the applicability of its basic concepts, will never be overthrown."

The subject of thermodynamics requires a thorough understanding of the various abstract quantities. This makes thermodynamics rather tricky and difficult to understand not only for students but also for academics outside of the engineering field. Vladimir I. Arnold, an eminent mathematician, once stated that, "every mathematician knows it is impossible to understand an elementary course in thermodynamics!"

In many engineering disciplines, thermodynamics is considered one of the basic engineering courses. This book is intended for use in such a course, usually taken in the sophomore year. I have provided the background material needed for students to solve practical problems related to thermodynamics. My approach is to stress the fundamentals rather than spoon-feed the subject matter. The material presented in this book can be covered in one semester.

Chapter 1 defines the terms commonly used in thermodynamics and introduces the concepts of heat and temperature. The concepts of "state/path function" and "reversible process" play a vital role in the solution of problems in the abstract world of thermodynamics. Chapter 2 talks about the calculation of work for reversible and irreversible processes. Chapter 3 examines the pressure-volume-temperature behavior of pure substances. The first and second laws of thermodynamics are extensively covered in Chapters 4 and 5, respectively. These two chapters can be regarded as the heart of this book. The last chapter, Chapter 6, is devoted to the thermodynamics of energy conversion.

My colleagues Tülay Özbelge, Önder Özbelge, Halil Kalıpçılar, and Harun Koku kindly read the entire manuscript and made many helpful suggestions. My thanks are also extended to Güngör Gündüz and Göknur Bayram for their advice and comments. I appreciate the help provided by my students Özgen Yalçın, Seval Gündüz, and Gökçe S. Avcıoğlu in proof reading and checking the numerical calculations. Finally, I am indebted to my wife, Ayşe, for her continuous support and encouragement during the writing of this book.

Suggestions and criticism from instructors and students using this book would be appreciated.

İSMAİL TOSUN
(itosun@metu.edu.tr)

Contents

Chapter 1

Introduction

1.1 Basic Concepts

Problems encountered in science and engineering can be solved by the application of the following *basic concepts*:

- Conservation of mass,
- Conservation of momentum,
- First law of thermodynamics (or conservation of energy),
- Second law of thermodynamics.

The first three basic concepts indicate that mass, momentum, and energy are all conserved quantities. A *conserved* quantity is one that can be transformed from one form to another. However, transformation does not alter the total amount of that quantity.

Note that half of the basic concepts come from the subject of thermodynamics. The word thermodynamics comes from the combination of the Greek words *"therme"*, meaning heat, and *"dunamis"*, meaning power. It is the study of energy and its transformations. Energy can be transformed into heat and work. Therefore, the subject of thermodynamics deals with how to convert energy efficiently into work.

The first law of thermodynamics is a statement of the conservation of energy, i.e., although energy can be transferred from one system to another in many forms, it can neither be created nor destroyed. Therefore, the total amount of energy available in the universe is constant. The second law of thermodynamics, on the other hand, tells us that as the energy is converted from one form to another its ability to produce useful work decreases. In other words, energy conversion leads to degradation of energy[1] as shown in

---

[1] While inflation degrades the buying power of money, conversion degrades energy!

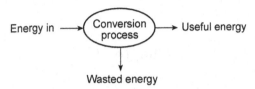

Fig. 1.1   Degradation of energy during energy conversion.

Fig. 1.1. The efficiency of a conversion process, η, defined by

$$\eta = \frac{\text{Useful energy output}}{\text{Energy input}} \tag{1.1-1}$$

is always less than 1. If a series of energy conversions is used, then the overall efficiency is the product of the efficiencies of individual processes, i.e.,

$$\text{Overall efficiency} = \prod_{i=1}^{k} \eta_i = \eta_1\eta_2\eta_3...\eta_k \tag{1.1-2}$$

Both the first and second laws of thermodynamics are in agreement with all human experience and, as a result, are also called *laws of nature*. Any feasible process in mother nature should satisfy both the first and second laws of thermodynamics.

1.2   Definitions

1.2.1   *System*

Any region that occupies a volume and has a boundary is called a *system*. The volume outside the boundary is called the *surroundings* of the system. The sum of the system and its surroundings is called the *universe*. Thermodynamics considers systems only at the macroscopic level. It is convenient to distinguish between three general types of systems:

• **Isolated system:** These are the set of systems that exchange neither mass nor energy with the surroundings. For example, the universe is an isolated system.

• **Closed system:** These are the set of systems that exchange energy (in the form of heat and work) but not mass with the surroundings.

• **Open system:** These are the set of systems that exchange both mass and energy with the surroundings.

The equations available to analyze closed and open systems are different from each other. Therefore, one should properly define the system before solving the problem.

1.2.2 *Property, state, and process*

To describe and analyze a system, some of the quantities that are characteristic of it must be known. These quantities are called *properties* and comprise volume, mass, temperature, pressure, etc. Thermodynamic properties are considered to be either *extensive* or *intensive*. When the property is proportional to the mass of the system, the property is *extensive*, i.e., volume, kinetic energy, potential energy. On the other hand, when the property is independent of the mass of the system, the property is *intensive*, i.e., viscosity, refractive index, density, temperature, pressure, mole fraction. An easy way to determine whether a property is intensive or extensive is to hypothetically divide the system into two equal parts with a partition. Each part will have the same value of intensive property, i.e., temperature, pressure, and density as the original system, but half the value of the extensive property, i.e., number of moles and volume. In other words, while the extensive properties are additive, intensive properties are not.

Specific (or molar) properties are extensive properties divided by the total mass (or total moles) of the system, i.e.,

$$\frac{\text{Specific}}{\text{property}} = \frac{\text{Extensive property}}{\text{Total mass}} \qquad \frac{\text{Molar}}{\text{property}} = \frac{\text{Extensive property}}{\text{Total moles}}$$

$$(1.2\text{-}1)$$

If φ represents any extensive property, then Eq. (1.2-1) is expressed as

$$\widehat{\varphi} = \frac{\varphi}{m} \qquad \text{and} \qquad \widetilde{\varphi} = \frac{\varphi}{n} \tag{1.2-2}$$

where m and n are the total mass and moles, respectively. Note that all specific (or molar) properties are intensive.

A complete list of the properties of a system describes its *state*. Consider a function

$$w = f(x, y) \tag{1.2-3}$$

in which there are three variables: w is dependent; x and y are independent. Obviously, once the values of x and y are specified, the value of w is automatically fixed. In thermodynamics we would say that "*the state of the system, w, is fixed when the thermodynamic properties x and y are* *specified*." Note that the mathematical term "point" is equivalent to the thermodynamic term "state".

The number of independent intensive properties needed to specify the state of a system is called the *degrees of freedom*. The *Gibbs phase rule* specifies the number of degrees of freedom, $\mathcal{F}$, for a given system at equilibrium and is expressed in the form

$$\boxed{\mathcal{P} + \mathcal{F} = \mathcal{C} + 2} \tag{1.2-4}$$

where $\mathcal{P}$ is the number of phases and $\mathcal{C}$ is the number of components. For example, for a single-phase and single-component system, such as water vapor, $\mathcal{P} = 1$ and $\mathcal{C} = 1$. Thus, from Eq. (1.2-4) $\mathcal{F} = 2$ and the state of such a system can be fixed by specifying temperature and pressure.

A change of state is called a *process*, and can occur in a number of ways. Work and heat can occur only during processes and only across the boundary of the system. The path followed in going from one state to another is known as the *process path*. If the final state is the same as the initial state, then the overall process is called a *cyclic process*.

1.2.3 *Equation of state*

Any mathematical relationship between temperature (T), pressure (P), and volume (V) is called an *equation of state*, i.e.,

$$f(T, P, V) = 0 \quad \begin{cases} T = T(P, V) \\ V = V(T, P) \\ P = P(T, V) \end{cases} \tag{1.2-5}$$

For example, an ideal gas equation of state is given by

$$PV - nRT = 0 \begin{cases} T = \dfrac{PV}{nR} \\[2mm] V = \dfrac{nRT}{P} \\[2mm] P = \dfrac{nRT}{V} \end{cases} \quad \text{or} \quad P\widetilde{V} - RT = 0 \begin{cases} T = \dfrac{P\widetilde{V}}{R} \\[2mm] \widetilde{V} = \dfrac{RT}{P} \\[2mm] P = \dfrac{RT}{\widetilde{V}} \end{cases} \tag{1.2-6}$$

where n is the number of moles and R is the ideal gas constant.

An ideal gas is composed of molecules occupying no volume and not interacting with each other. Any interaction of molecules in the form of attraction and/or repulsion tends to cause a deviation from ideal gas behavior. Monatomic (or inert) gases (He, Ne, Ar) are generally considered ideal. Simple gases, such as N_2, O_2, and CO_2, behave as an ideal gas at low pressures and high temperatures. Throughout this text, air will be considered an ideal gas.

1.2.4 *State and path functions*

The idea of a state function is one of the most important concepts to grasp in thermodynamics. If a system is caused to undergo a process, any property of the system whose value at the final state is the same no matter what path is used to carry out the process is called a *state function*[2]. On the other hand, if a property is a *path function*, its value at the final state will depend on what path is used and will be different for every path. For example, consider a system undergoing a process from an initial state (State 1) to the final state (State 2) by two different process paths (A and B) as shown in Fig. 1.2. The change in any thermodynamic property φ is denoted by $\Delta\varphi$ in which

$$\Delta\varphi = \varphi_{\text{final state}} - \varphi_{\text{initial state}} = \varphi_2 - \varphi_1 \qquad (1.2\text{-}7)$$

The property φ may be either a state function or a path function, depending on the following conditions:

$$\Delta\varphi \ (\text{Process A}) = \Delta\varphi \ (\text{Process B}) \quad \Rightarrow \quad \varphi \text{ is a state function} \qquad (1.2\text{-}8)$$

$$\Delta\varphi \ (\text{Process A}) \neq \Delta\varphi \ (\text{Process B}) \quad \Rightarrow \quad \varphi \text{ is a path function} \qquad (1.2\text{-}9)$$

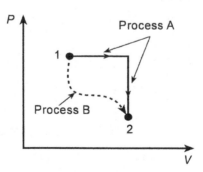

Fig. 1.2   A system undergoing a process by two different process paths.

---

[2]A property that is defined as the combination of other state functions is also a state function.

Thus, if a state function φ is caused to undergo a cyclic process, its initial and final values will be the same or $\Delta\varphi = 0$. On the other hand, if a path function φ is caused to undergo a cyclic process, its initial and final values will be different or $\Delta\varphi \neq 0$.

The quantities encountered in thermodynamics are all state functions except heat and work. The only exception to this statement is the work done by body forces, i.e., work done against a gravitational force. Body forces are conservative forces[3].

The term "state function" in thermodynamics corresponds to the term "exact differential" in mathematics. The expression $M(x,y)\,dx + N(x,y)\,dy$ is called an *exact differential* if there exists some $\varphi = \varphi(x,y)$ for which this expression is the *total differential* $d\varphi$, i.e.,

$$M(x,y)\,dx + N(x,y)\,dy = d\varphi(x,y) \tag{1.2-10}$$

A necessary and sufficient condition for the expression $M(x,y)\,dx + N(x,y)\,dy$ to be expressed as a total differential is that

$$\left(\frac{\partial M}{\partial y}\right)_x = \left(\frac{\partial N}{\partial x}\right)_y \tag{1.2-11}$$

The steps followed in the solution of a thermodynamic problem are shown in Fig. 1.3. The first step is to transform the real problem into the abstract world of thermodynamics in which necessary equations are provided for solving it. Once the problem is solved in the abstract world, then it is transformed back into the real world. In the abstract world of thermodynamics, actual processes are replaced by idealized processes since

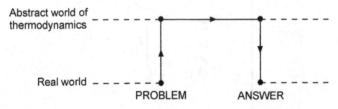

Fig. 1.3   The path followed in the solution of thermodynamics problems (Prausnitz, 1986).

---

[3]A force, $\mathbf{F}$, is called *conservative* if the work done by the force is a state function. In mathematical terms, $\nabla \times \mathbf{F} = 0$.

$$\Delta\varphi\,(\text{actual process}) = \Delta\varphi\,(\text{idealized process}) \qquad (1.2\text{-}12)$$

as long as φ is a state function.

1.2.5 *Equilibrium*

It is important to differentiate between the concepts of steady-state, uniform, and equilibrium:

• **Steady-state:** The term *steady-state* means that at a particular location in space the dependent variable φ does not change as a function of time, i.e.,

$$\left(\frac{\partial\varphi}{\partial t}\right)_{x,y,z} = 0 \qquad (1.2\text{-}13)$$

The partial derivative notation indicates that the dependent variable is a function of more than one independent variable. In this particular case, the independent variables are (x, y, z) and t. The specified location in space is indicated by the subscripts $(x,\ y,\ z)$ and Eq. (1.2-13) implies that φ is not a function of time, t. When an ordinary derivative is used, i.e., $d\varphi/dt = 0$, then this implies that φ is a constant. It is important to distinguish between partial and ordinary derivatives because the conclusions are very different.

• **Uniform:** The term *uniform* means that at a particular instant in time the dependent variable φ is not a function of position. This requires that all three of the partial derivatives with respect to position be zero, i.e.,

$$\left(\frac{\partial\varphi}{\partial x}\right)_{y,z,t} = \left(\frac{\partial\varphi}{\partial y}\right)_{x,z,t} = \left(\frac{\partial\varphi}{\partial z}\right)_{x,y,t} = 0 \qquad (1.2\text{-}14)$$

The variation of a physical quantity with respect to position is called *gradient*. Therefore, the gradient of a quantity must be zero for a uniform condition to exist with respect to that quantity.

• **Equilibrium:** A system is in *equilibrium* if both steady-state and uniform conditions are met simultaneously. This implies that the variables associated with the system, i.e., temperature, pressure, and density, are constant at all times and have the same magnitude at all positions within the system. A difference in any potential that causes a process to take place spontaneously is called a *driving force*. Driving force(s) turns out to be zero for a system in equilibrium. Thus, no work can be done by a system in equilibrium.

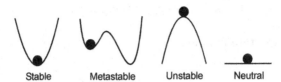

Stable   Metastable   Unstable   Neutral

Fig. 1.4   Classification of equilibrium states (Tester and Modell, 1997).

There are four classes of equilibrium states. For ease of conceptualization, they can be described by the mechanical analogy of a ball on a solid surface in a gravitational field as shown in Fig. 1.4. If the ball were pushed to the right or left and if it were to return to its original position, the state is *stable*. If the original position were *metastable*, the ball would return to the original position after a small perturbation, but there is the possibility that a large perturbation would displace the ball to a state of lower potential energy. If the original state were *unstable*, then even a minor perturbation would displace it to a position of lower potential energy. A system in a state of *neutral* equilibrium would be altered by any perturbation, but the potential energy would remain unchanged.

It should be kept in mind that thermodynamics considers systems only at equilibrium. By examining the system in its initial and final equilibrium states, it is possible to determine the heat and work interactions of the system with its surroundings during a process as shown in Fig. 1.5. Thermodynamics, however, does not consider the time it takes for the system to go from an initial equilibrium state to a final equilibrium state.

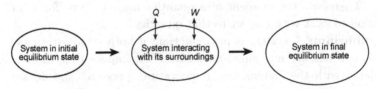

Fig. 1.5   A system undergoing a process.

1.3 The Zeroth Law of Thermodynamics

Hot bodies can be distinguished from cold ones by the sense of touch. In this way, it is possible to arrange bodies in the order of their hotness, deciding that A is hotter (or colder) than B, B than C, etc. The *degree of hotness* is our temperature sense. Take an object A that feels cold to the hand and an identical object B that feels hot. Place them in contact with each other. After a sufficient length of time, A and B give rise to the same temperature sensation. Then A and B are said to be in *thermal equilibrium*.

If A and B are in thermal equilibrium with a third body C, then A and B are in thermal equilibrium with each other. This statement is known as the *zeroth law of thermodynamics*. The zeroth law was formulated almost 100 years after the first and second laws. Since it must be stated before the first and second laws in a logical development of the subject, it has been agreed to designate it as the zeroth law[4]. The statement of the zeroth law, i.e., *A relates to C in the same way that B relates to C, then A and B must be related in the same way*, is by no means obvious or generally applicable. Consider, for example, the following example: If James loves Mary and Henry loves Mary, does that mean that James and Henry love each other? The third body is called a *thermometer*. Therefore, two systems at thermal equilibrium with each other must be at the same temperature. Temperature is an important property in the study of energy.

1.3.1 *Measurement of temperature*

Measurable physical properties that vary with temperature, such as the volume of a liquid, the electrical resistance of a wire, and the pressure of a gas in a fixed volume, are called *thermometric properties*. By choosing a particular thermometric substance and a particular thermometric property of this substance, it is possible to establish a temperature scale. For example, consider a thermometer consisting of a glass bulb and tube containing mercury. The length of mercury is the thermometric property, X, that will be used in setting up a temperature scale, T. How does X vary with T? To answer such a question, it is possible to propose various relationships between T and X such as

$$T = a X \qquad (1.3\text{-}1)$$

---

[4] A reader interested in the historical developments involving the laws of thermodynamics may refer to Sandler and Woodcock (2010).

$$T = a X + b \qquad (1.3\text{-}2)$$

$$T = a X^2 + b X + c \qquad (1.3\text{-}3)$$

It is known from experiments that certain processes always occur at fixed temperatures. For example, water boils at a fixed temperature under atmospheric pressure. Substances melt or solidify at fixed temperatures if the pressure is maintained at a constant value, and so on. The use of Eqs. (1.3-1), (1.3-2), and (1.3-3) requires, respectively, one, two, and three such points.

For example, two fixed points are needed when Eq. (1.3-2) is to be used. If the fixed points are chosen as

$$\begin{array}{l} \text{at } X = X_1 \ \ T = T_1 \\ \text{at } X = X_2 \ \ T = T_2 \end{array} \qquad (1.3\text{-}4)$$

then the use of Eq. (1.3-2) gives

$$T_1 = a X_1 + b \qquad (1.3\text{-}5)$$

$$T_2 = a X_2 + b \qquad (1.3\text{-}6)$$

Simultaneous solution of Eqs. (1.3-5) and (1.3-6) gives

$$a = \frac{T_2 - T_1}{X_2 - X_1} \quad \text{and} \quad b = \frac{X_2 T_1 - X_1 T_2}{X_2 - X_1} \qquad (1.3\text{-}7)$$

Therefore, the relationship between temperature and thermometric property is given by

$$T = \left(\frac{T_2 - T_1}{X_2 - X_1} \right) X + \frac{X_2 T_1 - X_1 T_2}{X_2 - X_1} \qquad (1.3\text{-}8)$$

A temperature scale established in accordance with the above rules is known as an *empirical temperature scale*[5].

In engineering practice Eq. (1.3-8) is used with the following two fixed points: the steam point (boiling point of water at a pressure of 1 atm) and the ice point (melting point of ice at a pressure of 1 atm). A temperature scale is obtained by dividing the distance between the ice point and the steam point. The ice and the steam points for the Celsius and Fahrenheit scales are given in Table 1.1.

---

[5]For a better understanding of various temperature scales, the reader should refer to the enlightening and entertaining article by Levenspiel (1975).

Table 1.1 The ice and steam points in different temperature scales.

Temperature Scale	Ice Point	Steam Point
Celsius scale	0°C	100°C
Fahrenheit scale	32°F	212°F

Denoting the values of the thermometric property measured at the steam and ice points by X_s and X_i, respectively, Eq. (1.3-8) becomes

$$T = 100 \left(\frac{X - X_i}{X_s - X_i} \right) \qquad \text{Celsius scale} \qquad (1.3\text{-}9)$$

$$T = 32 + 180 \left(\frac{X - X_i}{X_s - X_i} \right) \qquad \text{Fahrenheit scale} \qquad (1.3\text{-}10)$$

An empirical temperature scale has the following characteristics:

• The numbers on the scale are arbitrary.

• The degree on the thermometer is dependent on the thermometric property. However, all thermometers read the same at the ice and steam points.

Example 1.1 *A temperature scale of a thermometer is given by*

$$T = a \ln X + b$$

where a and b are constants. The thermometric properties at the ice and steam points are found as 1.03 *and* 3.78, *respectively. What is the temperature on the Celsius scale corresponding to the thermometric property of* 1.4?

Solution

To find out a relation between temperature and thermometric property, it is first necessary to evaluate the constants a and b. At the ice and steam points

$$0 = a \ln 1.03 + b$$

$$100 = a \ln 3.78 + b$$

Simultaneous solution of the above equations gives

$$a = 76.913 \qquad \text{and} \qquad b = -2.273$$

Table 1.2   The ideal gas constant, R.

$82.05\,\mathrm{cm^3.\,atm/\,mol.\,K}$	$8.314 \times 10^{-3}\,\mathrm{kPa.\,m^3/\,mol.\,K}$
$0.08205\,\mathrm{m^3.\,atm/\,kmol.\,K}$	$8.314 \times 10^{-5}\,\mathrm{bar.\,m^3/\,mol.\,K}$
$1.987\,\mathrm{cal/\,mol.\,K}$	$8.314 \times 10^{-2}\,\mathrm{bar.L/\,mol.\,K}$
$8.314\,\mathrm{J/\,mol.\,K}$	$8.314 \times 10^{-2}\,\mathrm{bar.\,m^3/\,kmol.\,K}$
$8.314 \times 10^{-6}\,\mathrm{MPa.\,m^3/\,mol.\,K}$	$83.14\,\mathrm{bar.\,cm^3/\,mol.\,K}$

Therefore, T is related to X as

$$T = 76.913 \ln X - 2.273$$

When $X = 1.4$, the temperature is

$$T = 76.913 \ln 1.4 - 2.273 = 23.6\,^\circ\mathrm{C}$$

A temperature scale that is independent of the thermometric substance is known as the *absolute (or thermodynamic) temperature scale*, the reference point of which is zero absolute temperature. It is the theoretical temperature at which all molecular motion virtually stops. All other temperatures are extrapolated linearly upwards from absolute zero.

The absolute scale related to the Celsius scale is the Kelvin scale and is designated K (without a degree symbol). The relation between these scales is given by

$$\mathrm{K} = {}^\circ\mathrm{C} + 273 \qquad (1.3\text{-}11)$$

On the other hand, the absolute scale related to the Fahrenheit scale is the Rankine scale and is designated $^\circ\mathrm{R}$. The relation between these scales is

$$^\circ\mathrm{R} = {}^\circ\mathrm{F} + 460 \qquad (1.3\text{-}12)$$

The temperature in the ideal gas equation of state, Eq. (1.2-6), must be an absolute temperature, i.e., K or $^\circ\mathrm{R}$. The values of the ideal gas constant, R, in different units are given in Table 1.2.

Example 1.2 *A rigid tank of $10\,\mathrm{m^3}$ volume contains air at $350\,\mathrm{K}$ and $900\,\mathrm{kPa}$. A relief valve is opened slightly, allowing air to escape to the atmosphere. The valve is closed when the pressure in the tank reaches $650\,\mathrm{kPa}$. If the temperature at this point is $298\,\mathrm{K}$, determine the mass of air that has escaped.*

Solution

Assuming air to be an ideal gas, the initial and final number of moles are calculated as

$$n_{initial} = \frac{(900)(10)}{(8.314 \times 10^{-3})(350)} = 3092.9 \, \text{mol}$$

$$n_{final} = \frac{(650)(10)}{(8.314 \times 10^{-3})(298)} = 2623.5 \, \text{mol}$$

The molecular weight of air is $29 \, \text{g/mol}$. *Thus, the mass of air that has escaped is*

$$m = \frac{(3092.9 - 2623.5)(29)}{1000} = 13.613 \, \text{kg}$$

The absolute temperature is a measure of the average kinetic energy of the molecules of a substance, i.e., the higher the temperature, the faster the speed of molecules. From the kinetic theory of gases, the average kinetic energy (KE) of a monatomic ideal gas molecule is given by

$$\text{Average KE of a molecule} = \frac{3}{2} kT \qquad (1.3\text{-}13)$$

where $k = 1.38 \times 10^{-23} \, \text{J/K}$ is the Boltzmann's constant. To find out the total kinetic energy, Eq. (1.3-13) should be multiplied by the total number of molecules calculated as

$$\begin{array}{c}
\text{Total number} \\
\text{of molecules}
\end{array} = \underbrace{\left(\begin{array}{c} \text{Number of} \\ \text{moles} \end{array} \right)}_{n} \underbrace{\left(\frac{\text{Number of molecules}}{\text{Mole}} \right)}_{\text{Avogadro's number } (\mathcal{N}=6.022\times 10^{23})} \qquad (1.3\text{-}14)$$

Thus

$$\text{Total KE} = \frac{3}{2} nRT \qquad (1.3\text{-}15)$$

where the ideal gas constant, R, is the product of the Boltzmann's constant and the Avogadro's number, i.e.,

$$R = k\mathcal{N} \qquad (1.3\text{-}16)$$

1.3.2 *Constant temperature (isothermal) process*

A *constant temperature process* is one in which the temperature of the system does not change throughout the process. It is also known as an *isothermal process*. Note that if the initial and final states are at the same temperature, this does not necessarily imply a constant temperature process. Since the phase change of a pure substance takes place at constant temperature, the melting of ice and the boiling of water are typical examples of isothermal processes.

For an ideal gas undergoing an isothermal process

$$PV = nRT = \text{constant} \tag{1.3-17}$$

On a *P-V* diagram, constant temperature lines are called *isotherms*. According to Eq. (1.3-17), isotherms are represented by equilateral hyperbolas as shown in Fig. 1.6.

1.4 Heat

Heat should not be confused with thermal energy, which is the total energy of all the molecules making up the substance. The total energy is the sum of the kinetic (translational, rotational, vibrational) and potential (stored energy of an interaction[6]) energies of molecules. Heat, on the other hand, is the transfer (or flow) of thermal energy from one system to another as a result of a temperature difference between them, the heat flowing from the higher to the lower temperature region.

Therefore, heat is an interaction between the system and its surroundings through the boundary of the system. It is neither a property of a

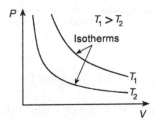

Fig. 1.6   Representation of isotherms in a *P-V* diagram for an ideal gas.

---

[6]Potential energy is converted into kinetic energy by the force associated with the interaction.

Fig. 1.7   Sign convention for heat.

system nor something a system has. All substances contain thermal energy, but it is meaningless to speak of a substance containing heat[7].

As a form of energy, heat has energy units. The relationship between different units is given by

$$1\,\text{Btu} = 1055\,\text{J} = 252\,\text{cal} = 778\,\text{ft. lb}_\text{f} \qquad (1.4\text{-}1)$$

According to the sign convention for heat, as shown in Fig. 1.7, if the system receives heat from the surroundings, it is positive. Otherwise, it is negative.

If the heat addition and/or removal results in a change in the temperature of a system, it is called *sensible heat*. When the heat addition and/or removal does not change the temperature of a system, it is called *latent heat*. Latent heat is associated with a phase change, i.e., melting (or freezing), evaporation (or condensation), and sublimation.

1.4.1 *Heat capacity*

The amount of sensible heat, Q, required to change the temperature of the system is found to be proportional to the mass, m, of the system and to the temperature change, ΔT. This implies

$$Q \propto \Delta T \qquad \Rightarrow \qquad Q = Km\Delta T \qquad (1.4\text{-}2)$$

where the proportionality constant K is defined as

$$K = \begin{cases} \widehat{C}_V & \text{Heat capacity at constant volume} \\ \widehat{C}_P & \text{Heat capacity at constant pressure} \end{cases} \qquad (1.4\text{-}3)$$

---

[7]According to the *caloric theory*, which was proposed in the 18th century, any object contained a certain amount of caloric; if more caloric flowed into the object, its temperature increased, and if caloric flowed out, the object's temperature decreased. When a material was broken apart, such as during burning, a great deal of caloric was believed to be released. Caloric was assumed to be massless, odorless, tasteless, and transparent. In spite of the mysterious nature of this fluid, the caloric theory did not explain many observations, such as the heating effect associated with friction.

Thus, Eq. (1.4-2) becomes

$$Q = \begin{cases} m\,\widehat{C}_V\,\Delta T & \text{constant volume process} \\ m\,\widehat{C}_P\,\Delta T & \text{constant pressure process} \end{cases} \tag{1.4-4}$$

Note that $\widehat{C}_V$ (or $\widehat{C}_P$) can be viewed as the energy required to raise the temperature of the unit mass of a substance by one degree as the volume (or pressure) is maintained constant. A common unit for heat capacity is kJ/kg.K or kJ/kg.°C. These two units are identical since

$$\Delta T\,(\text{K}) = \Delta T\,(^\circ\text{C}) \tag{1.4-5}$$

Heat capacities are sometimes given on a molar basis (especially for gases) and denoted by $\widetilde{C}_V$ and $\widetilde{C}_P$ with the units kJ/kmol.K or kJ/kmol.°C. Thus,

$$Q = \begin{cases} n\,\widetilde{C}_V\,\Delta T & \text{constant volume process} \\ n\,\widetilde{C}_P\,\Delta T & \text{constant pressure process} \end{cases} \tag{1.4-6}$$

The relationship between $\widetilde{C}_P$ and $\widehat{C}_P$ (or $\widetilde{C}_V$ and $\widehat{C}_V$) is given by

$$\widehat{C}_P = \frac{\widetilde{C}_P}{M} \tag{1.4-7}$$

where M is the molecular weight.

For solids and liquids $\widehat{C}_V \approx \widehat{C}_P$. Heat capacities for various solids and liquids are given in Table 1.3. Note that water has one of the highest heat capacities of all substances[8], which makes it an ideal substance for the cases that require a minimum ΔT for a given amount of heat transfer. For example, the heat required to increase the temperature of 10 kg of copper by 5°C (or 5 K) is

$$Q_{\text{Cu}} = (10)(0.39)(5) = 19.5\,\text{kJ}$$

In other words, if 19.5 kJ of heat is removed from 10 kg of copper, the decrease in temperature will be 5°C (or 5 K). If the same amount of heat is removed from 10 kg of water, the temperature drop will be

---

[8]Since water molecules form strong bonds with each other, it takes more energy to break them. Metals, on the other hand, form weaker bonds, resulting in lower heat capacities.

Table 1.3 Heat capacities of various substances at 293 K and 1 atm.

Substance	$\widehat{C}_V \approx \widehat{C}_P$ (kJ/kg. K)	Substance	$\widehat{C}_V \approx \widehat{C}_P$ (kJ/kg. K)
Aluminum	0.90	Rock	2.0
Copper	0.39	Sand (dry)	0.84
Ethyl alcohol	2.4	Snow	0.88
Glass	0.84	Vegetable oil	1.67
Human body	3.5	Water	4.2
Ice	2.0	Wood	1.8

$$\Delta T_{\mathrm{H_2O}} = \frac{19.5}{(10)(4.2)} = 0.46\,^\circ\mathrm{C} \ (\text{or } 0.46\,\mathrm{K})$$

which is much smaller than ΔT_{Cu}.

Consequences of this fact are observed in the following examples taken from everyday life:

• You burn your tongue when you take a bite from a hot pizza slice. The reason for this is the fact that you do not start eating from the crust! The temperature of the tomato sauce in the middle portion is rather high compared to crust temperature.
• The operating cost of a swimming pool is extremely high because of the amount of heat that must be supplied to keep the temperature at the desired level.
• Oceans retain energy longer than land does. Remember the phrase from the high school geography course, "land warms up and cools off faster than the oceans do."
• Sand can get much warmer than sea water. As a result, you burn your feet when you walk on a beach barefoot in the summer time.

For gases, however, $\widetilde{C}_P$ is always greater than $\widetilde{C}_V$ because at constant pressure the system is allowed to expand and the energy for this expansion work must also be supplied to the system. For monatomic and diatomic ideal gases, $\widetilde{C}_V$ and $\widetilde{C}_P$ values are given as

$$\widetilde{C}_V = \begin{cases} (3/2)\,R & \text{monatomic} \\ (5/2)\,R & \text{diatomic} \end{cases} \qquad \widetilde{C}_P = \begin{cases} (5/2)\,R & \text{monatomic} \\ (7/2)\,R & \text{diatomic} \end{cases} \tag{1.4-8}$$

Heat capacities of gases are only moderate functions of temperatures and thus for ordinary changes in temperature, i.e., up to several hundreds of degrees for air, the use of constant heat capacities is valid for engineering purposes. If heat capacities are dependent on temperature, then Eqs. (1.4-4) and (1.4-6) should be modified as

$$
Q = \begin{cases} m \displaystyle\int_{T_1}^{T_2} \widehat{C}_V(T)\, dT & \text{constant volume process} \\[3mm] m \displaystyle\int_{T_1}^{T_2} \widehat{C}_P(T)\, dT & \text{constant pressure process} \end{cases} \tag{1.4-9}
$$

and

$$
Q = \begin{cases} n \displaystyle\int_{T_1}^{T_2} \widetilde{C}_V(T)\, dT & \text{constant volume process} \\[3mm] n \displaystyle\int_{T_1}^{T_2} \widetilde{C}_P(T)\, dT & \text{constant pressure process} \end{cases} \tag{1.4-10}
$$

Example 1.3 *A meeting is held by* 25 *people in an insulated room with dimensions* 20 × 10 × 3 m. *Calculate the increase in air temperature after half an hour if there is no ventilation. Take the average metabolic rate, i.e., heat production per unit time, of a person as* 90 J/ s.

Solution

Assumptions

• *Each person occupies a volume of* $0.07\, \text{m}^3$.
• *Initial air pressure and temperature within the room are* 1 atm (1.013 bar) *and* 296 K, *respectively.*
• *Air is an ideal gas.*

System: *Air in the room*

Since air experiences a constant volume process, the increase in temperature can be calculated from Eq. (1.4-6) as

$$
\Delta T = \frac{Q}{n \widetilde{C}_V} \tag{1}
$$

where $\widetilde{C}_V = (5/2)R$. *To calculate the number of moles of air, it is first necessary to determine the volume of air contained in the room.*

$$
V_{room} = (20)(10)(3) = 600\, \text{m}^3 \quad \Rightarrow \quad V_{air} = 600 - (25)(0.07) = 598.25\, \text{m}^3
$$

Thus, the number of moles of air is

$$n = \frac{PV}{RT} = \frac{(1.013)(598.25)}{(8.314 \times 10^{-5})(296)} = 24,626 \, \text{mol}$$

Substitution of the numerical values into Eq. (1) gives the increase in temperature as

$$\Delta T = \frac{(25)(90)(1800)}{(24,626)(2.5 \times 8.314)} = 7.9 \, ^\circ\text{C}$$

1.4.2 *Adiabatic process*

An *adiabatic process* is one in which there is no exchange of heat between the system and its surroundings, i.e., $Q = 0$. An adiabatic process is usually confused with an isothermal process. Although there is no heat transfer during an adiabatic process, the temperature of a system can be changed by other means such as work. In the case of an isothermal process, however, temperature remains constant throughout the process. The adiabatic and isothermal processes are often opposite extremes and real processes fall in between.

There are three ways a process can be adiabatic:

• The system is well insulated so that heat transfer across the boundary of the system is negligible,

• Both the system and its surroundings are at the same temperature,

• The process occurs so rapidly that there is no time for the heat transfer to take place, i.e., heat transfer, being a "slow" process, is considered negligible in processes taking place very "rapidly".

Since *slow* and *rapid* are relative terms, there is no clear-cut recipe to differentiate slow processes from rapid ones. One should use "engineering judgement" in the analysis of a given problem. For example, consider the following two cases in which a rigid tank filled with a high-pressure gas at ambient temperature is evacuated by

(*i*) Punching a tiny hole in the surface of the tank,

(*ii*) Opening a large valve on the top of the tank.

Suppose that the tank is not insulated and it is required to find how the gas temperature within the tank changes with pressure. Since process (*i*) is rather slow, it allows heat transfer between the tank contents and ambient air to take place. Thus, the gas remaining in the tank may be considered to undergo an isothermal process. On the other hand, evacuation of the tank is very rapid in process (*ii*). Over the time scale of the evacuation process, heat transfer between the tank contents and ambient air is almost negligible. As a result, the gas remaining in the tank may be assumed to undergo an adiabatic process even though there is no insulation around the tank.

Problems

Problem related to Section 1.1

1.1 Posters for energy conservation remind us to use energy sensibly. Your friend, however, argues that since the total energy of the universe is constant according to the first law, we should not worry about conserving energy. Do you agree with your friend's argument?

Problems related to Section 1.2

1.2 For each of the following systems, indicate whether the system is isolated, closed, or open:

a) A closed bottle of wine,
b) A lake,
c) A pressure cooker,
d) A thermos flask,
e) A human body,
f) A car radiator.

1.3 A cylinder fitted with a piston contains a gas at a given temperature and pressure. The gas cools at constant pressure to a new volume by the transfer of heat. How would you define the system to determine the amount of work done and heat transferred? Also show the process path in a *P-V* diagram.

Problem related to Section 1.3

1.4 A mercury thermometer is calibrated at the ice and steam points. Using the following data, calculate the reading of this thermometer at 200 °C.

Indicate your assumptions.

$T\,(°C)$	$\rho\,(\mathrm{g/cm^3})$	$T\,(°C)$	$\rho\,(\mathrm{g/cm^3})$
0	13.5955	150	13.2330
50	13.4729	200	13.1148
100	13.3522	250	12.9975

(**Answer:** 201.15 °C)

Problems related to Section 1.4

1.5   Is temperature a measure of thermal energy? Discuss.

1.6   To drink hot coffee without scalding our lips and tongue, we prefer to take a sip. Why?

1.7   An electric current flows through a resistor that is immersed in running water. Considering the resistor as the system, is the heat flow positive, zero, or negative?

1.8   A block of copper and a block of aluminum, having equal mass, receive the same amount of energy by heat transfer. Which block would experience the greater temperature increase?

1.9   A rigid tank of $0.03\,\mathrm{m^3}$ volume contains air at atmospheric pressure and 150 °C. If the tank is allowed to cool to room temperature of 20 °C, calculate the amount of heat transferred. Assume air to be an ideal gas with $\widetilde{C}_V = (5/2)R$.

(**Answer:** $-2334.6\,\mathrm{J}$)

1.10[9]   Soldiers defending castles in the Middle Ages used to pour boiling oil down on the attacking enemy soldiers. Why did the defenders go all the trouble and expense of using oil, especially when the heat capacity of oil is less than half of that water? For quantitative comparison, use the following data:

	Heat Capacity $(\mathrm{kJ/kg.\,K})$	Boiling Point $(°C)$	Density $(\mathrm{kg/m^3})$
Water	4.20	100	958
Oil	2.00	300	800

---

[9]This problem is taken from Konak (1994).

1.11[10] Someone recommends a cold-water diet to lose weight. You are asked to drink ice cold water at $0\,°C$ to burn $8400\,kJ$ $(2000\,kcal)$ per day. Can you do it?

1.12 A vertical cylinder fitted with a frictionless piston, with a set of stops as shown in the figure below, initially contains air at $200\,kPa$ and $427\,°C$. It is then cooled as a result of heat transfer to the surroundings. The cross-sectional area of the piston is $0.05\,m^2$. Assume air to be an ideal gas with $\widetilde{C}_V = (5/2)R$ and $\widetilde{C}_P = (7/2)R$.

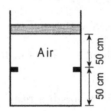

a) What is the temperature of the air at the moment the piston touches the stops?

b) What is the pressure of air in the cylinder if the air is further cooled until the temperature drops to $21\,°C$?

c) How much heat is transferred to the surroundings as a result of the overall cooling process?

(**Answer:** a) $77\,°C$, b) $168\,kPa$, c) $-19.5\,kJ$)

1.13 A rigid tank initially contains air at $60\,bar$. A valve is opened and the pressure in the tank falls rapidly to $35\,bar$. To calculate the air temperature at the final state, it is convenient to choose "air remaining in the tank when the pressure reaches $35\,bar$" as the system.

a) Is this an open or a closed system?
b) Is it plausible to assume that the system undergoes an adiabatic process? Why?

1.14 The temperature of a classroom with dimensions of $10 \times 8 \times 3\,m$ is to be maintained at $23\,°C$. There are 30 students and a teacher in the room. The metabolic rates of a student and a teacher are $320\,kJ/h$ and $490\,kJ/h$, respectively. Under steady conditions, the classroom loses heat to the outside at a rate of $350\,kJ/min$.

---

[10]This problem is taken from Konak (1994).

a) Estimate the power of a heater needed to keep the classroom temperature constant.

b) Over the weekend, the school is shut down and the classroom temperature drops to $15\,°C$. If the classroom is to be heated back to $23\,°C$ in 30 min on a Monday morning, estimate the power of a heater. Note that the classroom is now empty, but heat is lost to the outside at a rate of $350\,kJ/\,min$.

(**Answer:** a) $3\,kW$, b) $6.77\,kW$)

References

Konak, A.R., 1994, *Chem. Eng. Ed.*, **28** (3), 180-182.

Levenspiel, O., 1975, *Chem. Eng. Ed.*, **9** (3), 102-105, 137.

Prausnitz, J.M., 1986, *J. Non-Equil. Thermodyn.*, **11**, 49-66.

Sandler, S.I. and L.V. Woodcock, 2010, *J. Chem. Eng. Data*, **55**, 4485-4490.

Tester, J.F. and M. Modell, 1997, *Thermodynamics and Its Applications*, 3rd Ed., Prentice-Hall, Upper Saddle River, New Jersey.

Chapter 2

Calculation of Work in Reversible and Irreversible Processes

Calculation of work is extremely important in the study of thermodynamics since it deals with the relationship and conversions between heat and work. The equation to be used for the determination of work is dependent on the type of process, i.e., whether it is "reversible" or "irreversible". The concept of *reversibility* plays a vital role in the solution of thermodynamics problems. The purpose of this chapter is to explain what is meant by reversible and irreversible processes and to show how to calculate work interaction of a system during these processes with various examples.

2.1 Force

Force is defined by Newton's second law of motion. It states that the net force $\mathbf{F}_{net}$ acting on a body of mass m is proportional to the time rate of change of its momentum. If the mass is constant, the net force is proportional to the product of the mass of the body and its acceleration. Thus,

$$\sum \mathbf{F} = \mathbf{F}_{net} = K \frac{d}{dt}(m\mathbf{v}) = K\, m\mathbf{a} \qquad (2.1\text{-}1)$$

where $\mathbf{a} = d\mathbf{v}/dt$ is the acceleration of the body and K is a proportionality constant to be determined by the units used. In the SI system, $K = 1$ and force has the units of newton, N, or kg. m/ s^2.

The weight, $\mathbf{W}$, of a body is the force exerted on it by the earth's gravitational field, i.e.,

$$\mathbf{F}_{gravity} = \mathbf{W} = K\, m\mathbf{g} \qquad (2.1\text{-}2)$$

Near the earth's surface, $g = |\mathbf{g}| = 9.8\,\mathrm{m/s^2}$. At large distances from the earth's surface $g < 9.8\,\mathrm{m/s^2}$. The terms "mass" and "weight" are often confused with one another. Note that while weight is a force, mass is a property of a body.

In the English system, the mass unit is taken as the pound mass, lb_m, and the force unit, called the pound force (lb_f), is defined such that the weight in lb_f of a body at sea level will be numerically equal to its mass in lb_m. Since the magnitude of the acceleration of gravity at sea level is $32.2 \, ft/s^2$, Eq. (2.1-2) becomes

$$|\mathbf{W}| = K\,m(32.2) \qquad (2.1\text{-}3)$$

For $|\mathbf{W}|$ to be numerically equal to m, i.e., $|\mathbf{W}| = m$, the proportionality constant K must be

$$K = \frac{1}{32.2} \frac{lb_f.\,s^2}{lb_m.\,ft} \qquad (2.1\text{-}4)$$

It is customary to express K in the form

$$K = \frac{1}{g_c} \qquad (2.1\text{-}5)$$

where g_c is a conversion factor, equal to $32.2 \, lb_m.\,ft/(lb_f.\,s^2)$. It is important to note that while g_c has the magnitude of the acceleration of gravity at sea level, its units are not the same and it is not the acceleration due to gravity, or an acceleration of any kind. It is simply a conversion factor required by the selection of units. While g_c is a pure constant, the acceleration of gravity varies with distance from the earth.

Various units of force and proportionality constant are given in Table 2.1. The relation between a pound force and a newton is given by

$$1 \, lb_f = 4.448 \, N \qquad \text{and} \qquad 1 \, N = 0.2248 \, lb_f \qquad (2.1\text{-}6)$$

Table 2.1   Units of force and proportionality constant, K, in Newton's 2nd law.

Force	Mass	Acceleration	K
N	kg	m/s^2	1
dyn	g	cm/s^2	1
lb_f	lb_m	ft/s^2	$\dfrac{1}{32.2} \, lb_f.\,s^2/(lb_m.\,ft)$
Poundal	lb_m	ft/s^2	1

2.1.1 *Pressure and mechanical equilibrium*

The pressure, P, of a fluid on a surface is defined as the normal force exerted by the fluid per unit area of the surface, i.e.,

$$P = \frac{F}{A} \qquad (2.1\text{-}7)$$

Pressure has the units of Pa (N/m^2) in the SI system. However, the pressure in Pa is often too small for pressures encountered in practice. Therefore, the units kilopascal ($1\,kPa = 10^3\,Pa$) and megapascal ($1\,MPa = 10^6\,Pa$) are commonly used. Two other common pressure units are the bar and standard atmosphere:

$$1\,bar = 10^5\,Pa = 100\,kPa = 0.1\,MPa$$

$$1\,atm = 101,325\,Pa = 101.325\,kPa = 1.01325\,bar \qquad (2.1\text{-}8)$$

In the English system, the pressure unit is pound-force per square inch (lb_f/in^2 or psi), and $1\,atm = 14.696\,psi$.

Absolute pressure refers to the absolute value of the force per unit area exerted on the containing wall by a fluid. *Gauge pressure* represents the difference between the absolute pressure and the local atmospheric pressure. *Vacuum* represents the amount by which the atmospheric pressure exceeds the absolute pressure. The three pressure terms are schematically shown in Fig. 2.1. From these definitions one can easily conclude that (i) absolute pressure cannot be negative, and (ii) vacuum cannot be greater than the local atmospheric pressure.

A system is said to be in *mechanical equilibrium* with its surroundings when there is no pressure difference between them, i.e.,

$$\boxed{P_{sys} = P_{surr}} \quad \text{Condition of mechanical equilibrium} \qquad (2.1\text{-}9)$$

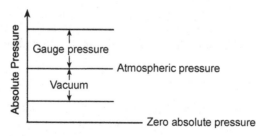

Fig. 2.1   Absolute, gauge, and vacuum pressures.

Example 2.1 *A vertical cylinder fitted with a frictionless piston contains carbon dioxide. The mass and the cross-sectional area of the piston are* 50 kg *and* 250 cm^2, *respectively. The atmospheric pressure outside the cylinder is* 100 kPa *and the local acceleration of gravity is* 9.8 m/s^2.

a) *What is the carbon dioxide pressure inside the cylinder if the piston is at rest?*

b) *If some heat is transferred to the gas and its volume doubles, do you expect the pressure inside the cylinder to change?*

Solution

System: *Piston*

a) *Since the piston is in equilibrium, the summation of forces acting on it must be zero, i.e.,*

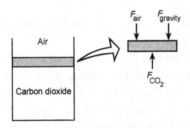

$$\sum F = 0 \quad \Rightarrow \quad F_{\text{gravity}} + F_{\text{air}} = F_{\text{CO}_2} \quad \Rightarrow \quad mg + P_{atm}A = P_{\text{CO}_2}A$$

Substitution of numerical values yields

$$(50)(9.8) + (1 \times 10^5)(250 \times 10^{-4}) = P_{\text{CO}_2}(250 \times 10^{-4})$$

Solving for P_{CO_2} *gives*

$$P_{\text{CO}_2} = 1.196 \times 10^5 \, \text{Pa} \, (\sim 1.2 \, \text{bar})$$

b) *At the final equilibrium state,* $\Sigma F = 0$ *again. Therefore, the pressure will be the same.*

Example 2.2 *A gas contained in two cylinders, A and B, is connected by a frictionless piston of two different diameters. The mass of the piston is* 10 kg *and the gas pressure inside cylinder A is* 200 kPa. *Calculate the pressure in cylinder B.*

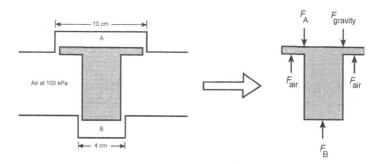

System: *Piston*

Since the piston is in equilibrium, the summation of forces acting on it must be zero:

$$F_A + F_{gravity} = F_B + F_{air} \quad \Rightarrow \quad P_A A_A + mg = P_B A_B + P_{air}(A_A - A_B)$$

Substitution of numerical values yields

$$(2 \times 10^5)\left[\frac{\pi(0.1)^2}{4}\right] + (10)(9.8)$$

$$= P_B\left[\frac{\pi(0.04)^2}{4}\right] + (1 \times 10^5)\left\{\frac{\pi\left[(0.1)^2 - (0.04)^2\right]}{4}\right\}$$

Solving for P_B gives

$$P_B = 802,986\,\text{Pa} \ (\sim 8\,\text{bar})$$

2.1.2   *Constant pressure (isobaric) process*

A *constant pressure process* is one in which the pressure of the system does not change throughout the process. It is also known as an *isobaric process*. It is important to note that if the initial and final states are at the same pressure, this does not necessarily imply a constant pressure process.

Consider an ideal gas confined in a cylinder fitted with a piston. If it undergoes a constant pressure process, then from the ideal gas equation of state

$$\frac{V}{T} = \frac{nR}{P} = \text{constant} \tag{2.1-10}$$

If the initial and final states are designated by 1 and 2, respectively, from Eq. (2.1-10)

$$\frac{V_1}{V_2} = \frac{T_1}{T_2} \qquad (2.1\text{-}11)$$

Consider a gas contained in a vertical cylinder fitted with a frictionless piston. When the system is in equilibrium, i.e., no movement of the piston, the pressure of the gas is equal to the summation of the ambient pressure and the pressure due to the weight of the piston. When the gas is heated, the piston moves up (gas volume increases) and reaches a new equilibrium state at which point the gas pressure remains the same. In other words, the gas experiences a constant pressure process.

2.2   Work

The concept of work in thermodynamics is an extension of the concept of work in mechanics. Since a complete understanding of the physical concept of work and of its mathematical nature is of utmost importance, it would appear opportune first to review the more familiar concept of work in mechanics.

2.2.1   *The concept of work in mechanics*

Work is a scalar quantity, although the two quantities involved in its definition, force and displacement, are vectors. It is defined by the scalar product (or dot product) of force, $\mathbf{F}$, and displacement, $\mathbf{x}$, i.e.,

$$W = \mathbf{F} \cdot \mathbf{x} \qquad (2.2\text{-}1)$$

Equation (2.2-1) can be expressed in the form

$$W = \begin{bmatrix} \text{Component of force in the} \\ \text{direction of displacement} \end{bmatrix} \begin{bmatrix} \text{Magnitude of} \\ \text{displacement} \end{bmatrix}$$

$$= \begin{bmatrix} \text{Magnitude} \\ \text{of force} \end{bmatrix} \begin{bmatrix} \text{Component of displacement} \\ \text{in the direction of force} \end{bmatrix} \qquad (2.2\text{-}2)$$

In the SI system, work is measured in newton-meter (N. m) or joule (J): $1\,\text{N. m} = 1\,\text{J}$. In the English system it is measured in foot-pounds force (ft. lb$_f$). These two units are related by

$$1\,\text{J} = 0.7376\,\text{ft. lb}_f \qquad 1\,\text{ft. lb}_f = 1.356\,\text{J} \qquad (2.2\text{-}3)$$

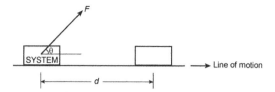

Fig. 2.2   Displacement of a system under the action of a force.

Power is the rate at which work is done in unit time and is given by

$$\dot{W} = \frac{dW}{dt} \tag{2.2-4}$$

In the SI system, the unit of power is J/s or watt (W). A familiar unit for power in the English system is the horsepower (hp), where

$$1\,hp = 550\,ft.\,lb_f/s = 745.7\,W \tag{2.2-5}$$

Consider a constant force F making an angle θ with the line of motion and acting on a particle whose displacement along the line of motion is d as shown in Fig. 2.2. According to Eq. (2.2-1)

$$W = Fd\,\cos\theta \tag{2.2-6}$$

When $\theta = 0°$, $W = Fd$. Thus, when a horizontal force draws a body horizontally, or when a vertical force lifts a body vertically, the work done by the force is simply Fd. When $\theta = 90°$, the force has no component in the direction of motion and no work is done on the body even if the body is moved horizontally over the ground. When $\theta > 90°$, work turns out to be negative, i.e., force and displacement vectors point in opposite directions.

The definition of work does not correspond to the common usage of the word. A person holding a weight may say that he/she is doing hard work in the physiological sense, i.e., he/she becomes tired. However, the work done by the person is zero since there is no displacement of the weight.

Work cannot be done if there is no opposing force. In other words, there must be some resistance against which the force operates. Otherwise, no work is done. For example, consider a rigid tank of known volume separated into two equal parts by a frictionless piston, which is held in place by a pin as shown in Fig. 2.3. While chamber A contains air at 10 bar, chamber B is evacuated.

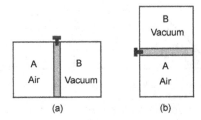

Fig. 2.3   A rigid tank separated by a piston into two chambers.

If the orientation of the tank is horizontal as shown in Fig. 2.3-a, the piston moves to the right when the pin is removed. The work done by air is zero since there is no resisting force. For the vertical orientation shown in Fig. 2.3-b, the piston moves up once the pin is removed. In this case the resisting force is the weight of the piston and work is done by air in raising the piston. However, if one chooses air and the piston together as the system, then the work done is again zero.

If the force acting on the system is a function of position, i.e., the force required to stretch a spring is proportional to its displacement, it is necessary to perform an integration to find the work done. If force acts in the direction of displacement, the expression for work becomes

$$W = \int_{x_1}^{x_2} F(x)\, dx \qquad (2.2\text{-}7)$$

Example 2.3 *A block of* 50 kg *is to be pushed up* 5 m *along a friction-less* 30° *inclined plane. Calculate the work required to push the block at a constant speed.*

Solution

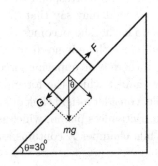

Since the motion is not accelerated, the resultant force parallel to the plane must be zero, i.e., $F = G$.

$$F = G = mg \sin \theta = (50)(9.8) \sin 30 = 245 \, \text{N}$$

Therefore, the work done by a person in pushing the block is

$$W = Fd = (245)(5) = 1225 \, \text{J}$$

Comment: *Note that the net work done on the block is zero since the net force on the block is zero. Therefore,*

$$W_{net} = W_{person} + W_{gravity} = 0$$

Then, the work done by the gravity is

$$W_{gravity} = -1225 \, \text{J}$$

Work is negative because the direction of the component of gravity and the displacement are in opposite directions.

2.2.2 *Kinetic energy and work*

Assume a displacement of a system in the direction of the force, and assume the force varies with position. In this case, the net work to carry out this displacement is given by

$$W_{net} = \int_{x_1}^{x_2} F_{net} \, dx \tag{2.2-8}$$

If the mass of the system is m and its acceleration is a, then according to Newton's second law of motion

$$F_{net} = ma \tag{2.2-9}$$

The acceleration can be expressed as

$$a = \frac{dv}{dt} = \frac{dv}{dx}\frac{dx}{dt} = v\frac{dv}{dx} \tag{2.2-10}$$

Substitution of Eqs. (2.2-9) and (2.2-10) into Eq. (2.2-8) gives

$$W_{net} = \int_{v_1}^{v_2} mv \, dv = \frac{1}{2} mv_2^2 - \frac{1}{2} mv_1^2 \tag{2.2-11}$$

or

$$\boxed{W_{net} = \Delta E_K} \tag{2.2-12}$$

where the kinetic energy, E_K, is defined by

$$E_K = \frac{1}{2}mv^2 \tag{2.2-13}$$

Note that the unit of kinetic energy is the same as the unit of work.

Equation (2.2-12) is known as the *work-energy theorem*. It simply states that the net work done to accelerate a body is independent of path and can be calculated by simply taking the difference in its kinetic energy. Therefore, it is not necessary to know how $F(x)$ varies as a function of position during the process.

Example 2.4 *A car having a mass of* 1500 kg *is traveling at* 70 km/ h.

a) *What is its kinetic energy?*

b) *How much work must be done to bring it to a stop?*

c) *How far does the car travel after the brakes are set before coming to rest? Assume a constant deceleration of* 7 m/ s^2.

Solution

a) *The kinetic energy of the car is calculated from Eq. (2.2-13) as*

$$E_K = \frac{1}{2}mv^2 = \frac{1}{2}(1500)\left[\frac{(70)(1000)}{3600}\right]^2 = 2.84 \times 10^5 \text{ J}$$

b) *Using Eq. (2.2-12)*

$$W_{net} = \Delta E_K = 0 - 2.84 \times 10^5 = -2.84 \times 10^5 \text{ J}$$

c) *Noting that the force and the displacement are in opposite directions*

$$W_{net} = Fd\cos 180 = \Delta E_K$$

Solving for the displacement gives

$$d = -\frac{\Delta E_K}{F} = -\frac{\Delta E_K}{ma} = \frac{2.84 \times 10^5}{(1500)(7)} = 27 \text{ m}$$

Comment: *The stopping distance increases with the square of the velocity. Therefore, drivers need to keep a safe distance between their vehicles to* avoid accidents. Besides, the distance calculated in part (c) is the stopping distance after the brakes were set. If the driver requires $2/3$ s to set the brakes, then the total stopping distance is

$$d = \left[\frac{(70)(1000)}{3600} \right] \left(\frac{2}{3} \right) + 27 \simeq 40 \, \text{m}$$

2.2.3   *Conservation of mechanical energy*

In general, forces acting on a particle can be classified as *surface forces* and *body forces*. Surface forces, such as normal stresses (pressure) and tangential stresses (shear stress), act by direct contact on a surface. Body forces, however, act on a volume remotely without any contact. Gravitational, electrical, and electromagnetic forces are examples of body forces.

Since the work done by body forces is a state function, body forces are also called *conservative forces*[1]. On the other hand, work done by surface forces is a path function and these forces are called *nonconservative forces*.

Consider one-dimensional motion of a body subjected to a conservative force, F_c. Conservative work, W_c, can be calculated from Eq. (2.2-7) as

$$W_c = \int_{x_1}^{x_2} F_c(x) \, dx \qquad (2.2\text{-}14)$$

Since the force is conservative, the work is independent of path. This is only possible if a function E_P is defined by

$$F_c(x) = - \frac{dE_P}{dx} \qquad (2.2\text{-}15)$$

in which the minus sign is used by convention. Hence, Eq. (2.2-14) becomes

$$W_c = - \int_{x_1}^{x_2} \frac{dE_P}{dx} \, dx = - \Big[E_P(x_2) - E_P(x_1) \Big] = - \Delta E_P \qquad (2.2\text{-}16)$$

The function E_P is called the *potential energy*.

Combination of Eqs. (2.2-12) and (2.2-16) gives

$$\boxed{\Delta E_K + \Delta E_P = 0} \qquad (2.2\text{-}17)$$

or

$$E_K + E_P = \text{constant} \qquad (2.2\text{-}18)$$

---

[1] Note that the work done by a conservative force in moving a system through a cyclic process is zero.

indicating that the sum of the kinetic and potential energies, i.e., total mechanical energy, is constant. This is the statement for the *conservation of mechanical energy*. However, note that the mechanical energy is conserved only when conservative forces exist[2].

Gravitational work

Consider a system located at an elevation of x above the earth's surface as shown in Fig. 2.4. Since the force of gravity is in the negative x-direction, then

$$F_{gravity} = -mg = \text{constant} \qquad (2.2\text{-}19)$$

Combination of Eqs. (2.2-15) and (2.2-19) yields

$$F_{gravity} = -\frac{dE_P}{dx} = -mg \qquad (2.2\text{-}20)$$

Integration of Eq. (2.2-20) gives

$$E_P = mgx + C \qquad (2.2\text{-}21)$$

where C is a constant of integration. Taking E_P to be equal to zero at $x = 0$ gives $C = 0$. Therefore, the *gravitational potential energy*[3] is defined by

$$\boxed{E_P = mgx} \qquad (2.2\text{-}22)$$

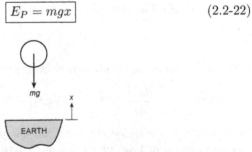

Fig. 2.4   A system located at an elevation x measured from the earth's surface.

---

[2]In around 1590 Galileo dropped objects off the *Leaning Tower of Pisa* and observed that, regardless of the weight of the two bodies, they reach the ground the same time with a velocity $v = \sqrt{2gh}$ as long as air resistance is the same on each body.

[3]It is the energy stored with the object by virtue of its position relative to the earth's surface.

If a stone of mass m is taken from the ground and raised to a height h, then the work done against the gravitational force is always a state function and, according to Eq. (2.2-16), is given by

$$W_{gravity} = -\Delta E_P = -mgh \qquad (2.2\text{-}23)$$

Example 2.5 *Resolve Example 2.3 by using the principle of conservation of mechanical energy.*

Solution

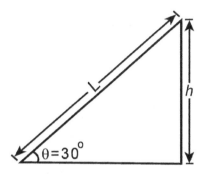

$$W_{gravity} = -\Delta E_P = -mgh = -mgL\sin\theta$$
$$= -(50)(9.8)(5)\sin 30 = -1225\,\text{J}$$

Spring work

For an ideal elastic spring, *Hooke's law* states that

$$F_{spring} = -kx \qquad (2.2\text{-}24)$$

where k is a spring constant. The minus sign indicates that the spring force is in the direction opposite to the direction of displacement. Combination of Eqs. (2.2-15) and (2.2-24) yields

$$F_{spring} = -\frac{dE_P}{dx} = -kx \qquad (2.2\text{-}25)$$

Integration of Eq. (2.2-25) gives

$$E_P = \frac{1}{2}kx^2 + C \qquad (2.2\text{-}26)$$

If E_P is taken as zero at $x = 0$, then the *elastic potential energy* is

$$E_P = \frac{1}{2} kx^2$$

(2.2-27)

Thus, according to Eq. (2.2-16), the work done by the spring is given by

$$W_{spring} = -\Delta E_P = -\frac{1}{2} k\left(x_2^2 - x_1^2\right)$$

(2.2-28)

Example 2.6 *An ideal spring* $(k = 100\,\mathrm{N/m})$ *is placed at the bottom of a frictionless inclined plane that makes an angle of* $\theta = 30°$ *with the horizontal. A* 10 kg *mass is released from rest at the top of the incline and comes to rest momentarily after compressing the spring* 2 m.

a) *What distance does the mass travel before coming to rest?*
b) *What is the speed of the mass just before it reaches the spring?*

Solution

a) System: *Spring and the mass*

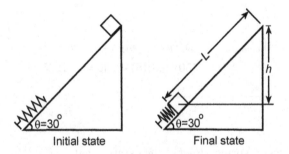

| Initial state | Final state |

The conservation of mechanical energy states that

$$\Delta E_K + \Delta E_P = 0 \qquad \Rightarrow \qquad (\Delta E_P)_{spring} + (\Delta E_P)_{mass} = 0 \qquad (1)$$

or

$$\left(\frac{1}{2} kx^2 - 0\right) + (0 - mgh) = 0 \qquad (2)$$

Since $h = L\sin\theta$, *we get*

$$L = \frac{kx^2}{2mg\sin\theta} = \frac{(100)(2)^2}{(2)(10)(9.8)\sin 30} = 4.08\,\mathrm{m} \qquad (3)$$

b) System: *The mass*

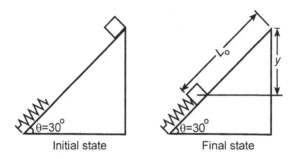

Initial state   Final state

The conservation of mechanical energy states that

$$\Delta E_K + \Delta E_P = 0 \tag{4}$$

or

$$\left(\frac{1}{2}mv^2 - 0\right) + (0 - mgy) = 0 \quad \Rightarrow \quad v = \sqrt{2gy} \tag{5}$$

The elevation y is given by

$$y = L_o \sin\theta = (4.08 - 2)\sin 30 = 1.04\,\text{m} \tag{6}$$

The use of Eq. (6) in Eq. (5) gives the velocity as

$$v = \sqrt{(2)(9.8)(1.04)} = 4.5\,\text{m/s} \tag{7}$$

2.2.4 *Work done by nonconservative forces*

Consider a system that consists of a hemispherical bowl and a steel ball as shown in Fig. 2.5. If the ball is released from a height h, it rolls backwards and forwards within the bowl. However, the height reached by the ball in each successive oscillation is reduced. Finally the ball comes to rest at the bottom of the bowl.

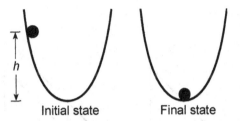

Initial state   **Final state**

Fig. 2.5   Initial and final states of a steel ball in a bowl.

The application of the conservation of mechanical energy gives

$$\Delta E_K + \Delta E_P = -mgh < 0 \qquad (2.2\text{-}29)$$

indicating that the total mechanical energy of the system diminishes, which manifests itself as an increase in the temperature of the ball and the bowl. In mechanics, this phenomenon is described by either of the following statements: "the mechanical energy has been dissipated", or "the mechanical energy has been converted into thermal energy."

In this particular example, a dissipative force called "friction" does negative work on the system and tends to diminish the total energy of the system. This can be shown by using the work-energy theorem, i.e., Eq. (2.2-12), expressed in the form

$$W_c + W_{nc} = \Delta E_K \qquad (2.2\text{-}30)$$

where W_c and W_{nc} represent the work done by conservative and nonconservative forces, respectively. Work done by conservative forces is given by Eq. (2.2-16). Substitution of this equation into Eq. (2.2-30) gives

$$\Delta E_K + \Delta E_P = W_{nc} \qquad (2.2\text{-}31)$$

Comparison of Eqs. (2.2-29) and (2.2-31) indicates that $W_{nc} < 0$. This comes from the fact that frictional force always acts in the direction opposite to the displacement.

To maintain the concept of conservation of energy, it is necessary to invent another form of energy besides potential and kinetic energies. Let the increase in energy due to frictional work be defined as *internal energy* and give it a symbol U, i.e.,

$$\Delta U = -W_{nc} \qquad (2.2\text{-}32)$$

Substitution of Eq. (2.2-32) into Eq. (2.2-31) gives

$$\Delta U + \Delta E_K + \Delta E_P = 0 \qquad (2.2\text{-}33)$$

or

$$U + E_K + E_P = \text{constant} \qquad (2.2\text{-}34)$$

Equation (2.2-34) states that there is no change in the sum of internal and mechanical energies when only conservative and frictional forces act on a system.

To understand the concept of internal energy consider the following two examples:

• When an astronaut on the space shuttle *Atlantis* looks at the earth, s(he) sees that the earth has an external kinetic energy due to its rotation and its motion around the sun. The earth also has an internal kinetic energy as a result of all the objects, i.e., people, cars, planes, trains, etc., moving on its surface that the astronaut cannot see and measure.

• Consider a watch on your arm. When you move your arm, the watch has a certain external kinetic energy and gravitational potential energy with respect to some reference plane, i.e., the floor. Note that the watch also has an internal energy as a result of the motion of the wheels located inside it. Again, we cannot measure this internal energy.

In both examples note that a physical object is composed of smaller objects, each of which can have a variety of internal and external energies. The sum of the internal and external energies of the smaller objects is apparent as internal energy of the larger objects.

Molecules have kinetic energy as a result of their vibration and random motion. They also have potential energy associated with their position and the attractive forces between them. At the microscopic level, the sum of these kinetic and potential energies, which cannot be seen, is referred to as the *internal energy*, U, in thermodynamics. In other words, internal energy is nothing but the thermal energy defined in Section 1.4.

Since thermodynamics considers systems at macroscopic level, it is not concerned with the origin of internal energy. Thermodynamics simply states that (i) internal energy is an extensive property, (ii) internal energy is a state function. The internal energy of a system is increased either by transferring heat to the system or by doing work on it. The relationship between heat, work, and change in internal energy will be explained in detail in Chapter 4.

2.3   The Concept of Reversibility

A process is called *reversible* (or *quasi-static)* if it is possible to restore the system and its surroundings to their original states and leave no other effects. For example, consider a closed system undergoing a process from state 1 to state 2 while receiving 1000 J of heat and 800 J of work. If the system can be returned to state 1 along the same path while rejecting 1000 J of heat to its surroundings and doing 800 J of work on the surroundings, then this process is called reversible. Note that a reversible process can be completely erased leaving no net effects at all on the physical world.

Real processes are all irreversible processes. For example, you can easily mix vodka and orange juice to prepare a cocktail. Now let us consider reversing the process, i.e., separating vodka from orange juice. Obviously, this requires the use of separation equipment. Some examples of irreversible processes are given below:

- Friction (solid-fluid and solid-solid),
- Conversion of energy from one form to another,
- Combustion,
- Flow through a constriction, such as a partially opened valve,
- Free expansion of a gas,
- Mixing of fluids,
- Flow of electric current through a resistance,
- Heat transfer over a finite temperature difference,
- Diffusion.

Although no real process is completely reversible, some processes can be approximated as reversible if they are executed very slowly, i.e., the driving force is small. For example, consider the heating of a metal block from T_1 to T_2 by contacting it with a series of heat reservoirs[4] at $T_1 + \Delta T$, $T_1 + 2\Delta T$, ..., $T_2 - \Delta T$, T_2 as shown in Fig. 2.6-a. Is this process reversible or irreversible? To answer this question, it is necessary to reverse the process as shown in Fig. 2.6-b, i.e., the block is cooled from T_2 to T_1 using the

---

[4]A heat reservoir is an idealization, like an ideal gas. It serves as a heat source or sink for another system and operates at constant temperature, i.e., transfer of heat does not appreciably alter the temperature of the reservoir. A lake is an example of a heat reservoir.

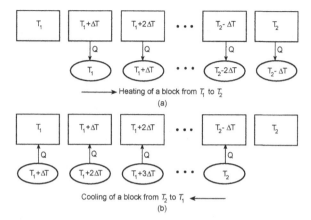

Fig. 2.6   Heating and cooling of a block by contacting it with a series of constant temperature reservoirs.

same heat reservoirs. When the block is restored to its initial temperature, note that all the heat reservoirs are also restored to their initial conditions, except for those at T_1 and T_2. The reservoir at T_1 will have gained a small amount of heat (enough to lower the block temperature from $T_1 + \Delta T$ to T_1) and the reservoir at T_2 will have lost a small amount of heat (to raise the block temperature from $T_2 - \Delta T$ to T_2). It is not possible to completely restore these two reservoirs. However, if ΔT is extremely small, these two reservoirs are *approximately* back to their initial states. Under these conditions the process is a reversible one. Although heat transfer through a finite temperature difference is always irreversible, by reducing the ΔT involved to very small values, it can be made approximately reversible.

Now consider a vertical cylinder fitted with a frictionless piston as shown in Fig. 2.7-a. The initial pressure of the gas contained in the cylinder is P_1. Let us increase the gas pressure to P_2 by putting a series of very small weights, each Δm, on the piston one at a time.

Is this process reversible or irreversible? To answer this question, it is necessary to reverse the process as shown in Fig. 2.7-b, i.e., the gas is expanded by removing the weights one by one. When the gas is restored to its initial pressure, each weight is restored to its initial height, except for a small Δz for the first and the fourth weights. Again, by making the Δm smaller and smaller, the first and the last weights are *approximately* restored to their initial heights. Under these conditions the process is considered a

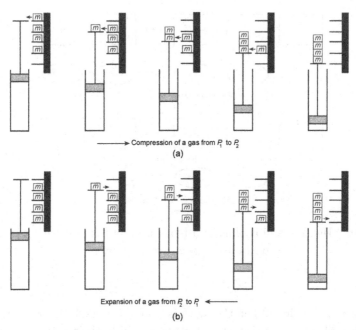

Fig. 2.7   Compression and expansion of a gas by putting a series of weights on a piston and removing them, respectively.

reversible one. Although compression or expansion of a gas under finite pressure difference is always irreversible, by reducing the pressure difference to very small values, it can be made nearly reversible. Since the driving force during a reversible process is infinitesimally small, the system is assumed to be in equilibrium at each and every stage of the process. Hence, a reversible process is also referred to as a *quasi-static process*.

2.3.1   *Reversible and irreversible work*

Consider a gas (the system) of volume V_{sys} and pressure P_{sys} contained in a cylinder fitted with a frictionless piston as shown in Fig. 2.8. An external force F_{ex} is applied to the piston. When the gas is in mechanical equilibrium with its surroundings, the external pressure counterbalances the force exerted by the gas on the piston:

$$P_{sys} = \frac{F_{ex}}{A} = P_{ex} \qquad (2.3\text{-}1)$$

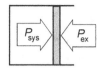

Fig. 2.8   Piston-cylinder device.

Now suppose that the external pressure decreases by a differential amount and the piston moves to the right as a result of this differential pressure difference. Thus, the gas volume within the cylinder increases by a differential amount dV_{sys}. The work done by the gas within the cylinder on the surroundings during this expansion process is given by

$$dW = F_{ex}\, dx = \frac{F_{ex}}{A}\, d(Ax) = P_{ex}\, dV_{sys} \qquad (2.3\text{-}2)$$

Since the pressure differential is very small, the motion of the piston takes place infinitesimally slowly. Under these circumstances, the process may be considered reversible and Eq. (2.3-1) holds at each and every stage of the process. Substitution of Eq. (2.3-1) into Eq. (2.3-2) gives

$$dW_{rev} = P_{sys}\, dV_{sys} \qquad (2.3\text{-}3)$$

In Eq. (2.3-3), P_{sys} is the absolute pressure of the system, which is always positive. However, the volume change dV_{sys} is positive during an expansion process and negative during a compression process.

According to the sign convention for work in thermodynamics, if the system receives work from the surroundings, it is positive. Otherwise, it is negative. This is illustrated in Fig. 2.9. Therefore, to be consistent with the sign convention used in thermodynamics, Eq. (2.3-3) is expressed as

$$dW_{rev} = -P_{sys}\, dV_{sys} \qquad (2.3\text{-}4)$$

Fig. 2.9   Sign convention for work.

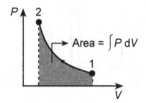

Fig. 2.10   The area under the reversible process path is equal to work.

The total work done by the gas on its surroundings may then be determined by

$$W_{rev} = -\int_{V_1}^{V_2} P_{sys} \, dV_{sys} \qquad (2.3\text{-}5)$$

where V_1 and V_2 represent the volumes occupied by the system at the initial and final states, respectively. Equation (2.3-5) indicates that the reversible work associated with the volume change is a function of the properties of a system. The evaluation of Eq. (2.3-5) requires a relationship between pressure and volume to be known:

• If the equation of state is given, first express either pressure in terms of volume or, volume in terms of pressure. Then, the integral may be evaluated analytically or numerically.

• If the relationship between P and V is given in terms of experimental data, the integral may be evaluated numerically or graphically. On a P-V diagram, as shown in Fig. 2.10, the area under the process path is equal, in magnitude, to the work done during a reversible expansion or compression process of a closed system.

Consider a system that changes its state from 1 to 2 by two different process paths, $\overline{1A2}$ and $\overline{1B2}$, as shown in Fig. 2.11. If the processes are reversible, the work done is equal to

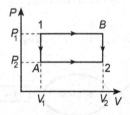

Fig. 2.11   A system undergoing a process from state 1 to state 2 by two different paths.

$$W_{1A2} = -\underbrace{\int_1^A P\,dV}_{dV=0} - \underbrace{\int_A^2 P\,dV}_{P=P_2} = -P_2\,(V_2 - V_1) \qquad (2.3\text{-}6)$$

$$W_{1B2} = -\underbrace{\int_1^B P\,dV}_{P=P_1} - \underbrace{\int_B^2 P\,dV}_{dV=0} = -P_1\,(V_2 - V_1) \qquad (2.3\text{-}7)$$

Equations (2.3-6) and (2.3-7) indicate that the work done by a system, in general, depends on the path of the process and hence it is a path function.

In the case of an irreversible process, $P_{sys} \neq P_{ex}$ and the integration of Eq. (2.3-2) gives

$$\boxed{W_{irrev} = -\int_{V_1}^{V_2} P_{ex}\,dV_{sys}} \qquad (2.3\text{-}8)$$

Example 2.7 *A cylinder fitted with a frictionless piston initially contains 2 moles of carbon dioxide at* 700 K. *Its volume is* 0.05 m³. *Calculate the work done by carbon dioxide if it undergoes a reversible isothermal expansion until the volume is doubled.*

Solution

Assumption

• CO_2 *is an ideal gas.*

System: CO_2 *in the piston-cylinder device*

Let 1 and 2 represent the initial and final states of carbon dioxide, respectively. Since the expansion process is reversible, the work done is given by

$$W = -\int_{V_1}^{V_2} P\,dV \qquad (1)$$

The pressure of CO_2, *P, is related to the volume, V, through the ideal gas equation of state as*

$$P = \frac{nRT}{V} \qquad (2)$$

Substitution of Eq. (2) into Eq. (1) and integration (note that the process is isothermal) yield

$$W = -nRT \int_{V_1}^{V_2} \frac{dV}{V} = nRT \ln \left(\frac{V_1}{V_2} \right) \tag{3}$$

Substitution of the numerical values leads to

$$W = (2)(8.314)(700) \ln \left(\frac{0.05}{0.1} \right) = -8,068 \text{ J} \tag{4}$$

Since work is done by the system, it is negative.

Alternative solution: *In the evaluation of the integral given by Eq. (1), it is also possible to relate volume to pressure as*

$$V = \frac{nRT}{P} \tag{5}$$

Thus,

$$dV = -\frac{nRT}{P^2} dP \tag{6}$$

The use of Eq. (6) in Eq. (1) gives

$$W = \int_{P_1}^{P_2} \frac{nRT}{P} dP = nRT \ln \left(\frac{P_2}{P_1} \right) \tag{7}$$

From the ideal gas equation of state

$$P_1 V_1 = P_2 V_2 \quad \Rightarrow \quad \frac{P_2}{P_1} = \frac{V_1}{V_2} \tag{8}$$

Substitution of Eq. (8) into Eq. (7) leads to Eq. (3).

Example 2.8 *The relation of pressure to volume for a closed system undergoing a cyclic reversible process is shown in the figure below. Calculate the net work for the cycle.*

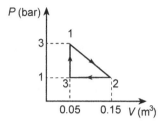

Solution

System: *Gas in the closed system*

• Process $1 \to 2$

The easiest way of calculating the work done is to find out the area under the process path on a P-V diagram. The sign of work is determined by inspection. Since this is an expansion process, work is done by the system and hence it is negative:

$$W = -(\text{Area of a trapezoid})$$

$$= -\frac{\left[(1+3) \times 10^5\right](0.15 - 0.05)}{2} = -2 \times 10^4 \, \text{J} \, (-20 \, \text{kJ})$$

Alternative solution: *Since the relationship between pressure and volume is linear, then*

$$P = mV + b \qquad\qquad (1)$$

where m and b are the slope and an intercept of a straight line. In this case,

$$m = \frac{(3-1) \times 10^5}{0.05 - 0.15} = -2 \times 10^6 \, \text{Pa}/\,\text{m}^3$$

To determine b, Eq. (1) must be evaluated either at state 1 or at state 2. Evaluation of Eq. (1) at state 1 gives

$$3 \times 10^5 = \left(-2 \times 10^6\right)(0.05) + b \qquad \Rightarrow \qquad b = 4 \times 10^5 \, \text{Pa}$$

Thus, from Eq. (2.3-5)

$$W = -\int_{V_1}^{V_2} P\,dV = -\int_{0.05}^{0.15} \left(-2 \times 10^6 \, V + 4 \times 10^5\right) dV$$

$$= \left(0.15^2 - 0.05^2\right) \times 10^6 - 4 \times 10^5 \left(0.15 - 0.05\right) = -2 \times 10^4 \, J \, (-20\,kJ)$$

• **Process 2 → 3**

This is a constant pressure process. From Eq. (2.3-5)

$$W = -\int_{V_2}^{V_3} P\,dV = -P\,(V_3 - V_2) = -(1 \times 10^5)(0.05 - 0.15) = 10,000\,J \,(10\,kJ)$$

• **Process 3 → 1**

Since the volume remains constant, i.e., $V_3 = V_1$, Eq. (2.3-5) gives

$$W = -\int_{V_3}^{V_1} P\,dV = 0$$

Note that, for the overall cycle

$$W_{net} = W_{12} + W_{23} + W_{31} = (-20) + (10) + (0) = -10\,kJ$$

Comment: *Since the work done by the system during the expansion process (W_{12}) is greater than the work done on the system during the compression process (W_{23}), the difference between these two is the net work done during the cycle. The net work is negative if the cycle is in the clockwise direction and positive if it is in the counterclockwise direction. In this case, the area of the triangle gives the same result, i.e.,*

$$W_{net} = -\frac{\left[(3-1) \times 10^5\right](0.15 - 0.05)}{2} = -10,000\,J \,(-10\,kJ)$$

Example 2.9 *A vertical cylinder fitted with a frictionless piston contains $0.003\,m^3$ of an ideal gas at 6.2 bar and 25 °C. The piston has a mass of 90 kg, cross-sectional area of $65\,cm^2$, and is held by a pin as shown in the figure below. The ambient pressure is 1 bar.*

The pin is now removed, allowing the piston to move. After a period of time, the system comes to equilibrium, with the final temperature being 25 °C.
a) *Determine the final pressure and volume of the gas.*
b) *Calculate the work done by the gas during this process.*

Solution

a) System: Piston

Let 1 and 2 represent the initial and final states of the gas within the cylinder. At the final equilibrium state, the piston is stationary and, according to Newton's second law of motion, summation of forces acting on the piston is zero, i.e.,

$$(F_{gas})_2 = F_{gravity} + F_{air}$$

or

$$P_2(65 \times 10^{-4}) = (90)(9.8) + (1 \times 10^5)(65 \times 10^{-4})$$

Solving for P_2 gives

$$P_2 = 2.4 \times 10^5 \, \text{Pa} \, (2.4 \, \text{bar})$$

The volume occupied by the gas can be calculated from the ideal gas equation of state as (note that $T_1 = T_2$)

$$V_2 = \frac{P_1 V_1}{P_2} = \frac{(6.2)(0.003)}{2.4} = 0.00775 \, \text{m}^3$$

b) System: *Gas contained within the cylinder*

Since the process is irreversible, the use of Eq. (2.3-8) gives

$$W_{irrev} = -\int_{V_1}^{V_2} P_{ex} \, dV_{sys} = -P_{ex} \, \Delta V$$

$$= -(2.4 \times 10^5)(0.00775 - 0.003) = -1140 \, \text{J}$$

This work is done against the piston and the surrounding atmosphere.

Comment: *A "frictionless piston" does not necessarily imply a reversible process. Mathematically speaking, it is the necessary but not the sufficient condition to make the process reversible. If a piston is frictionless, it simply implies that the pressure of the gas is counterbalanced by the pressure exerted by the piston and the ambient pressure when equilibrium is reached. It is the magnitude of the driving force that determines whether the process is reversible or not. In this example the process is irreversible since the initial driving force is quite high, i.e., $\Delta P = 6.2 - 2.4 = 3.8$ bar. Once the pin is removed, the piston shoots up and starts to oscillate up and down until the final equilibrium state is attained.*

Example 2.10 *A cylinder fitted with a frictionless piston initially contains $0.04\,\text{m}^3$ of an ideal gas at $200\,\text{kPa}$ as shown in the figure below. The weights placed on the piston are then removed at such a rate that the gas expands reversibly according to the relation*

$$PV^{1.2} = constant$$

until the total volume reaches $0.1\,\text{m}^3$. Determine the work done during this process.

Ideal Gas

System: *Ideal gas within the cylinder*

Let 1 and 2 represent the initial and final states of the gas within the cylinder. Since the expansion process is reversible, the work done by the gas is calculated from Eq. (2.3-5) as

$$W = -\int_{V_1}^{V_2} P\,dV \tag{1}$$

From the given relationship between pressure and volume, we have

$$PV^{1.2} = \text{constant} = k \qquad \Rightarrow \qquad P = \frac{k}{V^{1.2}} \tag{2}$$

Substitution of Eq. (2) into Eq. (1) gives

$$W = - \int_{V_1}^{V_2} \frac{k}{V^{1.2}} \, dV \tag{3}$$

Upon integration we get

$$W = \frac{k}{0.2} \left(\frac{1}{V_2^{0.2}} - \frac{1}{V_1^{0.2}} \right) \tag{4}$$

The constant k can be determined from the conditions of the initial state as

$$k = P_1 V_1^{1.2} = (200)(0.04)^{1.2} = 4.2 \, \text{kPa. m}^{3.6} \tag{5}$$

Substitution of the numerical values into Eq. (4) gives the work done by the system as

$$W = \frac{4.2}{0.2} \left(\frac{1}{0.1^{0.2}} - \frac{1}{0.04^{0.2}} \right) = -6.7 \, \text{kJ} \tag{6}$$

Alternative solution: *The integration of Eq. (3) can also be expressed in the form*

$$W = \frac{\left(kV^{-1.2} \right) V}{0.2} \bigg|_{V_1}^{V_2} = \frac{P_2 V_2 - P_1 V_1}{0.2} \tag{7}$$

The pressure at the final state can be determined from

$$P_1 V_1^{1.2} = P_2 V_2^{1.2} \quad \Rightarrow \quad P_2 = P_1 \left(\frac{V_1}{V_2} \right)^{1.2} \tag{8}$$

Substitution of the numerical values into Eq. (8) gives

$$P_2 = 200 \left(\frac{0.04}{0.1} \right)^{1.2} = 66.6 \, \text{kPa} \tag{9}$$

Therefore, the work done by the gas is

$$W = \frac{(66.6)(0.1) - (200)(0.04)}{0.2} = -6.7 \, \text{kJ}$$

Example 2.11 *A cylinder fitted with a frictionless piston initially contains $0.03 \, \text{m}^3$ of an ideal gas as shown in the figure below. At this state the gas pressure is $200 \, \text{kPa}$, which just balances the atmospheric pressure* *and the piston weight; an ideal spring touches but exerts no force on the piston. The gas is now heated until the volume is doubled and the gas pressure reaches 320 kPa. If the cross-sectional area of the piston is 0.3 m², determine:*

a) *The work done by the gas,*
b) *The fraction of this work done against the spring to compress it.*

Solution

a) System: *Gas within the cylinder*

Let 1 and 2 represent the initial and final states of the gas within the cylinder. Since heat transfer is a slow process and the spring is ideal, the process may be approximated as a reversible one and the work done by the system is calculated from Eq. (2.3-5), i.e.,

$$W = -\int_{V_1}^{V_2} P \, dV \tag{1}$$

The evaluation of the integral in Eq. (1) requires the variation of the gas pressure as a function of volume to be known. If V is the volume occupied by the gas at any given instant during the process, the displacement of the piston (or spring) from its original position is given by

$$x = \frac{V - V_1}{A} \tag{2}$$

The initial pressure of the gas, $P_1 = 200$ kPa, is due to the atmospheric pressure and the weight of the piston. As we start heating the gas, the pressure of the gas inside the cylinder is a result of the atmospheric pressure and the weight of the piston as well as the force of the spring, i.e.,

$$P = P_1 + \frac{|F_{spring}|}{A} = P_1 + \frac{kx}{A} \tag{3}$$

Substitution of Eq. (2) into Eq. (3) leads to

$$P = \left(P_1 - \frac{kV_1}{A^2}\right) + \left(\frac{k}{A^2}\right) V \tag{4}$$

indicating linear variation of gas pressure with volume. Thus, the process path can be represented as follows:

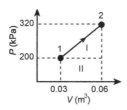

The easiest way of finding the work done is to find the area under the process path, which is the area of a trapezoid. The sign of the work is determined, by inspection, to be negative since it is done by the system:

$$W = -\frac{(200 + 320)(0.06 - 0.03)}{2} = -7.8\,\text{kJ}$$

b) *The rectangular area (region II) represents the work done against the piston and the atmosphere; the triangular area (region I) represents the work done against the spring. Hence, the work done by the gas on the spring is*

$$W = -\frac{(320 - 200)(0.06 - 0.03)}{2} = -1.8\,\text{kJ}$$

Therefore, 23% of the work done by the gas is against the spring.

Alternative solution: *The result in (b) could also be obtained as follows: The pressure exerted on the spring in the final state is*

$$(P_{spring})_2 = 320 - 200 = 120\,\text{kPa}$$

Therefore, the force applied on the spring is given by

$$(F_{spring})_2 = (120)(0.3) = 36\,\text{kN}$$

The final displacement of the piston (or the spring) is

$$x_2 = \frac{V_2 - V_1}{A} = \frac{0.06 - 0.03}{0.3} = 0.1\,\text{m}$$

Thus, the spring constant is given by the Hooke's law

$$k = \frac{F}{x} = \frac{36}{0.1} = 360\,\text{kN/m}$$

From Eq. (2.2-28)

$$W = -\frac{1}{2} k \left(x_2^2 - x_1^2\right) = -\frac{(360)\,(0.1)^2}{2} = -1.8\,\text{kJ}$$

Problems

Problems related to Section 2.1

2.1   Determine the mass and the weight of the air contained in a room whose dimensions are $10\,\text{m} \times 5\,\text{m} \times 3\,\text{m}$. Clearly state your assumptions.

2.2   At $45°$ latitude, the acceleration of gravity as a function of elevation z above sea level is given by

$$g = \alpha - \beta\,z$$

where

$$\alpha = 9.807\,\text{m}/\text{s}^2 \qquad \text{and} \qquad \beta = 3.32 \times 10^{-6}\,\text{s}^{-2}$$

Determine the height above sea level where the weight of a subject will decrease by 1%.

(**Answer:** $29,539\,\text{m}$)

2.3   A petcock is a small piece of mass placed on top of a small opening on the lid of a pressure cooker. It is used to reduce pressure within the cooker by releasing steam. Determine the mass of the petcock needed to maintain a pressure of $220\,\text{kPa}$ inside the pressure cooker. Take the opening cross-sectional area as $5\,\text{mm}^2$ and the atmospheric pressure as $100\,\text{kPa}$.

(**Answer:** $61\,\text{g}$)

2.4   A vertical cylinder fitted with a frictionless piston contains a gas as shown in the figure below. The piston has a mass of $5\,\text{kg}$ and cross-sectional area of $25\,\text{cm}^2$. A compressed spring above the piston exerts a force of $75\,\text{N}$ on the piston. If the atmospheric pressure is $100\,\text{kPa}$, determine the pressure inside the cylinder.

(**Answer:** 149.6 kPa)

2.5 A vertical cylinder fitted with a frictionless piston initially contains air at 300 kPa and 400 °C. The piston pushes against the stops as shown in the figure below. The piston has a mass of 20 kg and cross-sectional area of 50 cm². The atmospheric pressure is 100 kPa. The cylinder now cools as heat is transferred to the surroundings.

a) Determine the temperature when the piston begins to move down.
b) How far has the piston dropped when the temperature is decreased to 15 °C?

(**Answer:** a) 39.3 °C, b) 3.9 cm)

Problems related to Section 2.2

2.6 Determine the work required for a 2000 kg car to climb a 100 m long uphill road with a slope of 30° for the following cases:

a) At a constant velocity,
b) From rest to a final velocity of 30 m/s,
c) From 35 m/s to a final velocity of 5 m/s.

(**Answer:** a) 980 kJ, b) 1880 kJ, c) − 220 kJ)

2.7 Sergei Bubka, retired Ukrainian athlete, broke the world record for men's pole vaulting 35 times. If he weighs 75 kg, what should be the minimum speed of Bubka at take-off so that he clears the bar at 6 m high. Assume Bubka's center of mass[5] is initially 0.9 m off the ground and reaches its maximum height at the level of the bar itself.

(**Answer:** ∼ 10 m/s)

---

[5]The *center of mass* of a body is that point at which the entire mass of the body can be considered as concentrated for purposes of describing its translational motion.

2.8 The potential energy function for the force between two atoms in a diatomic (two-atom) molecule can be expressed in the form

$$E_P = \frac{a}{x^{12}} - \frac{b}{x^6}$$

where x is the distance between the two atoms and a and b are positive constants.

a) At what values of x is E_P a minimum?

b) At what values of x is E_P equal to zero?

c) Plot E_P as a function of x.

d) Determine the force between the atoms.

e) The energy needed to break up the molecule into separate atoms is called the *dissociation energy*. It is defined as the energy required to separate the two particles from their state of lowest energy to $x = \infty$. Determine the dissociation energy.

(**Answer:** a) $(2a/b)^{1/6}$, b) $(a/b)^{1/6}$, d) $12\,(a/x^{13}) - 6(b/x^7)$, e) $b^2/4a$)

2.9 Water flows over Niagara Falls at a rate of 5.5×10^6 kg/s and falls 50 m. How much power can a hydropower plant at Niagara Falls theoretically generate?

(**Answer:** 2695 MW)

2.10 In industry and everyday usage, we speak of energy in kW h. Daily kW h consumption of an appliance is calculated by the following formula:

$$\frac{P \times (\text{Hours used per day})}{1000}$$

where P is the power in W. The power consumption of a television is about 70 W. On average, if it is used for 5 hours a day, calculate its yearly energy expenditure in kW h.

(**Answer:** 127.75 kW h)

2.11 What is the minimum time to boil 0.5 L of water at $20\,^\circ$C with a 1000 kW heater?

(**Answer:** 2.8 min)

2.12 It takes 3 hammer blows to drive a 5 g steel nail completely into wood. If the speed of the hammer just before it strikes the nail is 3 m/s, estimate the increase in the nail's temperature. The mass of the hammer is 1.5 kg and the heat capacity of steel is 460 J/kg. K. Indicate your assumptions.

(**Answer:** $8.8\,^\circ$C)

Problems related to Section 2.3

2.13 An ideal gas at a given state expands to a fixed volume by two types of reversible processes: (*i*) constant pressure, (*ii*) constant temperature. For which case is the work done greater?

2.14 A vertical cylinder fitted with a frictionless piston contains of 0.5 kg of air at 300 kPa and 35 °C. Heat is slowly added and the piston is moved so that the air temperature remains constant. The final pressure is 100 kPa.

a) Show the process in a *P-V* diagram.
b) Calculate the work done during this process.
c) If $Q = n\tilde{C}_V \Delta T$ or $Q = n\tilde{C}_P \Delta T$, then it would appear that no heat is transferred since $\Delta T = 0$. Explain this paradox.

(**Answer:** b) $- 48,503$ J)

2.15 Carbon dioxide contained in a piston-cylinder device is compressed reversibly from 0.3 to 0.1 m^3. Calculate the work done on the carbon dioxide if the pressure and volume during this compression process are related by

$$PV^2 = 8$$

where P is in kPa and V is in m^3.

(**Answer:** 53.3 kJ)

2.16 A cylinder fitted with a frictionless piston contains 0.2 m^3 of gas at 5 bar. The gas expands reversibly until the final volume becomes 0.6 m^3. Calculate the work done by the gas if:

a) Pressure remains constant,
b) Pressure is directly proportional to volume,
c) Pressure is proportional to the square of volume.

(**Answer:** a) $- 200$ kJ, b) $- 400$ kJ, c) $- 866.7$ kJ)

2.17 An existing two-stage process consists of the following steps:

a) One mole of air at 900 K and 3 bar is cooled at constant volume to 300 K.
b) The air is then heated at constant pressure until its temperature reaches 900 K.

It is proposed to replace this two-stage process by a single isothermal expansion of the air from 900 K and 3 bar to some final pressure P^*. If all processes take place reversibly, determine the value of P^* that makes the work of the proposed process equal to that of the existing one?

(**Answer:** 1.54 bar)

2.18 A rigid $0.4\,\mathrm{m^3}$ tank A and cylinder B are connected as shown in the figure below. Tank A initially contains air at $250\,\mathrm{kPa}$ and $30\,^\circ\mathrm{C}$. Cylinder B contains a frictionless piston of a mass such that a pressure of $150\,\mathrm{kPa}$ inside the cylinder is required to raise the piston. Initially the piston is at the bottom of the cylinder. The valve is opened and air flows into B and eventually reaches a uniform state of $150\,\mathrm{kPa}$ and $30\,^\circ\mathrm{C}$ throughout. Calculate the work done by air.

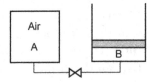

(**Answer:** $-40,054\,\mathrm{J}$)

2.19 A weather balloon is constructed of a nonstretchable plastic material to produce an inflated spherical shape $10\,\mathrm{m}$ in diameter. Initially the balloon is collapsed and is to be filled with helium at a location where the pressure is $1\,\mathrm{atm}$.

a) Calculate the work done by helium during the inflation process.
b) What will be the work done by helium if the plastic material is elastic and stretches during the filling process?

(**Answer:** a) $-53,000\,\mathrm{kJ}$)

2.20 A vertical cylinder fitted with a frictionless piston contains $1\,\mathrm{m^3}$ of nitrogen at $100\,\mathrm{kPa}$ as shown in the figure below. At this state, an ideal spring ($k = 200\,\mathrm{kN/m}$) is touching the piston but exerts no force on it. The cross-sectional area of the piston is $0.8\,\mathrm{m^2}$. Heat is transferred to the nitrogen, causing it to expand until its volume doubles. Determine:

a) The final pressure,
b) The total work done by the nitrogen,
c) The fraction of the total work done against the spring.

(**Answer:** a) $412.5\,\mathrm{kPa}$, b) $-256.25\,\mathrm{kJ}$, c) 61%)

2.21   A vertical cylinder fitted with a piston contains one mole of an ideal gas at 100 bar and 21 °C. The piston is initially held in place by a pin. The pin is removed, allowing the gas to expand isothermally against an external pressure of 10 bar and come to rest.

a) Calculate the work done by the gas during the expansion.

b) Compare the result you find in part (a) with the work done by an isothermal reversible expansion from 100 bar to 10 bar at 21 °C.

(**Answer:** a) -2200 J, b) -5628.2 J)

2.22   A vertical cylinder fitted with a frictionless piston contains 2800 cm^3 of ideal gas at 3.5 bar and 27 °C. The piston is initially held in place by a pin. The piston has a mass of 90 kg and a cross-sectional area of 65 cm^2. The ambient pressure is 1 bar. The pin is removed, allowing the piston to move. After a sufficient period of time, the system comes to equilibrium, with the final temperature being 27 °C.

a) Determine the final pressure and volume of the ideal gas.

b) Calculate the work done by the gas.

(**Answer:** a) 2.357 bar, 4158 cm^3, b) -320 J)

2.23   An elastic sphere initially has a diameter of 1 m and contains a gas at 1 bar. Due to heat transfer, the diameter of the sphere increases to 1.1 m. During the process the gas pressure inside the sphere is proportional to the sphere diameter. Calculate the work done by the gas.

(**Answer:** $-18,225$ J)

2.24   A rectangular chamber of 40 cm width contains water with a free surface. One end of the chamber is fitted with a sliding piston as shown in the figure below. Initially, $x = 60$ cm and $y = 50$ cm. Calculate the work done by the water on the piston when the chamber length is increased slowly from 60 cm to 80 cm.

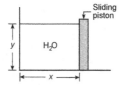

Hint: The hydrostatic pressure distribution is given by

$$P = P_{atm} + \rho g y$$

where $P_{atm} = 100\,\text{kPa}$, and y is the distance measured from the surface of the water.

(**Answer:** 3525.7 J)

2.25   Consider a rigid tank with the top open to the atmosphere as shown in the figure below. A frictionless piston of mass 10 kg and a cross-sectional area of $50\,\text{cm}^2$ divides the tank into chambers A and B. Initially chamber A contains liquid water and chamber B contains 0.036 mol air at 128 kPa and 300 K. Air is heated so that the piston moves upward. Calculate the work done by air when the piston touches the top of the tank.

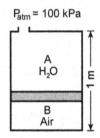

(**Answer:** -532.4 J)

Chapter 3

Pressure-Volume-Temperature (PVT) Relations for Pure Substances

A *pure substance* is one that has a fixed chemical composition throughout. It may exist in more than one phase, but the chemical composition of all phases must be the same, i.e., ice and liquid water mixture; liquid water and steam mixture. For a pure substance, the relations among pressure-volume-temperature (PVT) are expressed in the form of equations (known as equation of state), diagrams (T-V, P-V, and P-T), and property tables. The purpose of this chapter is to explain how to use PVT diagrams and property tables to identify the state of a system. Specification of the initial and final states is a prerequisite in the calculation of heat and work interactions of a system undergoing a process.

3.1 Phase Change of a Pure Subtance

Consider a certain amount of liquid water placed in a cylinder fitted with a frictionless piston as shown in Fig. 3.1-a. Suppose that the ambient pressure together with the piston weight maintains a pressure of 0.125 MPa in the cylinder, and that the initial temperature is 25 °C. At 0.125 MPa and 25 °C (State 1), water is called *subcooled liquid* (or *compressed liquid*), implying that it is not about to vaporize if heat is transferred to the system. As heat is transferred to the water, its temperature increases appreciably, the specific volume increases slightly, and the pressure remains constant. When the temperature reaches 105.99 °C, additional heat transfer results in boiling, i.e., phase change. A liquid that is about to boil is called a *saturated liquid* (State 2). As heat is supplied to vaporize the liquid, its temperature and pressure remain constant while its specific volume increases. During this stage, liquid and vapor phases are in equilibrium with each other (State 3). When all the liquid vaporizes, the entire cylinder is filled with vapor,

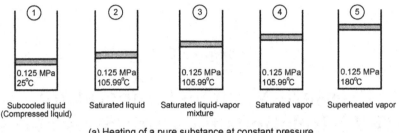

(a) Heating of a pure substance at constant pressure.

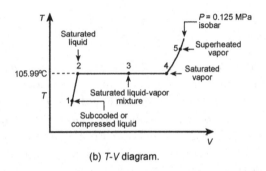

(b) *T-V* diagram.

Fig 3.1   Different states of a pure substance during heating under
constant pressure.

called *saturated vapor* (State 4). Note that any heat loss from a saturated
vapor results in condensation. Further transfer of heat to a saturated vapor
causes an increase in both the temperature and specific volume of the vapor
and it is called a *superheated vapor* (State 5).

The heating process of water at constant pressure is represented in a
T-V diagram in Fig. 3.1-b. State 1 is the initial subcooled liquid (or
compressed liquid) state. A *saturated liquid* is represented by state 2. A
saturated liquid is ready to boil, i.e., addition of heat causes liquid to
boil, and $T = 105.99\,°C$ represents the *boiling point temperature*. State 4
represents a saturated vapor. A saturated vapor is ready to condense, i.e.,
removal of heat causes vapor to condense, and $T = 105.99\,°C$ represents
the *dew point temperature*[1]. The horizontal line joining states 2 and 4
represents a process (both isobaric and isothermal) in which phase change

---

[1]For a pure substance, the boiling point and dew point temperatures are the same.
However, for mixtures these two temperatures are different from each other.

from liquid to vapor, or vice versa, takes place. During phase change the liquid and vapor phases are in equilibrium with each other and, as a result, both temperature and pressure remain constant. The line joining states 4 and 5 represents the process in which the steam is superheated at constant pressure.

At a given pressure, the temperature at which a pure substance boils is called the *saturation temperature*, T^{sat}. On the other hand, at a given temperature, the pressure at which a pure substance boils is known as the *vapor* (or *saturation*) *pressure*, P^{vap}. The boiling of a pure component starts when its vapor pressure equals the ambient pressure. This is the reason why water boils at a temperature less than $100\,^{\circ}$C at the top of a mountain.

3.2 PVT Diagrams for Pure Substances

3.2.1 *Temperature-volume (T-V) diagram*

To develop a T-V diagram for water, we repeat the heating process described in Section 3.1 at different pressures and the resulting curves are given in Fig. 3.2-a. Note that as the pressure increases while the specific volume of saturated liquid increases, the specific volume of the saturated vapor decreases. Thus, the horizontal line connecting the saturated liquid and saturated vapor states becomes shorter as pressure increases. At a pressure of 22.09 MPa, however, the horizontal line between the saturated liquid and vapor states shrinks to a point at which the constant pressure line has a point of inflection with a zero slope. This point is called the *critical point*.

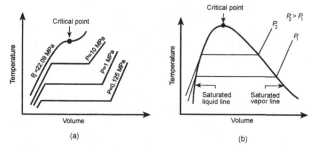

Fig. 3.2   Temperature-volume diagram for water.

At the critical point, the saturated liquid and saturated vapor states shrink to a point, i.e., they are identical. The temperature, pressure, and specific volume at the critical point are called the critical temperature, critical pressure, and critical volume, respectively. Above the critical point, liquid and vapor are indistinguishable from each other.

The saturated liquid states in Fig. 3.2-a can be connected by a line called the *saturated liquid line*. Saturated vapor states can also be connected by a line called the *saturated vapor line*. These two lines meet at the critical point, forming a dome as shown in Fig. 3.2-b.

Note that all the subcooled (or compressed) liquid states are located in the region to the left of the saturated liquid line, and this is called the *subcooled (or compressed) liquid region*. All the superheated vapor states are located to the right of the saturated vapor line, and this is called the *superheated vapor region*. In these two regions, which are located outside of the dome, a pure substance exists in a single phase, either liquid or vapor. The region located under the dome is called the *saturated liquid-vapor mixture region* (or the *wet region*) in which the liquid and vapor phases are in equilibrium.

3.2.2   *Pressure-volume (P-V) diagram*

The general shape of the P-V diagram of a pure substance is very much like that of the T-V diagram, but the constant temperature lines, i.e., isotherms, on this diagram have a downward trend as shown in Fig. 3.3.

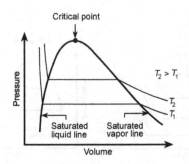

Fig. 3.3   Pressure-volume diagram of a pure substance.

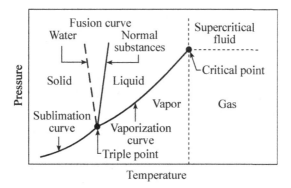

Fig. 3.4   Phase diagram of a pure substance.

3.2.3   *Pressure-temperature (P-T) diagram*

A pure substance may exist as a solid, liquid, or vapor. The P-T diagram (or the *phase diagram*) is a graphical way of showing the effects of pressure and temperature on the phase of a pure substance as shown in Fig. 3.4. Note that the three phases are separated from each other by three lines.

The curve separating the solid and vapor phases is called the *sublimation curve*. Along the sublimation curve the solid and vapor phases are in equilibrium. The slope of the sublimation curve gives the rate of change of sublimation (or vapor) pressure of a solid with temperature.

The curve separating the solid and liquid phases is called the *fusion* (or *melting*) *curve*. Along the fusion curve the solid and liquid phases are in equilibrium. The slope of the fusion curve gives the rate of change of the melting (or freezing) pressure of a solid with temperature. While the fusion curve has a positive slope for most substances, the slope becomes negative for water.

The curve separating the liquid and vapor phases is called the *vaporization curve*. Along the vaporization curve the vapor and liquid phases are in equilibrium. The slope of the vaporization curve gives the rate of change of vapor pressure of liquid with temperature. The vaporization curve ends at the *critical temperature and pressure* of the substance. At temperatures and pressures higher than the critical values, substances exist in the fluid (or supercritical) region and are called *supercritical fluids*. They possess both the gaseous properties (viscosity, diffusivity, surface tension) of being able to diffuse into substances easily, and the liquid property (density) of being able to dissolve substances.

When $P < P_c$, a substance in the gaseous state is called either a *gas* ($T > T_c$) or a *vapor* ($T < T_c$). Under isothermal conditions, while a vapor can be liquefied by exerting pressure, a gas cannot be liquefied no matter what pressure is applied to it. In other words, a pure gas cannot be liquefied at temperatures above its critical temperature no matter what pressure is imposed on it. This is the reason why N_2 ($T_c = -147\,°C$) cannot be liquefied at room temperature.

The *triple point* is the only point on the phase diagram where the solid, liquid, and vapor phases coexist in equilibrium[2]. In other words, it is the intersection of the liquid-vapor (vapor pressure curve), solid-liquid (fusion or melting curve), and solid-vapor (sublimation pressure curve) coexistence curves. Note that the number of degrees of freedom, $\mathcal{F}$, is zero at the triple point. If the triple point pressure is less than 1 atm, as in the case of water ($P_t = 0.006\,atm$), it is possible to have all three phases of a substance under atmospheric pressure depending on temperature. If the triple point pressure is higher than 1 atm, as in the case of carbon dioxide ($P_t = 5.1172\,atm$), then a substance cannot exist in the liquid form under atmospheric pressure, and the transition from the solid to vapor form, i.e., sublimation, takes place with an increase in temperature.

Since the fusion curve generally has a very steep slope, the triple point temperature for most substances is close to their melting (or freezing) temperature at atmospheric pressure, known as the *normal melting (or freezing) point*.

Example 3.1 *Sketch the following processes in the P-V and P-T diagrams:*

a) *A superheated vapor is cooled at constant pressure until liquid just begins to form,*
b) *A superheated vapor is cooled at constant pressure until all the vapor is gone,*
c) *A liquid-vapor two-phase mixture is heated at constant volume until all the liquid is vaporized.*

---

[2]In general, the triple point is the point of intersection of three different phases. If a substance exists in different forms of solid, e.g., graphite and diamond for carbon, it can have more than one triple point.

Solution

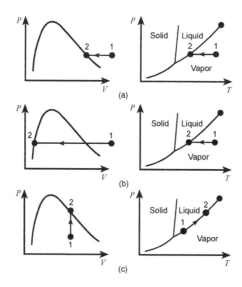

3.3 Property Tables

Some thermodynamic properties can be measured easily, but others cannot be measured directly and are calculated by using the relations that relate them to measurable properties. The results of these measurements and calculations are generally presented in the form of tables. For example, the "Steam Tables" given in Appendix A are a tabulation of the thermodynamic properties for water in liquid and vapor phases.

Some of the websites providing property tables for steam and other substances are given below:

- **Thermophysical properties of fluid systems**
  webbook.nist.gov/chemistry/fluid

- **Ammonia**
  www.ammonia-properties.com
  www.peacesoftware.de/einigewerte/nh3_e.html

- **Carbon dioxide**
  www.carbon-dioxide-properties.com
  www.peacesoftware.de/einigewerte/co2_e.html

- **Methane**

  www.peacesoftware.de/einigewerte/methan_e.html

- **Nitrogen**

  www.peacesoftware.de/einigewerte/stickstoff_e.html

- **Oxygen**

  www.peacesoftware.de/einigewerte/o2_e.html

- **Steam**

  www.steamtablesonline.com

  www.wolframalpha.com/examples/SteamTables.html

  www.efunda.com/materials/water/steamtable_general.cfm

  www.peacesoftware.de/einigewerte/wasser_dampf_e.html

Before going into the discussion of how to use property tables to specify the state of a system, definition of a new thermodynamic property, called *enthalpy*, is necessary.

3.3.1 *Enthalpy*

Enthalpy, H, is defined by

$$\boxed{H = U + PV}$$
(3.3-1)

In an open system, enthalpy can be interpreted as the amount of energy transferred across a system boundary by a moving flow. The term PV represents the flow work[3] and has the units of energy. Thus, H also has the units of energy. Since U, P, and V are all state functions, any combination of them must also be a state function. Therefore, enthalpy is a state function. It is also an extensive property. Note that enthalpy per unit mass, $\widehat{H}$, and enthalpy per unit mole, $\widetilde{H}$, are defined as

$$\widehat{H} = \widehat{U} + P\widehat{V}$$
(3.3-2)

$$\widetilde{H} = \widetilde{U} + P\widetilde{V}$$
(3.3-3)

3.3.2 *Saturated liquid and saturated vapor*

The properties of saturated liquid and saturated vapor are listed in Appendix A in two different tables: *Table A.1 - Saturated Water: Temperature Table* and *Table A.2 - Saturated Water: Pressure Table*. Both tables give the same information. The only difference is that the properties are listed

---

[3]Flow work will be explained in Section 4.2.

under temperature in Table A.1 and under pressure in Table A.2. Therefore, it is more convenient to use Table A.1 when temperature is given and Table A.2 when pressure is given.

The superscript L will be used to denote the properties of a saturated liquid and the superscript V to denote the properties of saturated vapor. The difference between the saturated vapor and saturated liquid states are designated by Δ, i.e.,

$$\Delta\widehat{\varphi} = \widehat{\varphi}^V - \widehat{\varphi}^L \qquad \text{where} \qquad \widehat{\varphi} = \widehat{V}, \widehat{U}, \widehat{H} \qquad (3.3\text{-}4)$$

Example 3.2 *A rigid tank of* $0.1\,\mathrm{m}^3$ *volume contains saturated water vapor at* $7\,\mathrm{MPa}$. *Determine the temperature and mass of vapor in the tank.*

Solution

System: *Contents of the tank*

From Table A.2 in Appendix A, $T^{sat} = 285.88\,°\mathrm{C}$ *and* $\widehat{V}^V = 0.02737\,\mathrm{m}^3/\mathrm{kg}$. *The mass of saturated water vapor is calculated from*

$$m = \frac{V}{\widehat{V}} = \frac{0.1}{0.02737} = 3.654\,\mathrm{kg}$$

While saturated liquid is in equilibrium with its own vapor, saturated vapor is in equilibrium with its own liquid. Thus, application of Gibbs phase rule gives

$$2 + \mathcal{F} = 1 + 2 \qquad \Rightarrow \qquad \mathcal{F} = 1$$

Therefore, to specify the state of a saturated liquid or vapor, it is necessary to know only one independent intensive property.

3.3.3 Vapor-liquid mixture

During a vaporization or condensation process, a substance exists as partly liquid and partly vapor. To analyze this mixture properly, one needs to know the proportions of the liquid and vapor phases in the mixture. The term *quality*, x, is defined as the mass fraction of a vapor in a vapor-liquid mixture. The specific property of any quantity, $\widehat{\varphi}$, is defined as

$$\widehat{\varphi} = (1 - x)\,\widehat{\varphi}^L + x\,\widehat{\varphi}^V = \widehat{\varphi}^L + x\left(\widehat{\varphi}^V - \widehat{\varphi}^L\right) \qquad (3.3\text{-}5)$$

or

$$\boxed{\widehat{\varphi} = \widehat{\varphi}^L + x \, \Delta\widehat{\varphi}} \qquad \text{Two-phase mixture} \qquad (3.3\text{-}6)$$

Since $0 \le x \le 1$, for any specific property of a two-phase mixture we have

$$\boxed{\widehat{\varphi}^L \le \widehat{\varphi} \le \widehat{\varphi}^V} \qquad \text{Two-phase mixture} \qquad (3.3\text{-}7)$$

While the quality of a saturated liquid is 0, the quality of a saturated vapor is 1. It is important to note that quality has no meaning in the subcooled (or compressed) liquid and superheated vapor regions.

In the two-phase region, the liquid and vapor phases are in equilibrium with each other, i.e., the phases are at the same temperature and pressure. However, the other specific properties ($\widehat{V}$, $\widehat{U}$, $\widehat{H}$) of the liquid and vapor phases are different from each other. As the quality changes from 0 to 1, the specific properties of the mixture will also change. As a result, specification of a two-phase system requires one more independent intensive property besides T (or P). In other words, specification of any two of the $\widehat{V}$, $\widehat{U}$, $\widehat{H}$, x, and T (or P) is necessary to specify the state of a two-phase mixture.

Example 3.3 *Calculate the quality of steam with an internal energy of* 1200 kJ/ kg *at* 50 kPa.

Solution

From Table A.2 in Appendix A

$$P = 50\,\text{kPa} \left\}\begin{array}{l} \widehat{U}^L = 340.44\,\text{kJ/ kg} \\ \Delta\widehat{U} = 2143.4\,\text{kJ/ kg} \end{array}\right.$$

The use of Eq. (3.3-6) gives

$$x = \frac{\widehat{U} - \widehat{U}^L}{\Delta\widehat{U}} = \frac{1200 - 340.44}{2143.4} = 0.4$$

Comment: *A mixture of steam and liquid water is also known as "wet steam".*

Example 3.4 *A rigid tank of* 0.3 m³ *volume contains a liquid water and water vapor mixture in equilibrium at* 700 kPa. *If the mass of the mixture is* 1.5 kg, *calculate:*

a) *The mass and volume of liquid,*
b) *The mass and volume of vapor,*
c) *The total internal energy of the mixture.*

Solution

From Table A.2 in Appendix A

$$P = 700\,\text{kPa} \left.\begin{array}{l}\\ \\ \\ \end{array}\right\} \begin{array}{ll} \widehat{V}^L = 0.001108\,\text{m}^3/\text{kg} & \widehat{U}^L = 696.44\,\text{kJ}/\text{kg} \\ \widehat{V}^V = 0.2729\,\text{m}^3/\text{kg} & \Delta\widehat{U} = 1876.1\,\text{kJ}/\text{kg} \\ & \widehat{U}^V = 2572.5\,\text{kJ}/\text{kg} \end{array}$$

System: *Two-phase mixture at* 700 kPa

a) *Specific volume of the mixture is given by*

$$\widehat{V} = \frac{\text{Total volume}}{\text{Total mass}} = \frac{0.3}{1.5} = 0.2\,\text{m}^3/\text{kg}$$

The use of Eq. (3.3-6) gives the quality of the mixture as

$$x = \frac{\widehat{V} - \widehat{V}^L}{\Delta\widehat{V}} = \frac{0.2 - 0.001108}{0.2729 - 0.001108} = 0.732$$

Thus,

$$\text{Mass of liquid} = (1 - 0.732)(1.5) = 0.402\,\text{kg}$$
$$\text{Volume of liquid} = (0.402)(0.001108) = 0.0004\,\text{m}^3$$

b) *The mass and the volume of vapor are*

$$\text{Mass of vapor} = (0.732)(1.5) = 1.098\,\text{kg}$$
$$\text{Volume of vapor} = (1.098)(0.2729) = 0.2996\,\text{m}^3$$

Alternatively,

$$\text{Mass of vapor} = 1.5 - 0.402 = 1.098\,\text{kg}$$
$$\text{Volume of vapor} = 0.3 - 0.0004 = 0.2996\,\text{m}^3$$

c) *The use of Eq. (3.3-6) gives*

$$\widehat{U} = \widehat{U}^L + x\,\Delta\widehat{U} = 696.44 + (0.732)(1876.1) = 2069.75\,\text{kJ}/\text{kg}$$

Thus, the total internal energy is

$$U = m\,\widehat{U} = (1.5)(2069.75) = 3104.63\,\text{kJ}$$

Alternatively,

$$U = m_L \widehat{U}^L + m_V \widehat{U}^V = (0.402)\,(696.43) + (1.098)(2572.5) = 3104.57\,\text{kJ}$$

3.3.4 *Superheated vapor*

In the region to the right of the saturated vapor line, a pure substance exists as superheated vapor. In this region

$$P < P^{vap} \text{ at a given temperature}$$
$$T > T^{sat} \text{ at a given pressure}$$

as shown in Fig. 3.5. The *degrees of superheat* is defined as the temperature in excess of the saturation temperature at a given pressure, i.e.,

$$\boxed{\text{Degrees of superheat} = T - T^{sat}} \qquad (3.3\text{-}8)$$

The properties of superheated vapor are tabulated in Appendix A as *Table A.3 - Superheated Water Vapor.* For a superheated vapor, application of Gibbs phase rule gives the number of degrees of freedom as

$$1 + \mathcal{F} = 1 + 2 \qquad \Rightarrow \qquad \mathcal{F} = 2$$

Therefore, two independent intensive properties are needed to specify the state of a superheated vapor.

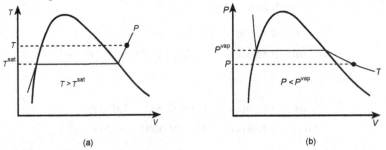

Fig.3.5 Representation of a superheated vapor in T-V and P-V diagrams.

Example 3.5 *Calculate the degrees of superheat and the internal energy for steam at* 3 MPa *and* 300 °C.

Solution

From Table A.2 in Appendix A, $T^{sat} = 233.90\,°C$, and from Table A.3 in Appendix A, $\widehat{U} = 2750.1\,kJ/\,kg$. The use of Eq. (3.3-8) gives the degrees of superheat as

$$\text{Degrees of superheat} = T - T^{sat} = 300 - 233.9 = 66.1\,°C$$

3.3.5 Compressed (subcooled) liquid

In the region to the left of the saturated liquid line, a pure substance exists as compressed (or subcooled) liquid. In this region

$$P > P^{vap} \text{ at a given temperature (compressed liquid)}$$
$$T < T^{sat} \text{ at a given pressure (subcooled liquid)}$$

as shown in Fig. 3.6. *Compressed liquid* implies that the pressure is greater than the vapor pressure for a given temperature. On the other hand, *subcooled liquid* implies that the temperature is lower than the saturation temperature for the given pressure.

The properties of compressed liquid are tabulated in Appendix A as *Table A.4 - Compressed Liquid Water.* As an approximation, properties of compressed liquid are equal to those of a saturated liquid at the given temperature, i.e.,

$$\boxed{\widehat{\varphi}(T, P) \simeq \widehat{\varphi}^{L}(T, P^{vap})}$$ For a compressed liquid   (3.3-9)

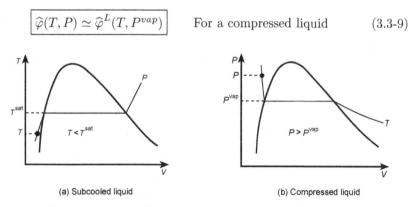

(a) Subcooled liquid   (b) Compressed liquid

Fig. 3.6   Representation of a subcooled or compressed liquid in T-V and P-V diagrams.

For a compressed liquid, application of Gibbs phase rule, Eq. (1.2-4), gives the number of degrees of freedom as

$$1 + \mathcal{F} = 1 + 2 \qquad \Rightarrow \qquad \mathcal{F} = 2$$

Therefore, two independent intensive properties are required to specify the state of a compressed (or subcooled) liquid.

Example 3.6 *Determine the specific volume and internal energy of a subcooled liquid at* $15\,\mathrm{MPa}$ *and* $100\,^\circ\mathrm{C}$.

Solution

From Table A.4 in Appendix A, $\widehat{V} = 0.0010361\,\mathrm{m}^3/\mathrm{kg}$ *and* $\widehat{U} = 414.74\,\mathrm{kJ}/\mathrm{kg}$.

As an approximation, we can also read the values of $\widehat{V}^L$ *and* $\widehat{U}^L$ *at* $100\,^\circ\mathrm{C}$. *From Table A.1 in Appendix A,* $\widehat{V}^L = 0.001044\,\mathrm{m}^3/\mathrm{kg}$ *and* $\widehat{U}^L = 418.94\,\mathrm{kJ}/\mathrm{kg}$.

3.3.6 Interpolation of the values in property tables

When the state of a system does not fall exactly at a value reported in Steam Tables, then interpolation is required between the two adjacent properties in a given table. For example, let us assume that the specific properties $\widehat{\varphi}_1$ and $\widehat{\varphi}_2$ are given at temperatures T_1 and T_2, respectively. To determine the value of $\widehat{\varphi}^*$ at any temperature between T_1 and T_2, we assume linear variation of $\widehat{\varphi}$ with respect to temperature between T_1 and T_2 as shown in Fig. 3.7.

The slope of the straight line is

$$\frac{d\widehat{\varphi}}{dT} = \frac{\widehat{\varphi}_2 - \widehat{\varphi}_1}{T_2 - T_1} = \text{constant } (C) \tag{3.3-10}$$

Integration of Eq. (3.3-10) from $\widehat{\varphi}_1$ to $\widehat{\varphi}^*$ gives

$$\int_{\widehat{\varphi}_1}^{\widehat{\varphi}^*} d\widehat{\varphi} = C \int_{T_1}^{T^*} dT \qquad \Rightarrow \qquad \widehat{\varphi}^* = \widehat{\varphi}_1 + C\,(T^* - T_1) \tag{3.3-11}$$

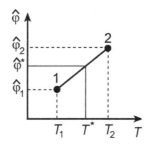

Fig. 3.7 Linear variation of a specific property $\widehat{\varphi}$ with respect to temperature between T_1 and T_2.

It is also possible to integrate Eq. (3.3-10) from $\widehat{\varphi}^*$ to $\widehat{\varphi}_2$ with the result

$$\int_{\widehat{\varphi}^*}^{\widehat{\varphi}_2} d\widehat{\varphi} = C \int_{T^*}^{T_2} dT \quad \Rightarrow \quad \widehat{\varphi}^* = \widehat{\varphi}_2 - C\,(T_2 - T^*) \qquad (3.3\text{-}12)$$

Example 3.7 *Calculate $\widehat{U}$ of steam at 800 kPa and 237 °C.*

Solution

From Table A.3 in Appendix A, at 800 kPa (0.8 MPa)

$T\ (°C)$	$\widehat{U}\ (kJ/kg)$
200	2630.6
250	2715.5

Assuming a linear variation of $\widehat{U}$ between 200 °C and 250 °C, the change in $\widehat{U}$ with respect to temperature is

$$\frac{d\widehat{U}}{dT} = \frac{2715.5 - 2630.6}{250 - 200} = 1.698\,\text{kJ/kg.°C}$$

Upon integration we obtain the value of $\widehat{U}$ at the desired temperature:

$$\int_{2630.6}^{\widehat{U}} d\widehat{U} = 1.698 \int_{200}^{237} dT \Rightarrow \widehat{U} = 2630.6 + 1.698\,(237 - 200) = 2693.4\,\text{kJ/kg}$$

or

$$\int_{\widehat{U}}^{2715.5} d\widehat{U} = 1.698 \int_{237}^{250} dT \Rightarrow \widehat{U} = 2715.5 - 1.698\,(250 - 237) = 2693.4\,\text{kJ/kg}$$

Example 3.8 *Calculate the specific enthalpy, $\widehat{H}$, of steam at 1.08 MPa and 315 °C.*

Solution

Since neither T nor P falls exactly at a value reported in the Steam Table, double interpolation is necessary in this specific case. From Table A.3 in Appendix A

$$\text{at } 1\,\text{MPa} \begin{cases} T = 300\,°\text{C} \quad \widehat{H} = 3051.2\,\text{kJ/kg} \\ T = 350\,°\text{C} \quad \widehat{H} = 3157.7\,\text{kJ/kg} \end{cases}$$

$$\text{at } 1.2\,\text{MPa} \begin{cases} T = 300\,°\text{C} \quad \widehat{H} = 3045.8\,\text{kJ/kg} \\ T = 350\,°\text{C} \quad \widehat{H} = 3153.6\,\text{kJ/kg} \end{cases}$$

The value of the specific enthalpy can be calculated by two different methods.

Method I: *First keep T constant and calculate $\widehat{H}$ values at the desired pressure of 1.08 MPa:*

$$P = 1.08\,\text{MPa} \,\&\, T = 300\,°\text{C} \quad \widehat{H} = 3051.2 + \left(\frac{3045.8 - 3051.2}{1.2 - 1} \right)(1.08 - 1)$$
$$= 3049.04\,\text{kJ/kg}$$

$$P = 1.08\,\text{MPa} \,\&\, T = 350\,°\text{C} \quad \widehat{H} = 3157.7 + \left(\frac{3153.6 - 3157.7}{1.2 - 1} \right)(1.08 - 1)$$
$$= 3156.06\,\text{kJ/kg}$$

Then calculate $\widehat{H}$ at 1.08 MPa and 315 °C:

$$\widehat{H} = 3049.04 + \left(\frac{3156.06 - 3049.04}{350 - 300} \right)(315 - 300) = 3081.15\,\text{kJ/kg}$$

Method II: *First keep P constant and calculate $\widehat{H}$ values at the desired temperature of 315 °C:*

$$T = 315\,°\text{C} \,\&\, P = 1\,\text{MPa} \quad \widehat{H} = 3051.2 + \left(\frac{3157.7 - 3051.2}{350 - 300} \right)(315 - 300)$$
$$= 3083.15\,\text{kJ/kg}$$

$$T = 315\,°\text{C} \,\&\, P = 1.2\,\text{MPa} \quad \widehat{H} = 3045.8 + \left(\frac{3153.6 - 3045.8}{350 - 300} \right)(315 - 300)$$
$$= 3078.14\,\text{kJ/kg}$$

Then calculate $\widehat{H}$ *at* 315 °C *and* 1.08 MPa:

$$\widehat{H} = 3083.15 + \left(\frac{3078.14 - 3083.15}{1.2 - 1}\right)(1.08 - 1) = 3081.15 \, \text{kJ/kg}$$

Example 3.9 *The specific volume and the specific enthalpy of steam are* 0.0509 m³/kg *and* 3593.8 kJ/kg, *respectively. Estimate its temperature and pressure.*

Solution

From Table A.3 in Appendix A, at 7 MPa

T (°C)	$\widehat{V}$ (m³/kg)	$\widehat{H}$ (kJ/kg)
500	0.04814	3410.3
550	0.05195	3530.9
600	0.05565	3650.3

At 8 MPa

T (°C)	$\widehat{V}$ (m³/kg)	$\widehat{H}$ (kJ/kg)
550	0.04516	3521.0
600	0.04845	3642.0
700	0.05481	3882.4

$\widehat{H}$ *versus* $\widehat{V}$ *plots at two different pressures are shown in the following graph.*

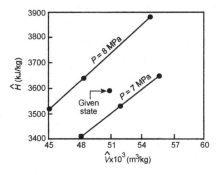

First, let us calculate T *and* $\widehat{V}$ *values corresponding to* $\widehat{H} = 3593.8 \, \text{kJ/kg}$ *at two different pressures:*

At 7 MPa

$$T = 550 + \left(\frac{600 - 550}{3650.3 - 3530.9} \right) (3593.8 - 3530.9) = 576.3\,^\circ\mathrm{C}$$

$$\widehat{V} = 0.05195 + \left(\frac{0.05565 - 0.05195}{3650.3 - 3530.9} \right) (3593.8 - 3530.9) = 0.0539\,\mathrm{m}^3/\mathrm{kg}$$

At 8 MPa

$$T = 550 + \left(\frac{600 - 550}{3642 - 3521} \right) (3593.8 - 3521) = 580.1\,^\circ\mathrm{C}$$

$$\widehat{V} = 0.04516 + \left(\frac{0.04845 - 0.04516}{3642 - 3521} \right) (3593.8 - 3521) = 0.04714\,\mathrm{m}^3/\mathrm{kg}$$

Therefore, the values of P *and* T *corresponding to* $\widehat{V} = 0.0509\,\mathrm{m}^3/\mathrm{kg}$ *are*

$$P = 7 + \left(\frac{8 - 7}{0.04714 - 0.0539} \right) (0.0509 - 0.0539) = 7.44\,\mathrm{MPa}$$

$$T = 576.3 + \left(\frac{580.1 - 576.3}{0.04714 - 0.0539} \right) (0.0509 - 0.0539) = 578\,^\circ\mathrm{C}$$

3.4 Examples

The following examples demonstrate the use of steam tables in the solution of thermodynamics problems. Always keep in mind that while one independent intensive property is needed to specify the state of saturated liquid and saturated vapor, two independent intensive properties are required to specify the state of two-phase mixture, compressed (or subcooled) liquid, and superheated vapor.

Example 3.10 *A rigid tank contains water vapor at* 200 °C *and at an unknown pressure. When the tank is cooled to* 150 °C, *the vapor starts condensing. Estimate the initial pressure in the tank.*

Solution

System: *Contents of the rigid tank*

Let 1 and 2 be the initial and final states, respectively. Since the vapor starts condensing at $150\,^\circ$C, this implies that the vapor is saturated at state 2. The process is shown in the following P-V diagram:

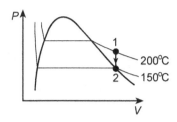

From Table A.1 in Appendix A

$$\text{at } 150\,^\circ\text{C} \qquad \widehat{V}^V = 0.3928\,\text{m}^3/\text{kg}$$

From Table A.3 in Appendix A

$$\text{at } 200\,^\circ\text{C} \left\{ \begin{array}{l} P = 0.5\,\text{MPa} \quad \widehat{V} = 0.4249\,\text{m}^3/\text{kg} \\ P = 0.6\,\text{MPa} \quad \widehat{V} = 0.3520\,\text{m}^3/\text{kg} \end{array} \right.$$

Using linear interpolation

$$P = 0.5 + \left(\frac{0.6 - 0.5}{0.3520 - 0.4249} \right) (0.3928 - 0.4249) = 0.544\,\text{MPa}$$

Example 3.11 *The cylinder shown in the figure below contains 0.085 kg of wet steam at $37\,^\circ$C. The frictionless piston has a mass of 100 kg and a cross-sectional area of $400\,\text{cm}^2$, and is resting on the stops. The volume at this point is $0.018\,\text{m}^3$. Atmospheric pressure outside is 100 kPa. Heat is now transferred to the system until the cylinder contains saturated vapor.*

a) *What is the temperature of the water when the piston first rises from the stops?*

b) *Calculate the work done for the overall process.*

Solution

a) System: *Contents of the cylinder*

The pressure, P, exerted by the surrounding atmosphere and the piston weight is

$$P = 1 \times 10^5 + \frac{(100)(9.8)}{400 \times 10^{-4}} = 124,500 \, \text{Pa} \, (0.125 \, \text{MPa})$$

To locate the state of the system, the specific volume should be compared with $\widehat{V}^L$ and $\widehat{V}^V$. From Table A.2 in Appendix A, at 0.125 MPa

$$T^{sat} = 105.99\,^\circ\text{C} \qquad \widehat{V}^L = 0.001048 \, \text{m}^3/\text{kg} \qquad \widehat{V}^V = 1.3749 \, \text{m}^3/\text{kg}$$

When the piston first rises from the stops, the specific volume is

$$\widehat{V} = \frac{\text{Total volume}}{\text{Total mass}} = \frac{0.018}{0.085} = 0.212 \, \text{m}^3/\text{kg}$$

Since $\widehat{V}^L \leq \widehat{V} \leq \widehat{V}^V$, both liquid and vapor are present. Thus,

$$T = T^{sat} \simeq 106\,^\circ\text{C}$$

b) *The volume at the final state is*

$$V = (0.085)(1.3749) = 0.1169 \, \text{m}^3$$

When the piston starts to rise, the process is a reversible isobaric process. Therefore the work done by the system is

$$W = -\int_{V_1}^{V_2} P \, dV = -P \, \Delta V = -(125)(0.1169 - 0.018) = -12.4 \, \text{kJ}$$

The overall process path is shown in the figure below:

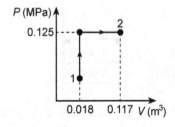

Problems

Problems related to Section 3.3

3.1   Using the Steam Tables, calculate the following:

a) Pressure at $900\,°C$ for which the specific volume is the same as at $12.5\,MPa$ and $550\,°C$.

b) Temperature at $0.3\,MPa$ for which the internal energy is the same as at $5\,MPa$ and $450\,°C$.

c) Temperature at $30\,kPa$ for which the internal energy is the same as at $30\,MPa$ and $425\,°C$.

(**Answer:** a) $18.99\,MPa$, b) $420.7\,°C$, c) $69.10\,°C$)

3.2   Determine the state of water for the following conditions:

a) $T = 70\,°C$ and $\widehat{V} = 0.015\,m^3/kg$
b) $T = 70\,°C$ and $\widehat{V} = 0.80\,m^3/kg$
c) $P = 0.5\,MPa$ and $\widehat{V} = 0.40\,m^3/kg$
d) $P = 3\,MPa$ and $\widehat{U} = 1854.2\,kJ/kg$
e) $\widehat{U} = 4015.4\,kJ/kg$ and $\widehat{H} = 4596.6\,kJ/kg$
f) $\widehat{U} = 1585.6\,kJ/kg$ and $\widehat{H} = 1610.5\,kJ/kg$

3.3   The specific volume and the specific enthalpy of steam are $0.11913\,m^3/kg$ and $4289.4\,kJ/kg$, respectively. Estimate its temperature and pressure.

(**Answer:** $4.38\,MPa$, $862\,°C$)

3.4   A rigid tank contains a substance at its critical point. What is the phase of this substance when the tank is cooled?

3.5   Steam at $400\,kPa$ and $200\,°C$ is compressed isothermally until its quality becomes 90%. Calculate the change in enthalpy.

(**Answer:** $-261.42\,kJ/kg$)

3.6   A vertical cylinder fitted with a frictionless piston contains $0.2\,kg$ of water at $150\,°C$ with a volume of $0.123\,m^3$. It is then cooled to $30\,°C$.

a) What is the initial pressure?
b) What is the condition of water at the final state?

(**Answer:** a) $312\,kPa$)

3.7   A cylindrical tank fitted with a sight glass for the measurement of the liquid level contains a mixture of liquid water and water vapor in equilibrium with each other at $80\,°C$. Liquid water is slowly withdrawn from a valve placed at the bottom until the liquid level drops 20 cm. The temperature remains constant during the process. If the cross-sectional area of the tank is $0.1 \, m^2$, determine the mass of water withdrawn from the tank.

(**Answer:** 19.43048 kg)

3.8 A vertical cylinder fitted with a frictionless piston contains 1 kg of liquid water and 2 kg of water vapor. The mass of the piston is such that it maintains a constant pressure of 600 kPa inside the cylinder.

a) What is the initial temperature within the cylinder?
b) Calculate the change in volume when the contents of the tank are heated to 500 °C.

(**Answer:** b) $1.1435 \, m^3$)

3.9 A rigid tank of $0.04 \, m^3$ volume contains saturated water vapor at 375 kPa. Cooling coils placed in the tank reduce the temperature of the tank contents to 40 °C. Calculate:

a) The initial temperature and the mass of water in the tank,
b) The final pressure,
c) The mass of liquid water in the tank at the final conditions.

(**Answer:** a) 141.32 °C, 0.0814 kg, b) 7.384 kPa, c) 0.0812 kg)

3.10 A rigid tank of $0.8 \, m^3$ volume contains steam at 1.6 MPa and 250 °C. Cooling coils placed in the tank reduce the temperature of the tank contents.

a) Determine the temperature at which condensation starts.
b) Calculate the quality of steam when the pressure drops to 0.8 MPa.

(**Answer:** a) 195 °C b) 0.588)

3.11 Consider a cylinder fitted with a frictionless piston as shown in the figure below. The total volume of the cylinder is $0.8 \, m^3$ when the piston touches the stops. Initially, the cylinder contains wet steam at 150 °C and 20% quality. Under these conditions, the piston is at rest and the enclosed volume is $0.3 \, m^3$. The contents of the cylinder are then heated until the temperature reaches 400 °C. Determine:

a) Final pressure,
b) The work done by steam.

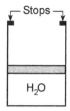

(**Answer:** a) 1.443 MPa, b) − 237.9 kJ)

3.12 A rigid tank of $0.5\,\text{m}^3$ volume contains wet steam at $80\,^\circ\text{C}$ and 1% quality. The tank is then heated until the water inside becomes saturated vapor. Determine the final temperature and pressure within the tank.

(**Answer:** $271\,^\circ\text{C}$, 5.588 MPa)

3.13 A vertical cylinder fitted with a frictionless piston initially contains steam at $200\,^\circ\text{C}$ with a volume of $0.05\,\text{m}^3$. The mass of the piston is such that it maintains a constant pressure of 0.8 MPa inside the cylinder. As a result of heat transfer from an external source, the temperature of steam is increased to $700\,^\circ\text{C}$. Determine the work for this process.

(**Answer:** − 45.9 kJ)

3.14 A vertical cylinder fitted with a piston contains $0.02\,\text{m}^3$ of wet steam at $180\,^\circ\text{C}$ and 90% quality. The piston is initially held in place by a pin. The piston has a mass of 100 kg and a cross-sectional area of $49\,\text{cm}^2$. The ambient pressure is 100 kPa. The pin is removed, allowing the piston to move. After a sufficient period of time, the system comes to equilibrium, with the final temperature being $180\,^\circ\text{C}$.

a) Determine the final pressure and volume of the steam,
b) Calculate the work done by the steam during this process.

(**Answer:** a) 300 kPa, $0.0782\,\text{m}^3$, b) − 17.46 kJ)

3.15 The cylinder shown in the figure below contains 5 kg of water at $80\,^\circ\text{C}$. The frictionless piston has a cross-sectional area of $0.3\,\text{m}^2$ and is resting on the stops. At this state, an ideal spring touches but exerts no force on the piston. Heat is now transferred to the water, causing the pressure to increase to 350 kPa, at which point the piston first rises from the stops. The water at this state is a saturated liquid. More heat is transferred to the water until its pressure and temperature reach 1.6 MPa and $450\,^\circ\text{C}$, respectively. Determine the work done by the water.

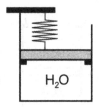

(**Answer:** $-995\,\mathrm{kJ}$)

3.16   The cylinder shown in the figure below contains 1 kg of wet steam at 950 kPa. The frictionless piston has a cross-sectional area of $0.5\,\mathrm{m}^2$. At this state, an ideal spring with a spring constant of $k = 120\,\mathrm{kN/m}$ touches but exerts no force on the piston. The water is then heated to expand to a final volume of $1.4\,\mathrm{m}^3$. Determine the work done by the water.

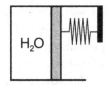

(**Answer:** $-418.4\,\mathrm{kJ}$)

3.17   A cylinder fitted with a piston contains 3 kg of steam at 10 MPa and 400 °C. It expands against a constant pressure of 2 MPa, until the forces balance. During the process, the piston generates $748,740\,\mathrm{J}$ of work. Determine the final temperature.

(**Answer:** $400\,°\mathrm{C}$)

3.18   A cylinder fitted with a piston contains 749 g of wet steam at 200 kPa. It is heated at constant pressure until the volume reaches $0.8\,\mathrm{m}^3$. The steam is then compressed isothermally until the first liquid droplet appears. Estimate the final steam pressure in the cylinder.

(**Answer:** $1.4\,\mathrm{MPa}$)

Chapter 4

The First Law of Thermodynamics

According to the first law of thermodynamics, energy is converted from one form to another and transferred from one system to another but its total is conserved. The purpose of this chapter is first to show how to formulate this statement mathematically for different types of systems, and then apply these equations to solve energy related engineering problems.

4.1 The First Law for a Closed System

As stated in Section 2.2.4, the energy associated with microscopic motions and forces is called an *internal energy*. Since this energy cannot be seen, it is usually separated from the macroscopic, i.e., measurable, mechanical energy so as to express the total energy of the system as

$$E = U + E_K + E_P \tag{4.1-1}$$

Consider a closed system receiving energy, in the form of heat and work, from its surroundings as shown in Fig. 4.1. The first law of thermodynamics states that the total energy of the universe is constant, i.e.,

$$E_{universe} = \text{constant} \tag{4.1-2}$$

In other words, the change in the total energy of the universe is zero:

$$\Delta E_{universe} = 0 \tag{4.1-3}$$

Since the universe is composed of the system and its surroundings, then Eq. (4.1-3) can be expressed in the form

$$\Delta E_{sys} + \Delta E_{surr} = 0 \tag{4.1-4}$$

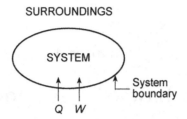

Fig. 4.1   A closed system receiving heat and work from the surroundings.

The increase in the total energy of the system is given by

$$\Delta E_{sys} = \Delta U + \Delta E_K + \Delta E_P \qquad (4.1\text{-}5)$$

On the other hand, the decrease in the total energy of the surroundings is given by

$$\Delta E_{surr} = -Q - W \qquad (4.1\text{-}6)$$

Substitution of Eqs. (4.1-5) and (4.1-6) into Eq. (4.1-4) gives

$$\boxed{\Delta U + \Delta E_K + \Delta E_P = Q + W} \qquad (4.1\text{-}7)$$

which is known as the *first law of thermodynamics for a closed system*. In differential form, Eq. (4.1-7) is expressed as[1]

$$dU + dE_K + dE_P = \delta Q + \delta W \qquad (4.1\text{-}8)$$

If the changes in kinetic and potential energies are negligible, then Eq. (4.1-7) reduces to

$$\boxed{\Delta U = Q + W} \qquad \text{Negligible changes in } E_K \text{ and } E_P \qquad (4.1\text{-}9)$$

The term W in Eqs. (4.1-7) and (4.1-9) includes expansion and non-expansion types of work. Expansion (or contraction) work is related to the change in the volume of the system. Non-expansion work, on the other hand, includes shaft work (work done on the system by a rotating mechanical device), chemical work, electrical work, etc.

Example 4.1 *When a system is taken from state 1 to state 2 along the path $\overline{1A2}$, it is found that $Q_{1A2} = 200\,\text{kJ}$ and $W_{1A2} = -70\,\text{kJ}$.*

---

[1]Since Q and W are path functions, these quantities in differential form are expressed as δQ and δW.

a) *If* $Q_{1B2} = 140\,\text{kJ}$ *along the path* $\overline{1B2}$, *what is* W_{1B2} *along the same path?*

b) *If* $W_{21} = 25\,\text{kJ}$ *for the diagonal return path* $\overline{21}$, *what is* Q_{21} *for this path?*

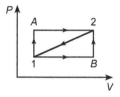

Solution

a)

$$\Delta U_{1A2} = Q_{1A2} + W_{1A2} = 200 - 70 = 130\,\text{kJ}$$

Since U is a state function, then

$$\Delta U_{1B2} = \Delta U_{1A2} = 130\,\text{kJ}$$

Application of Eq. (4.1-9) for the process path $\overline{1B2}$ *gives*

$$W_{1B2} = \Delta U_{1B2} - Q_{1B2} = 130 - 140 = -10\,\text{kJ}$$

b) *Note that*

$$\Delta U_{21} = -\Delta U_{1A2} = -130\,\text{kJ}$$

Application of Eq. (4.1-9) for the process path $\overline{21}$ *gives*

$$Q_{21} = \Delta U_{21} - W_{21} = -130 - 25 = -155\,\text{kJ}$$

Alternative solution: *Note that the process* $1 \to A \to 2 \to 1$ *is a cyclic process and Eq. (4.1-9) is written as*

$$\underbrace{\Delta U_{cycle}}_{0} = Q_{cycle} + W_{cycle}$$

or

$$0 = Q_{1A2} + Q_{21} + W_{1A2} + W_{21}$$

Therefore, Q_{21} *is calculated as*

$$Q_{21} = -Q_{1A2} - W_{1A2} - W_{21} = -200 + 70 - 25 = -155\,\text{kJ}$$

4.1.1 *Reversible processes in a closed system*

If changes in kinetic and potential energies are negligible, the first law of thermodynamics is expressed in differential form as

$$dU = \delta Q + \delta W \qquad (4.1\text{-}10)$$

For a reversible process, in the absence of shaft work, Eq. (4.1-10) becomes

$$dU = \delta Q - P\,dV \qquad (4.1\text{-}11)$$

Constant volume (isometric or isochoric) process

In this case Eq. (4.1-11) simplifies to

$$dU = \delta Q \qquad (4.1\text{-}12)$$

Integration of Eq. (4.1-12) gives

$$\boxed{Q = \Delta U} \qquad \text{Reversible constant volume process} \qquad (4.1\text{-}13)$$

For a constant volume process, Eq. (1.4-4) states that

$$\delta Q = m\widehat{C}_V\,dT \qquad (4.1\text{-}14)$$

Combination of Eqs. (4.1-12) and (4.1-14) gives

$$dU = m\widehat{C}_V\,dT \qquad (4.1\text{-}15)$$

Equation (4.1-15) implies that

$$\boxed{\widehat{C}_V = \left(\frac{\partial \widehat{U}}{\partial T}\right)_{\widehat{V}}} \qquad (4.1\text{-}16)$$

Integration of Eq. (4.1-15) results in

$$\boxed{\Delta U = U_{final} - U_{initial} = m\int_{T_{initial}}^{T_{final}} \widehat{C}_V\,dT} \qquad (4.1\text{-}17)$$

Equation (4.1-17) is developed for a closed system undergoing a reversible constant volume process. However, since U is a state function, this equation is applicable to all processes in which $V_1 = V_2$ as shown in Fig. 4.2-a.

If $\widehat{C}_V$ is independent of temperature, Eq. (4.1-17) reduces to

$$\boxed{\Delta U = m\widehat{C}_V\,\Delta T} \qquad \widehat{C}_V \neq \widehat{C}_V(T) \qquad (4.1\text{-}18)$$

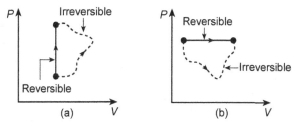

Fig. 4.2 (a) Eq. (4.1-17) applies to all processes in which $V_1 = V_2$.
(b) Eq. (4.1-25) applies to all processes in which $P_1 = P_2$.

Constant pressure (isobaric) process

Since pressure remains constant, Eq. (4.1-11) is expressed as

$$dU = \delta Q - d(PV) \tag{4.1-19}$$

or

$$\delta Q = dH \tag{4.1-20}$$

Integration of Eq. (4.1-20) gives

$$\boxed{Q = \Delta H} \qquad \text{Reversible isobaric process} \tag{4.1-21}$$

For a constant pressure process, Eq. (1.4-4) states that

$$\delta Q = m\widehat{C}_P \, dT \tag{4.1-22}$$

Combination of Eqs. (4.1-20) and (4.1-22) gives

$$dH = m\widehat{C}_P \, dT \tag{4.1-23}$$

Equation (4.1-23) implies that

$$\boxed{\widehat{C}_P = \left(\frac{\partial \widehat{H}}{\partial T}\right)_P} \tag{4.1-24}$$

Integration of Eq. (4.1-23) results in

$$\boxed{\Delta H = H_{final} - H_{initial} = m \int_{T_{initial}}^{T_{final}} \widehat{C}_P \, dT} \tag{4.1-25}$$

Equation (4.1-25) is developed for a closed system undergoing a reversible isobaric process. However, since H is a state function, this equation is applicable to all processes in which $P_1 = P_2$ as shown in Figure 4.2-b.

If $\widehat{C}_P$ is independent of temperature, Eq. (4.1-25) reduces to

$$\boxed{\Delta H = m\,\widehat{C}_P\,\Delta T} \qquad \widehat{C}_P \neq \widehat{C}_P(T) \tag{4.1-26}$$

4.1.2 *Internal energy and enthalpy change for liquids and solids*

Solids and liquids are generally considered *incompressible* substances, i.e., volume change with temperature and pressure is negligible. Thus, the change in internal energy can be calculated from Eq. (4.1-17), i.e.,

$$\boxed{\Delta U = m \int_{T_{initial}}^{T_{final}} \widehat{C}_V\,dT} \qquad \text{Liquids and solids} \tag{4.1-27}$$

Using the definition of enthalpy given by Eq. (3.3-1)

$$\Delta H = \Delta U + \Delta(PV) \tag{4.1-28}$$

Since V does not change, Eq. (4.1-28) becomes

$$\Delta H = \Delta U + V\,\Delta P \tag{4.1-29}$$

Substitution of Eq. (4.1-27) into Eq. (4.1-29) gives

$$\Delta H = m\left(\int_{T_{initial}}^{T_{final}} \widehat{C}_V\,dT + \widehat{V}\,\Delta P \right) \tag{4.1-30}$$

As stated in Section 1.4.1, $\widehat{C}_V \simeq \widehat{C}_P$ for liquids and solids. Moreover, the term $\widehat{V}\,\Delta P$ is insignificant for liquids and solids. Under these circumstances, one can conclude that for liquids and solids

$$\boxed{\Delta U = \Delta H = m \int_{T_{initial}}^{T_{final}} \widehat{C}_V\,dT = m \int_{T_{initial}}^{T_{final}} \widehat{C}_P\,dT} \tag{4.1-31}$$

For example, consider the following values for saturated liquid water from Table A.1 in Appendix A:

T (°C)	P (kPa)	$\widehat{V}^L$ (m^3/ kg)	$\widehat{U}^L$ (kJ/ kg)	$\widehat{H}^L$ (kJ/ kg)
20	2.339	0.001002	83.95	83.96
80	47.39	0.001029	334.86	334.91

The changes in internal energy and enthalpy per unit mass are

$$\Delta\widehat{U} = 334.86 - 83.95 = 250.91 \text{ kJ/ kg}$$

$$\Delta\widehat{H} = 334.91 - 83.96 = 250.95 \text{ kJ/ kg}$$

implying that $\Delta\widehat{U} \approx \Delta\widehat{H}$. Also note that the term $\Delta(P\widehat{V})$ is

$$\Delta(P\widehat{V}) = (47.39)(0.001029) - (2.339)(0.001002) = 0.046421 \text{ kJ/ kg}$$

indicating that its contribution is almost insignificant.

It is also possible to calculate $\Delta\widehat{U}$ and $\Delta\widehat{H}$ from Eq. (4.1-31). Taking $\widehat{C}_V = \widehat{C}_P = 4.2 \text{ kJ/ kg. K}$,

$$\Delta\widehat{U} = \Delta\widehat{H} = \widehat{C}_P \, \Delta T = (4.2)(80 - 20) = 252 \text{ kJ/ kg}$$

which is almost identical with the previously calculated values from the Steam Table[2].

Example 4.2 *A block of aluminum with a mass of* 3 kg *is cooled from* 75 °C *to* 30 °C. *Calculate the change in internal energy. For aluminum, heat capacity is given as* 0.9 kJ/ kg. K.

Solution

From Eq. (4.1-31)

$$\Delta U = (3)(0.9)(30 - 75) = -121.5 \text{ kJ}$$

Example 4.3 *An insulated copper tank initially contains* 4 kg *of water at* 300 K. *The mass of the tank is* 0.5 kg. *Also consider a* 0.2 kg *copper block with an initial temperature of* 400 K. *The heat capacities of water and copper are* 4200 J/ kg. K *and* 380 J/ kg. K, *respectively.*

---

[2]In engineering calculations, keep in mind that $1000 + 1$ or $1000 - 1$ is equal to 1000!

a) *If the copper block is immersed in the water and allowed to come to equilibrium, calculate the change in internal energy of the water.*

b) *If the copper block is dropped into the water from a height of 50 m and allowed to come to equilibrium, calculate the change in internal energy of the water. Assume no loss of water from the tank due to splashing.*

Solution

a) System: *Copper block + water + tank*

From Eq. (4.1-9)

$$\Delta U = \underbrace{Q}_{0} + \underbrace{W}_{0} \qquad \Rightarrow \qquad \Delta U_{Cu} + \Delta U_{H_2O} + \Delta U_{tank} = 0$$

Expressing internal energy changes in terms of temperature difference gives

$$(0.2)(380)(T_{final} - 400) + (4)(4200)(T_{final} - 300)$$
$$+ (0.5)(380)(T_{final} - 300) = 0$$

Solving for the final temperature yields $T_{final} = 300.445$ K. *Thus, the change in internal energy of the water is*

$$\Delta U_{H_2O} = (4)(4200)(300.445 - 300) = 7476 \text{ J}$$

b) System: *Copper block + water + tank*

From Eq. (4.1-7)

$$\Delta U + \underbrace{\Delta E_K}_{0} + \Delta E_P = \underbrace{Q}_{0} + \underbrace{W}_{0} \Rightarrow \Delta U_{Cu} + \Delta U_{H_2O} + \Delta U_{tank} + \Delta E_P = 0$$

Expressing internal energy changes in terms of temperature difference gives

$$(0.2)(380)(T_{final} - 400) + (4)(4200)(T_{final} - 300)$$
$$+ (0.5)(380)(T_{final} - 300) - (0.2)(9.8)(50) = 0$$

Solving for the final temperature yields $T_{final} = 300.451$ K. *Thus, the change in internal energy of the water is*

$$\Delta U_{H_2O} = (4)(4200)(300.451 - 300) = 7576.8 \text{ J}$$

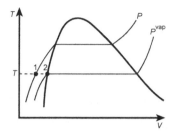

Fig. 4.3   Estimation of the enthalpy of a compressed liquid.

As stated in Section 3.3.5, the enthalpy of a compressed liquid can be approximated as the enthalpy of a saturated liquid at the given temperature, i.e., $\widehat{H}(T, P) \simeq \widehat{H}^L(T, P^{vap})$. We are now in a position to prove this statement. Consider a compressed liquid at a given temperature and pressure as shown in Fig. 4.3 (state 1). State 2 represents the saturated liquid at the given temperature. If an isothermal process path is followed in going from state 2 to state 1, from Eq. (4.1-31)

$$\Delta \widehat{H} = 0 \qquad \Rightarrow \qquad \widehat{H}(T, P) - \widehat{H}^L(T, P^{vap}) = 0 \qquad (4.1\text{-}32)$$

or

$$\boxed{\widehat{H}(T, P) \simeq \widehat{H}^L(T, P^{vap})} \quad \text{Compressed (subcooled) liquid} \qquad (4.1\text{-}33)$$

For example, from Table A.4 in Appendix A, the enthalpy of a compressed liquid at 5 MPa and 120 °C is 507.09 kJ/ kg. On the other hand, the enthalpy of a saturated liquid at 120 °C is 503.71 kJ/ kg from Table A.1 in Appendix A. As far as the accuracy of the engineering calculations is concerned, these two values are close enough to each other.

4.1.3   *Reversible processes involving an ideal gas*

An ideal gas is one that satisfies the following conditions:

• As stated in Section 1.2.3, the equation of state is given by

$$P\widetilde{V} = RT \qquad (4.1\text{-}34)$$

• The internal energy is dependent only on temperature[3].

---

[3]In Section 2.2.4, the internal energy is defined as the sum of kinetic and potential energies of molecules making up the substance. Potential energy is the energy of molecules interacting with other molecules, i.e., to make a chemical bond, or to attract/repel each

The enthalpy of an ideal gas is defined by

$$\widetilde{H}^{IG} = \widetilde{U}^{IG} + P\widetilde{V} \tag{4.1-35}$$

The use of Eq. (4.1-34) in Eq. (4.1-35) leads to

$$\boxed{\widetilde{H}^{IG} = \widetilde{U}^{IG} + RT} \tag{4.1-36}$$

indicating that enthalpy is also dependent only on temperature.

For a single-phase and single-component system, Gibbs phase rule, Eq. (1.2-4), gives the number of degrees of freedom as two. Hence, the state of such a system is specified by two independent intensive properties. The dependence of internal energy on temperature and volume, i.e.,

$$\widetilde{U} = \widetilde{U}(T, \widetilde{V}) \tag{4.1-37}$$

leads to a convenient relationship to use in calculating internal energy change[4]. The total differential of $\widetilde{U}$ is given by

$$d\widetilde{U} = \underbrace{\left(\frac{\partial \widetilde{U}}{\partial T}\right)_{\widetilde{V}}}_{\widetilde{C}_V} dT + \left(\frac{\partial \widetilde{U}}{\partial \widetilde{V}}\right)_T d\widetilde{V} \tag{4.1-38}$$

For an ideal gas, the internal energy is dependent only on temperature. Hence, $\widetilde{U} \neq \widetilde{U}(\widetilde{V})$ and Eq. (4.1-38) simplifies to

$$d\widetilde{U}^{IG} = \widetilde{C}_V^* \, dT \qquad \Rightarrow \qquad \boxed{\Delta\widetilde{U}^{IG} = \int_{T_{initial}}^{T_{final}} \widetilde{C}_V^* \, dT} \tag{4.1-39}$$

where $\widetilde{C}_V^*$ represents the *ideal gas heat capacity at constant volume*, i.e., $\widetilde{C}_V^* = d\widetilde{U}^{IG}/dT$. If $\widetilde{C}_V^*$ is independent of temperature, Eq. (4.1-39) reduces to

$$\boxed{\Delta\widetilde{U}^{IG} = \widetilde{C}_V^* \, \Delta T} \qquad \widehat{C}_V^* \neq \widehat{C}_V^*(T) \tag{4.1-40}$$

Keep in mind that Eqs. (4.1-39) and (4.1-40) are valid for an ideal gas regardless of what kind of process is considered.

---

other. For an ideal gas there are no interactions between molecules. As a result, potential energy is always zero for an ideal gas and the internal energy is simply the kinetic energy of molecules. Kinetic energy of an ideal monatomic gas, given by Eq. (1.3-15), indicates that $U = U(T)$.

[4]The functional form of internal energy can also be expressed as $\widetilde{U} = \widetilde{U}(P, T)$ or $\widetilde{U} = \widetilde{U}(P, \widetilde{V})$.

The dependence of enthalpy on temperature and pressure, i.e.,

$$\widetilde{H} = \widetilde{H}(T, P) \tag{4.1-41}$$

leads to a convenient relationship to use in calculating enthalpy change[5]. The total differential of $\widetilde{H}$ is given by

$$d\widetilde{H} = \underbrace{\left(\frac{\partial \widetilde{H}}{\partial T}\right)_P}_{\widetilde{C}_P} dT + \left(\frac{\partial \widetilde{H}}{\partial P}\right)_T dP \tag{4.1-42}$$

For an ideal gas, enthalpy is independent of pressure, i.e., $\widetilde{H} = \widetilde{H}(T)$ only, and Eq. (4.1-42) simplifies to

$$d\widetilde{H}^{IG} = \widetilde{C}_P^* \, dT \qquad \Rightarrow \qquad \boxed{\Delta \widetilde{H}^{IG} = \int_{T_{initial}}^{T_{final}} \widetilde{C}_P^* \, dT} \tag{4.1-43}$$

where $\widetilde{C}_P^*$ represents the *ideal gas heat capacity at constant pressure*, i.e., $\widetilde{C}_P^* = d\widetilde{H}^{IG}/dT$. If $\widetilde{C}_P^*$ is independent of temperature, Eq. (4.1-43) reduces to

$$\boxed{\Delta \widetilde{H}^{IG} = \widetilde{C}_P^* \, \Delta T} \qquad \widehat{C}_P^* \neq \widehat{C}_P^*(T) \tag{4.1-44}$$

which is valid for an ideal gas regardless of what kind of process is considered.

For an ideal gas, $\widetilde{C}_V^*$ and $\widetilde{C}_P^*$ values are not independent of each other. The relationship between these two heat capacities can be found by differentiation of Eq. (4.1-36) with respect to temperature as

$$\frac{d\widetilde{H}^{IG}}{dT} = \frac{d\widetilde{U}^{IG}}{dT} + R \tag{4.1-45}$$

or

$$\boxed{\widetilde{C}_P^* = \widetilde{C}_V^* + R} \tag{4.1-46}$$

---

[5]The functional form of enthalpy can also be expressed as $\widetilde{H} = \widetilde{H}(T, \widetilde{V})$ or $\widetilde{H} = \widetilde{H}(P, \widetilde{V})$.

The ratio of heat capacities, i.e., $\widetilde{C}_P^*/\widetilde{C}_V^*$, is denoted by γ and is a useful quantity in calculations involving an ideal gas. Note that

$$\boxed{\gamma = \frac{\widetilde{C}_P^*}{\widetilde{C}_V^*} = 1 + \frac{R}{\widetilde{C}_V^*}} \tag{4.1-47}$$

In the analysis of the following processes, heat capacities are considered to be independent of temperature.

Constant volume (isometric or isochoric) process

Changes in internal energy and enthalpy are given by

$$\Delta \widetilde{U}^{IG} = \widetilde{C}_V^* \, \Delta T \tag{4.1-48}$$

$$\Delta \widetilde{H}^{IG} = \widetilde{C}_P^* \, \Delta T \tag{4.1-49}$$

For a constant volume process, the work associated with the displacement of system boundaries is zero. Besides, if shaft work is zero or negligible, then

$$\widetilde{W} = 0 \tag{4.1-50}$$

The amount of heat transferred can be found from the application of the first law

$$\Delta \widetilde{U}^{IG} = \widetilde{Q} + \underbrace{\widetilde{W}}_{0} \tag{4.1-51}$$

or

$$\widetilde{Q} = \widetilde{C}_V^* \, \Delta T \tag{4.1-52}$$

Constant pressure (isobaric) process

Changes in internal energy and enthalpy are given by

$$\Delta \widetilde{U}^{IG} = \widetilde{C}_V^* \, \Delta T \tag{4.1-53}$$

$$\Delta \widetilde{H}^{IG} = \widetilde{C}_P^* \, \Delta T \tag{4.1-54}$$

The heat transferred is

$$\widetilde{Q} = \Delta \widetilde{H}^{IG} = \widetilde{C}_P^* \, \Delta T \tag{4.1-55}$$

The work done in a reversible process is given by

$$\widetilde{W} = -\int_{\tilde{V}_1}^{\tilde{V}_2} P\,d\tilde{V} = -P\,\Delta\tilde{V} \qquad (4.1\text{-}56)$$

Note that it is also possible to calculate $\widetilde{W}$ from the first law

$$\Delta\tilde{U}^{IG} = \tilde{Q} + \widetilde{W} \qquad (4.1\text{-}57)$$

or

$$\widetilde{W} = \tilde{C}_V^* \,\Delta T - \tilde{C}_P^* \,\Delta T = -R\,\Delta T \qquad (4.1\text{-}58)$$

which is identical with Eq. (4.1-56).

Constant temperature (isothermal) process

Since temperature remains constant, then

$$\Delta\tilde{U}^{IG} = 0 \qquad (4.1\text{-}59)$$

$$\Delta\tilde{H}^{IG} = 0 \qquad (4.1\text{-}60)$$

The work done in a reversible process is given by

$$\widetilde{W} = -\int_{\tilde{V}_1}^{\tilde{V}_2} P\,d\tilde{V} = -RT\int_{\tilde{V}_1}^{\tilde{V}_2} \frac{1}{\tilde{V}}\,d\tilde{V}$$

$$= -RT\ln\left(\frac{\tilde{V}_2}{\tilde{V}_1}\right) = -RT\ln\left(\frac{P_1}{P_2}\right) \qquad (4.1\text{-}61)$$

The amount of heat transferred can be found from the application of the first law

$$\underbrace{\Delta\tilde{U}^{IG}}_{0} = \tilde{Q} + \widetilde{W} \qquad (4.1\text{-}62)$$

or

$$\tilde{Q} = RT\ln\left(\frac{\tilde{V}_2}{\tilde{V}_1}\right) = RT\ln\left(\frac{P_1}{P_2}\right) \qquad (4.1\text{-}63)$$

Adiabatic process

An adiabatic process is one in which the amount of heat transferred between the system and its surroundings is zero, i.e.,

$$\widetilde{Q} = 0 \tag{4.1-64}$$

The changes in internal energy and enthalpy are given by

$$\Delta \widetilde{U}^{IG} = \widetilde{C}_V^* \, \Delta T \tag{4.1-65}$$

$$\Delta \widetilde{H}^{IG} = \widetilde{C}_P^* \, \Delta T \tag{4.1-66}$$

The work done in a reversible process can be calculated from

$$\widetilde{W} = -\int_{\widetilde{V}_1}^{\widetilde{V}_2} P \, d\widetilde{V} = -\int_{\widetilde{V}_1}^{\widetilde{V}_2} \frac{RT}{\widetilde{V}} \, d\widetilde{V} \tag{4.1-67}$$

Since temperature changes in an adiabatic process, evaluation of this integral is only possible numerically. Therefore, it is much more convenient to calculate work by the application of the first law, i.e.,

$$\Delta \widetilde{U}^{IG} = \underbrace{\widetilde{Q}}_{0} + \widetilde{W} \qquad \Rightarrow \qquad \widetilde{W} = \widetilde{C}_V^* \, \Delta T \tag{4.1-68}$$

The use of Eq. (4.1-47) to express $\widetilde{C}_V^*$ in terms of γ leads to

$$\widetilde{W} = \frac{R \, \Delta T}{\gamma - 1} \tag{4.1-69}$$

or

$$\widetilde{W} = \frac{\Delta(P\widetilde{V})}{\gamma - 1} \tag{4.1-70}$$

Note that the calculations of $\Delta \widetilde{U}$, $\Delta \widetilde{H}$, and $\widetilde{W}$ require the value of the temperature to be known at the final state. In some problems, we may know the values of either pressure or volume at the final state. Thus, it is necessary to develop equations relating T to P, T to V, and P to V. To do this, consider Eq. (4.1-68) in differential form

$$d\widetilde{U}^{IG} = \delta \widetilde{W} \tag{4.1-71}$$

or

$$\widetilde{C}_V^* \, dT = - P \, d\widetilde{V} \tag{4.1-72}$$

The use of the ideal gas equation of state to express P in terms of T and $\widetilde{V}$, and rearrangement of Eq. (4.1-72) lead to

$$\frac{dT}{T} = -\frac{R}{\widetilde{C}_V^*}\frac{d\widetilde{V}}{\widetilde{V}} \tag{4.1-73}$$

With the help of Eq. (4.1-47), Eq. (4.1-73) can be expressed in the form

$$\frac{dT}{T} = -(\gamma - 1)\frac{d\widetilde{V}}{\widetilde{V}} \tag{4.1-74}$$

Integration of Eq. (4.1-74) leads to

$$\boxed{\frac{T_2}{T_1} = \left(\frac{\widetilde{V}_1}{\widetilde{V}_2}\right)^{\gamma-1}} \tag{4.1-75}$$

From the ideal gas equation of state

$$\frac{P_1}{P_2}\frac{\widetilde{V}_1}{\widetilde{V}_2} = \frac{T_1}{T_2} \tag{4.1-76}$$

Elimination of $\widetilde{V}$ between Eqs. (4.1-75) and (4.1-76) gives

$$\boxed{\frac{T_2}{T_1} = \left(\frac{P_2}{P_1}\right)^{(\gamma-1)/\gamma}} \tag{4.1-77}$$

Finally, elimination of T between Eqs. (4.1-75) and (4.1-77) results in

$$\boxed{\left(\frac{\widetilde{V}_1}{\widetilde{V}_2}\right)^{\gamma} = \left(\frac{P_2}{P_1}\right)} \tag{4.1-78}$$

From Eq. (4.1-78), it is possible to conclude that

$$\boxed{P\widetilde{V}^{\gamma} = \text{constant}} \qquad \text{reversible and adiabatic process} \tag{4.1-79}$$

How one can represent a reversible and adiabatic process in a P-$\widetilde{V}$ diagram? To answer this question, first consider a reversible and isothermal process that is represented by an equilateral hyperbola in a P-$\widetilde{V}$ diagram with an equation

$$P\widetilde{V} = RT = \text{constant} \tag{4.1-80}$$

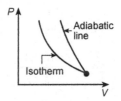

Fig. 4.4   At a given point, an adiabatic line has a steeper slope than an isotherm.

The slope of a hyperbola (or isotherm) is given by

$$\frac{dP}{d\widetilde{V}} = -\frac{P}{\widetilde{V}} \qquad (4.1\text{-}81)$$

which is negative. For a reversible and adiabatic process, on the other hand, the relationship between P and $\widetilde{V}$ is given by Eq. (4.1-79). The slope of an adiabatic curve is given by

$$\frac{dP}{d\widetilde{V}} = -\gamma \left(\frac{P}{\widetilde{V}} \right) \qquad (4.1\text{-}82)$$

Since P, $\widetilde{V}$, and γ are all positive, $dP/d\widetilde{V}$ is again negative. Since $\gamma > 1$, comparison of Eqs. (4.1-81) and (4.1-82) indicates that the reversible and adiabatic curve has a steeper negative slope than the reversible and isothermal curve passing through the same point as shown in Fig. 4.4.

▶ **Adiabatic reversible expansion**

Consider expansion of air from 50 atm and 75 °C to atmospheric pressure by an adiabatic and reversible process. Taking $\gamma = \widetilde{C}_P^* / \widetilde{C}_V^* = 7/5$ for air, the final temperature can be calculated from Eq. (4.1-77) as

$$T_2 = (75 + 273) \left(\frac{1}{50} \right)^{2/7} = 113.8\,\text{K}\ (-159.2\,^\circ\text{C})$$

Since $Q = 0$, work done by air during expansion, which is negative, leads to a decrease in internal energy and hence in the temperature of air. It is important to keep in mind that all adiabatic expansion processes, reversible or not, cause a substantial decrease in a fluid's temperature.

For example, when you open a champagne bottle you observe the formation of white fog or smoke in the neck and just above the mouth of the bottle. As a result of the expansion of pressurized gas in the bottle, which is a mixture of air and carbon dioxide, temperature drops considerably, leading to condensation of moisture in the air to form the observed fog.

Mechanical explosions are generally considered adiabatic since they occur very quickly. To estimate the maximum damage caused by an explosion, the adiabatic expansion process is also considered reversible.

▶ **Adiabatic reversible compression**

Consider compression of air from atmospheric pressure and 25 °C to 30 atm by an adiabatic and reversible process. The final temperature can be calculated from Eq. (4.1-77) as

$$T_2 = (25 + 273) \left(\frac{30}{1}\right)^{2/7} = 787.5 \, \text{K} \, (514.5 \, ^\circ\text{C})$$

Since $Q = 0$, work done on the air during compression causes an increase in internal energy and hence in the temperature of the system. One should be extremely careful when increasing the pressure of flammable vapors. An enormous increase in temperature may result in autoignition, leading to plant accidents.

Polytropic process

The relationship between pressure and volume during compression and expansion of gases is sometimes expressed as

$$\boxed{P\widetilde{V}^n = \text{constant}} \tag{4.1-83}$$

where n is a constant for any given change. Such processes are called *polytropic* and the equations available for adiabatic processes can also be used for polytropic ones by replacing γ with n.

Among the various processes described above, note that the most general process is the polytropic process given by Eq. (4.1-83). Depending on the process, the exponent n takes the following values:

$$n = \begin{cases} \infty & \text{Constant volume process} \\ 0 & \text{Constant pressure process} \\ 1 & \text{Constant temperature process} \\ \gamma & \text{Adiabatic process} \end{cases} \tag{4.1-84}$$

Example 4.4 *An insulated horizontal cylinder fitted with a piston contains 1.6 mol of helium at 800 kPa. Its volume is 0.005 m³ and the ambient pressure is 100 kPa. The piston does not conduct heat.*

a) *Calculate the work done when the gas is expanded reversibly until the internal pressure is equal to the ambient pressure.*

b) *Calculate the work done when the gas is expanded very suddenly until the internal pressure is equal to the ambient pressure.*

Solution

System: *Contents of the cylinder*

a) *The process is reversible and adiabatic (no heat loss between the contents of the cylinder and the surroundings). Assuming helium as an ideal gas with $\widetilde{C}_P^* = (5/2)R$, application of the first law gives*

$$\Delta U = \underbrace{Q}_{0} + W \qquad \Rightarrow \qquad W = \Delta U = n\widetilde{C}_V^*(T_2 - T_1) \tag{1}$$

The initial temperature can be calculated from the ideal gas equation of state

$$T_1 = \frac{P_1 V_1}{nR} = \frac{(800)(0.005)}{(1.6)(8.314 \times 10^{-3})} = 301 \, \text{K}$$

The final temperature can be calculated from Eq. (4.1-77), i.e.,

$$T_2 = T_1 \left(\frac{P_2}{P_1}\right)^{(\gamma-1)/\gamma}$$

where

$$\gamma = \frac{\widetilde{C}_P^*}{\widetilde{C}_V^*} = \frac{5}{3} \qquad \Rightarrow \qquad \frac{\gamma-1}{\gamma} = \frac{2}{5}$$

Thus,

$$T_2 = (301) \left(\frac{100}{800}\right)^{2/5} = 131 \, \text{K}$$

Substitution of the numerical values into Eq. (1) gives

$$W = (1.6)(1.5 \times 8.314)(131 - 301) = -3392 \, \text{J}$$

b) *In this case, the process is irreversible and adiabatic. Application of the first law gives*

$$\Delta U = \underbrace{Q}_{0} + W$$

or

$$n\,\widetilde{C}_V^*(T_2 - T_1) = -P_{ex}(V_2 - V_1)$$

Substitution of the numerical values gives

$$(1.6)(1.5{\times}8.314)(T_2{-}301) = - \left(100 \times 10^3\right)\left[\frac{(1.6)(8.314 \times 10^{-3})T_2}{100} - 0.005\right]$$

Solving for T_2 results in

$$T_2 = 195.7\,\mathrm{K}$$

Thus,

$$W = \Delta U = (1.6)(1.5 \times 8.314)(195.7 - 301) = -2101\,\mathrm{J}$$

Example 4.5 *One mole of air undergoes the following reversible changes in a series of nonflow processes:*

i) From an initial state of $20\,°\mathrm{C}$ and $100\,\mathrm{kPa}$ (state 1), it is compressed adiabatically to $500\,\mathrm{kPa}$ (state 2),
ii) It is then cooled to $20\,°\mathrm{C}$ at a constant pressure of $500\,\mathrm{kPa}$ (state 3),
iii) Finally, the gas is expanded isothermally to its original state.

a) *Sketch the paths followed in each process in a single P-V diagram,*
b) *Calculate the work done, the heat transferred, and the changes in internal energy and enthalpy for each of the three processes and for the entire cycle.*

Solution

a) *Assuming air as an ideal gas, the paths followed in each process are shown below.*

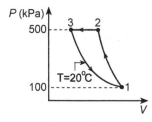

b) System: *Air in a closed system*

Process $1 \rightarrow 2$ (Reversible adiabatic)

Application of the first law, Eq. (4.1-9), gives

$$\Delta \tilde{U}_{12} = \underbrace{\tilde{Q}_{12}}_{0} + \tilde{W}_{12} \tag{1}$$

The temperature at state 2 is determined from Eq. (4.1-77) as

$$T_2 = T_1 \left(\frac{P_2}{P_1} \right)^{(\gamma-1)/\gamma} = (20 + 273) \left(\frac{500}{100} \right)^{2/7} = 464 \, \text{K} \tag{2}$$

Thus, the changes in internal energy and enthalpy are

$$\Delta \tilde{U}_{12} = \tilde{C}_V^*(T_2 - T_1) = (2.5 \times 8.314)(464 - 293) = 3554.2 \, \text{J/mol} \tag{3}$$

$$\Delta \tilde{H}_{12} = \tilde{C}_P^*(T_2 - T_1) = (3.5 \times 8.314)(464 - 293) = 4975.9 \, \text{J/mol} \tag{4}$$

The work done on the system is determined from Eq. (1) as

$$\tilde{W}_{12} = \Delta \tilde{U}_{12} = 3554.2 \, \text{J/mol} \tag{5}$$

Process $2 \to 3$ (Reversible isobaric)

The changes in internal energy and enthalpy are

$$\Delta \tilde{U}_{23} = \tilde{C}_V^*(T_3 - T_2) = (2.5 \times 8.314)(293 - 464) = -3554.2 \, \text{J/mol} \tag{6}$$

$$\Delta \tilde{H}_{23} = \tilde{C}_P^*(T_3 - T_2) = (3.5 \times 8.314)(293 - 464) = -4975.9 \, \text{J/mol} \tag{7}$$

For an isobaric process, heat transferred is calculated from

$$\tilde{Q}_{23} = \Delta \tilde{H}_{23} = -4975.9 \, \text{J/mol} \tag{8}$$

The work done on the system is calculated from the first law, Eq. (4.1-9), as

$$\tilde{W}_{23} = \Delta \tilde{U}_{23} - \tilde{Q}_{23} = -3554.2 + 4975.9 = 1421.7 \, \text{J/mol} \tag{9}$$

Process $3 \to 1$ (Reversible isothermal)

Since temperature remains constant, then

$$\Delta \tilde{U}_{31} = \Delta \tilde{H}_{31} = 0 \tag{10}$$

The work done by the system is

$$\tilde{W}_{31} = -RT \ln \left(\frac{P_3}{P_1} \right) = -(8.314)(293) \ln \left(\frac{500}{100} \right) = -3920.6 \, \text{J/mol} \tag{11}$$

The heat transferred is calculated from the first law, Eq. (4.1-9), as

$$\widetilde{Q}_{31} = -\widetilde{W}_{31} = 3920.6 \, \text{J/mol} \tag{12}$$

The following table summarizes the values of $\widetilde{Q}$, $\widetilde{W}$, $\Delta\widetilde{U}$, and $\Delta\widetilde{H}$ for each process as well as for the entire process:

Process	$\widetilde{Q}$ (J/ mol)	$\widetilde{W}$ (J/ mol)	$\Delta\widetilde{U}$ (J/ mol)	$\Delta\widetilde{H}$ (J/ mol)
$1 \to 2$	0	3554.2	3554.2	4975.9
$2 \to 3$	-4975.9	1421.7	-3554.2	-4975.9
$3 \to 1$	3920.6	-3920.6	0	0
$\sum$	-1055.3	1055.3	0	0

Comment: *Since internal energy and enthalpy are state functions, for the overall cyclic process $\Delta\widetilde{U} = \Delta\widetilde{H} = 0$. When $\Delta\widetilde{U}_{cycle} = 0$, from the first law $\widetilde{Q}_{cycle} = -\widetilde{W}_{cycle}$.*

Example 4.6 *An insulated rigid tank of $0.15 \, \text{m}^3$ volume initially contains $0.2 \, \text{kg}$ of wet steam at $65\,^\circ\text{C}$. The steam is stirred with a paddle wheel until the pressure reaches $500 \, \text{kPa}$.*

a) *Determine the final temperature in the tank,*
b) *Calculate the work done by the paddle wheel.*

Solution

System: *Contents of the tank*

a) *To determine the properties at the final state, one more independent intensive variable must be known besides pressure ($500 \, \text{kPa}$). Since neither the volume of the tank nor its total mass changes, the specific volume of the system remains constant, i.e.,*

$$\widehat{V}_2 = \widehat{V}_1 = \frac{\text{Total volume}}{\text{Total mass}} = \frac{0.15}{0.2} = 0.75 \, \text{m}^3/\text{kg}$$

From Table A.2 in Appendix A

$$\text{at } 500 \, \text{kPa} \qquad \widehat{V}^V = 0.3749 \, \text{m}^3/\text{kg}$$

Since $\widehat{V}_2 > 0.3749 \, \text{m}^3/\text{kg}$, the system exists as superheated steam at state 2. From Table A.3 in Appendix A, at $500 \, \text{kPa}$

$T\ (°C)$	$\widehat{V}\ (\mathrm{m^3/kg})$
500	0.7109
600	0.8041

The temperature at which $\widehat{V} = 0.75\,\mathrm{m^3/kg}$ can be determined by interpolation:

$$T = 500 + \left(\frac{600 - 500}{0.8041 - 0.7109}\right)(0.75 - 0.7109) \simeq 542\,°C$$

The process path in a T-V diagram is shown in the figure below:

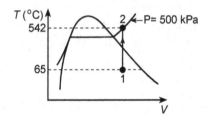

b) *Application of the first law gives*

$$\Delta U = \underbrace{Q}_{0} + W$$

To calculate the shaft work done on the system, internal energy values at the initial and final states must be determined.

- **Initial state**

$$T = 65\,°C \begin{cases} \widehat{V}^L = 0.001020\,\mathrm{m^3/kg} & \widehat{U}^L = 272.02\,\mathrm{kJ/kg} \\ \widehat{V}^V = 6.197\,\mathrm{m^3/kg} & \Delta\widehat{U} = 2191.1\,\mathrm{kJ/kg} \end{cases}$$

$$0.75 = 0.001020 + x\,(6.197 - 0.001020) \qquad \Rightarrow \qquad x = 0.12$$

$$\widehat{U}_1 = 272.02 + (0.12)(2191.1) \simeq 535\,\mathrm{kJ/kg}$$

- **Final state**

From Table A.3 in Appendix A, at 500 kPa

T (°C)	$\widehat{U}$ (kJ/kg)
500	3128.4
600	3299.6

By interpolation

$$\widehat{U}_2 = 3128.4 + \left(\frac{3299.6 - 3128.4}{600 - 500}\right)(542 - 500) = 3200.3\,\text{kJ/kg}$$

Hence, the work done is

$$W = m\,\Delta\widehat{U} = (0.2)(3200.3 - 535) = 533\,\text{kJ}$$

Example 4.7 *A rigid tank of* $0.3\,\text{m}^3$ *volume initially contains saturated steam at* 600 kPa. *The tank contents are cooled until the quality reaches* 50%.

a) *Determine the final temperature and pressure in the tank,*
b) *Calculate the amount of heat that must be removed.*

Solution

System: *Contents of the tank*

a) *The process path in a P-V diagram is shown in the figure below:*

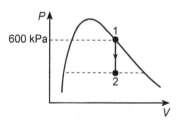

Since neither the volume of the tank nor its total mass changes, the system undergoes a constant volume process leading to wet steam at state 2.

The properties at the initial state (state 1) are given as

$$\begin{aligned} P_1 &= 600\,\text{kPa} \\ (T^{sat} &= 158.85\,°\text{C}) \end{aligned}\left.\right\} \quad \begin{aligned} \widehat{V}_1 &= 0.3157\,\text{m}^3/\text{kg} \\ \widehat{U}_1 &= 2567.4\,\text{kJ/kg} \end{aligned}$$

Since the specific volume of the system remains constant

$$\widehat{V}_2 = \widehat{V}_1 = 0.3157\,\text{m}^3/\text{kg} = \widehat{V}_2^L + 0.5(\widehat{V}_2^V - \widehat{V}_2^L) \tag{1}$$

or

$$\widehat{V}_2^L + \widehat{V}_2^V = 0.6314\,\text{m}^3/\text{kg} \tag{2}$$

One more independent intensive variable is needed to specify the final state (state 2). The temperature (or pressure) at the final state can be determined by a trial-and-error procedure as follows:

- *Assume T_2,*
- *Using Table A.1 in Appendix A read the values of $\widehat{V}_2^L$ and $\widehat{V}_2^V$,*
- *Check whether the summation of $\widehat{V}_2^L$ and $\widehat{V}_2^V$ is equal to $0.6314\,\text{m}^3/\text{kg}$.*

The next question is how to assume T_2. Since $\widehat{V}_2^V \gg \widehat{V}_2^L$, the dominant term in Eq. (2) is $\widehat{V}_2^V$. From Table A.1 in Appendix A

T (°C)	$\widehat{V}^V$ (m³/kg)
130	0.6685
135	0.5822

Therefore $130\,°\text{C} < T_2 < 135\,°\text{C}$. Let us assume that the temperature of the final state is $132.2\,°\text{C}$. From Table A.1 in Appendix A, by interpolation

$$\begin{array}{l} T = 132.2\,°\text{C} \\ (P^{vap} = 0.289\,\text{MPa}) \end{array} \Bigg\} \quad \begin{array}{l} \widehat{V}^L = 0.001072\,\text{m}^3/\text{kg} \quad \widehat{U}^L = 554.41\,\text{kJ/kg} \\ \widehat{V}^V = 0.6305\,\text{m}^3/\text{kg} \quad \widehat{U}^V = 2542.1\,\text{kJ/kg} \end{array}$$

Now, it is necessary to check whether Eq. (2) is satisfied:

$$\widehat{V}_2^L + \widehat{V}_2^V = 0.001072 + 0.6305 = 0.6316 \quad \Rightarrow \quad \text{Checks!}$$

Therefore, the temperature is $\sim 132\,°\text{C}$ and the pressure is $289\,\text{kPa}$.

b) *Application of the first law gives*

$$\Delta U = Q + \underbrace{W}_{0} \tag{3}$$

To calculate the heat transferred, internal energy of the final state must be determined. The internal energy at the final state is

$$\widehat{U}_2 = \widehat{U}_2^L + 0.5(\widehat{U}_2^V - \widehat{U}_2^L) = \frac{\widehat{U}_2^L + \widehat{U}_2^V}{2} = \frac{554.41 + 2542.1}{2}$$

$$= 1548.3\,\text{kJ/kg} \tag{4}$$

The mass of the tank contents is

$$m = \frac{V}{\widehat{V}} = \frac{0.3}{0.3157} = 0.95\,\text{kg} \tag{5}$$

From Eq. (3), the amount of heat removed from the system is

$$Q = m\left(\widehat{U}_2 - \widehat{U}_1\right) = (0.95)(1548.3 - 2567.4) = -968.1\,\text{kJ}$$

Example 4.8 *An insulated rigid cylinder of* 0.06 m³ *volume is divided into two equal chambers by a frictionless piston. The piston conducts heat and is initially held in place by a pin as shown in the figure below. Chamber* A *contains air at* 15 MPa *and* 350 °C, *and chamber* B *contains air at* 1.8 MPa *and* 350 °C. *What are the final temperature and pressure when the pin is removed?*

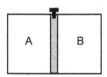

Solution

System: *Contents of chambers* A *and* B

Application of the first law gives

$$\Delta U_A + \Delta U_B = \underbrace{Q}_{0} + \underbrace{W}_{0} \tag{1}$$

Since the piston is conducting heat, temperatures in both chambers are equal to each other at the final equilibrium state. Expressing ΔU *in terms of* ΔT *gives*

$$n_A \widetilde{C}_V^*(T_2 - 350) + n_B \widetilde{C}_V^*(T_2 - 350) = 0 \tag{2}$$

or

$$(n_A + n_B)(T_2 - 350) = 0 \qquad \Rightarrow \qquad T_2 = 350\,°\text{C} \tag{3}$$

The final pressure can be determined from the ideal gas law, i.e.,

$$P_2 = \frac{(n_A + n_B)RT_2}{V_A + V_B} \tag{4}$$

The number of moles of air in each chamber can be determined from the initial state as

$$n_A = \frac{P_{A_1} V_{A_1}}{R T_{A_1}} = \frac{(15)(0.03)}{(8.314 \times 10^{-3})(350 + 273)} = 0.0869\,\text{kmol} \qquad (5)$$

$$n_B = \frac{P_{B_1} V_{B_1}}{R T_{B_1}} = \frac{(1.8)(0.03)}{(8.314 \times 10^{-3})(350 + 273)} = 0.0104\,\text{kmol} \qquad (6)$$

Substitution of the values into Eq. (4) gives the final pressure as

$$P_2 = \frac{(0.0869 + 0.0104)(8.314 \times 10^{-3})(350 + 273)}{0.06} = 8.4\,\text{MPa}$$

Alternative solution: *Since* $\Delta T = 0$, ΔH *is also zero. From the definition of enthalpy one can write*

$$\underbrace{\Delta H}_{0} = \underbrace{\Delta U}_{0} + \Delta(PV) \qquad (7)$$

Expanding the $\Delta(PV)$ *term gives*

$$P_2(V_{A_2} + V_{B_2}) - (P_{A_1} V_{A_1} + P_{B_1} V_{B_1}) = 0 \qquad (8)$$

Solving for P_2 *results in*

$$P_2 = \frac{P_{A_1} V_{A_1} + P_{B_1} V_{B_1}}{V_{A_2} + V_{B_2}} \qquad (9)$$

Since $V_{A_1} = V_{B_1} = (V_{A_2} + V_{B_2})/2$, *Eq. (9) simplifies to*

$$P_2 = \frac{P_{A_1} + P_{B_1}}{2} = \frac{15 + 1.8}{2} = 8.4\,\text{MPa} \qquad (10)$$

Example 4.9 *Repeat Example 4.8 if chamber* A *initially contains steam at* 15 MPa *and* 350 °C, *and chamber* B *contains steam at* 1.8 MPa *and* 350 °C.

Solution

System: *Contents of chambers* A *and* B

Application of the first law gives

$$\Delta U_A + \Delta U_B = \underbrace{Q}_{0} + \underbrace{W}_{0} \qquad (1)$$

Let 1 and 2 represent the initial and final states, respectively. In the final state, the temperature and pressure in each chamber are equal to each other, i.e., $\widehat{U}_{A_2} = \widehat{U}_{B_2} = \widehat{U}_2$. Thus, Eq. (1) is expressed as

$$m_A\left(\widehat{U}_2 - \widehat{U}_{A_1}\right) + m_B\left(\widehat{U}_2 - \widehat{U}_{B_1}\right) = 0 \tag{2}$$

From Table A.3 in Appendix A

$$\left.\begin{array}{l} P_{A_1} = 15\,\text{MPa} \\ T_{A_1} = 350\,^\circ\text{C} \end{array}\right\} \begin{array}{l} \widehat{V}_{A_1} = 0.01147\,\text{m}^3/\text{kg} \\ \widehat{U}_{A_1} = 2520.4\,\text{kJ}/\text{kg} \end{array}$$

and

$$\left.\begin{array}{l} P_{B_1} = 1.8\,\text{MPa} \\ T_{B_1} = 350\,^\circ\text{C} \end{array}\right\} \begin{array}{l} \widehat{V}_{B_1} = 0.15457\,\text{m}^3/\text{kg} \\ \widehat{U}_{B_1} = 2863.0\,\text{kJ}/\text{kg} \end{array}$$

The amounts of steam in chambers A and B are

$$m_A = \frac{V_{A_1}}{\widehat{V}_{A_1}} = \frac{0.03}{0.01147} = 2.6155\,\text{kg} \qquad m_B = \frac{V_{B_1}}{\widehat{V}_{B_1}} = \frac{0.03}{0.15457} = 0.1941\,\text{kg}$$

Substitution of numerical values into Eq. (2) gives

$$(2.6155)(\widehat{U}_2 - 2520.4) + (0.1941)(\widehat{U}_2 - 2863.0) = 0$$

Solving for $\widehat{U}_2$ gives

$$\widehat{U}_2 = 2544\,\text{kJ}/\text{kg}$$

To determine the properties at the final state, one more independent intensive property is needed besides specific internal energy. Since the total volume and total mass do not change, the specific volume of the system at the final state is

$$\widehat{V}_2 = \frac{\text{Total volume}}{\text{Total mass}} = \frac{0.06}{2.6155 + 0.1941} = 0.02136\,\text{m}^3/\text{kg}$$

The temperature and pressure at the final state should be determined by a trial-and-error procedure so as to give $\widehat{U}_2 = 2544\,\text{kJ}/\text{kg}$ and $\widehat{V}_2 = 0.02136\,\text{m}^3/\text{kg}$. The next question is how to assume T_2 (or P_2). The final pressure will be between 1.8 MPa and 15 MPa. If both chambers contain equal amounts, it is plausible to assume that the final pressure is $(1.8 + 15)/2 = 8.4\,\text{MPa}$. Since chamber A contains more steam, then the final pressure will be between 8.4 MPa and 15 MPa. The possibilities for the condition of steam at the final state are (i) superheated vapor, (ii) all vapor, (iii) wet steam. Examination of $\widehat{V}$ and $\widehat{U}$ values in Table A.3 in

Appendix A indicates that steam cannot exist as a superheated steam at the final state. From Table A.1 in Appendix A

$$\text{at } T = 135\,^{\circ}\text{C} \qquad \widehat{V}^V = 0.5822\,\text{m}^3/\text{kg and } \widehat{U}^V = 2545.0\,\text{kJ}/\text{kg}$$

Although 2545.0 kJ/ kg *is very close to* 2544 kJ/ kg, 0.5822 m³/ kg *is very far from the required value of* 0.02136 m³/ kg. *Therefore, steam also cannot exist as a saturated vapor at the final state. This leaves us with the possibility of wet steam at the final state. From Table A.1 in Appendix A*

$$\left.\begin{array}{l} T = 300\,^{\circ}\text{C} \\ (P^{vap} = 8.581\,\text{MPa}) \end{array}\right\} \begin{array}{l} \widehat{V}^L = 0.001404\,\text{m}^3/\text{kg} \quad \widehat{U}^L = 1332.0\,\text{kJ}/\text{kg} \\ \qquad\qquad\qquad\qquad \Delta\widehat{U} = 1231.0\,\text{kJ}/\text{kg} \\ \widehat{V}^V = 0.02167\,\text{m}^3/\text{kg} \quad \widehat{U}^V = 2563.0\,\text{kJ}/\text{kg} \end{array}$$

This value is chosen such that $\widehat{V}^V > \widehat{V}_2$ but not very far from 0.02136 m³/ kg. *The quality of the wet steam is*

$$2544 = 1332 + x\,(1231) \qquad \Rightarrow \qquad x = 0.985$$

The specific volume is

$$\widehat{V}_2 = 0.001404 + (0.985)(0.02167 - 0.001404) = 0.02137\,\text{m}^3/\text{kg}$$

which is almost equal to 0.02136 m³/ kg. *Therefore, the final temperature and pressure are* 300 °C *and* 8.581 MPa, *respectively.*

Example 4.10 *A rigid cylinder of* 0.6 m³ *volume is divided into two equal chambers by a frictionless piston as shown in the figure below. The piston conducts heat. Initially chamber A contains air at* 700 kPa, *and chamber B contains wet steam at* 700 kPa *and 15% quality. Heat is transferred to both chambers until all the liquid in B evaporates. Determine the total heat transfer during the process.*

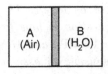

Solution

System: *Contents of chambers A and B*

Application of the first law gives

$$\Delta U = Q + \underbrace{W}_{0} \tag{1}$$

or

$$Q = \Delta U_A + \Delta U_B = n_A (\widetilde{C}_V^*)_A (T_{A_2} - T_{A_1}) + m_B (\widehat{U}_{B_2} - \widehat{U}_{B_1}) \tag{2}$$

• Initial state

Chamber B

From Table A.2 in Appendix A

$$P = 700\,\text{kPa} \left.\right\} \quad \widehat{V}^L = 0.001108\,\text{m}^3/\text{kg} \quad \widehat{U}^L = 696.44\,\text{kJ}/\text{kg}$$
$$(T^{sat} \simeq 165\,^\circ\text{C}) \quad \widehat{V}^V = 0.2729\,\text{m}^3/\text{kg} \quad \Delta\widehat{U} = 1876.1\,\text{kJ}/\text{kg}$$

The specific volume of the wet steam is

$$\widehat{V}_{B_1} = 0.001108 + (0.15)(0.2729 - 0.001108) = 4.1877 \times 10^{-2}\,\text{m}^3/\text{kg}$$

The total mass is

$$m_B = \frac{0.3}{4.1877 \times 10^{-2}} = 7.164\,\text{kg}$$

The initial internal energy is

$$U_{B_1} = (7.164) \left[696.44 + (0.15)(1876.1) \right] = 7005.4\,\text{kJ}$$

Chamber A

Since the piston is conducting, the initial temperatures in both chambers are equal to each other, i.e., $T_{A_1} = T_{B_1} = 165\,^\circ\text{C}$. Thus, the number of moles of air in chamber A is

$$n_A = \frac{P_{A_1} V_{A_1}}{R T_{A_1}} = \frac{(700)(0.3)}{(8.314)(165 + 273)} = 0.0577\,\text{kmol}$$

• Final state

Since the piston is conducting heat, final temperature and pressure in both chambers are the same, i.e., $T_{A_2} = T_{B_2} = T_2$ and $P_{A_2} = P_{B_2} = P_2$. It is also given that the steam in chamber B is a saturated vapor. The final pressure is determined by a trial-and-error procedure as follows:

- *Assume P_2*
- *Calculate V_{A_2} and V_{B_2}*

• *Check whether the summation of V_{A_2} and V_{B_2} is equal to $0.6\,m^3$.*

The next question is how to assume P_2. If saturated steam occupies the whole cylinder, i.e., the piston moves to the extreme left, the specific volume is $\widehat{V} = 0.6/7.164 = 0.0838\,m^3/kg$. From Table A.2 in Appendix A, the pressure at which $\widehat{V}^V = 0.0838\,m^3/kg$ is calculated by interpolation as $2.39\,MPa$. Since the actual value of $\widehat{V}^V$ will be less than $0.0838\,m^3/kg$, the final pressure must be greater than $2.39\,MPa$.

Let us assume that $P_2 = 2.81\,MPa$. From Table A.2 in Appendix A by interpolation

$$T^{sat} \simeq 230.1\,°C \qquad \widehat{V}^V = 0.071734\,m^3/kg \qquad \widehat{U}^V = 2603.7\,kJ/kg$$

The volumes of each chamber are

$$V_{A_2} = \frac{n_A R T_{A_2}}{P_{A_2}} = \frac{(0.0577)(8.314)(230.1 + 273)}{2810} = 0.0859\,m^3$$

$$V_{B_2} = m_B \widehat{V}_{B_2} = (7.164)(0.071734) = 0.5139\,m^3$$

Since $V_{A_2} + V_{B_2} = 0.5998\,m^3 \simeq 0.6\,m^3$, the assumed pressure of $2.81\,MPa$ is correct. The heat transferred is then calculated from Eq. (2) as

$$Q = (0.0577)(2.5 \times 8.314)(230.1 - 165) + [(7.164)(2603.7) - 7005.4]$$
$$= 11,726\,kJ$$

4.2   Generalized Development of the First Law

As stated in Section 1.1, the concepts or ideas that are the basis of science and engineering are *mass*, *momentum*, and *energy*. These are all conserved quantities. For any quantity φ that is conserved, an *inventory rate equation* can be written to describe the transformation of the conserved quantity. Inventory of the conserved quantity is based on a specified unit of time, which is reflected in the term, *rate*. For a single component system, rate equations for mass and energy are expressed in the form

$$\begin{bmatrix} \text{Rate of} \\ \text{mass in} \end{bmatrix} - \begin{bmatrix} \text{Rate of} \\ \text{mass out} \end{bmatrix} = \begin{bmatrix} \text{Rate of mass} \\ \text{accumulation} \end{bmatrix} \qquad (4.2\text{-}1)$$

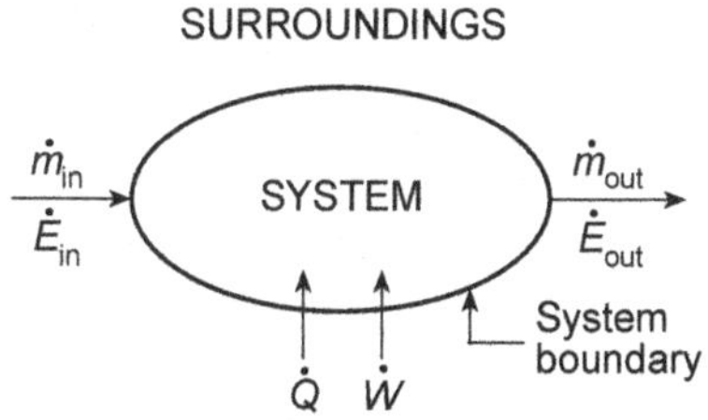

SURROUNDINGS

Fig. 4.5 Unsteady-state flow system exchanging energy in the form of heat and work with the surroundings.

$$\begin{bmatrix} \text{Rate of} \\ \text{energy in} \end{bmatrix} - \begin{bmatrix} \text{Rate of} \\ \text{energy out} \end{bmatrix} = \begin{bmatrix} \text{Rate of energy} \\ \text{accumulation} \end{bmatrix} \tag{4.2-2}$$

Consider an unsteady-state flow system as shown in Fig. 4.5. The conservation statement for mass, Eq. (4.2-1), is expressed as

$$\boxed{\dot{m}_{in} - \dot{m}_{out} = \frac{dm_{sys}}{dt}} \tag{4.2-3}$$

where the mass flow rate, $\dot{m}$, is the product of density, ρ, average velocity, v, and cross-sectional area of duct, A, such that

$$\dot{m} = \rho\, v\, A \tag{4.2-4}$$

An average velocity is calculated by dividing volumetric flow rate, $\dot{Q}$, by cross-sectional area, i.e.,

$$v = \frac{\dot{Q}}{A} \tag{4.2-5}$$

Density, ρ, is constant for an incompressible fluid. For gases, however, it is dependent on temperature and pressure. For example, for an ideal gas

$$\rho = \frac{PM}{RT} \tag{4.2-6}$$

where M is the molecular weight.

Multiplication of Eq. (4.2-3) by dt leads to

$$\boxed{dm_{in} - dm_{out} = dm_{sys}} \tag{4.2-7}$$

Dividing Eq. (4.2-7) by the molecular weight results in[6]

$$\boxed{dn_{in} - dn_{out} = dn_{sys}} \qquad (4.2\text{-}8)$$

The rate of energy entering or leaving the system by a flowing stream is given by

$$\dot{E} = \widehat{E}\dot{m} = (\widehat{U} + \widehat{E}_K + \widehat{E}_P)\dot{m} \qquad (4.2\text{-}9)$$

For a system shown in Figure 4.5, the statement of the conservation of energy, given by Eq. (4.2-2), is mathematically expressed as

$$\left[(\widehat{U} + \widehat{E}_K + \widehat{E}_P)\dot{m}\right]_{in} + \dot{Q} + \dot{W} - \left[(\widehat{U} + \widehat{E}_K + \widehat{E}_P)\dot{m}\right]_{out}$$
$$= \frac{d}{dt}\left[(\widehat{U} + \widehat{E}_K + \widehat{E}_P)m\right]_{sys} \qquad (4.2\text{-}10)$$

So far two types of work are considered: (*i*) work associated with the expansion or contraction of the system boundaries, (*ii*) shaft work, i.e., work done by/on the system through an either rotating or reciprocating shaft that crosses the boundaries of the system. The shaft work can be either calculated indirectly by the application of the first law of thermodynamics or can be calculated directly if the torque and the rate of rotation of the shaft are known. A mathematical expression for the calculation of shaft work will be developed in Chapter 5. For flow systems, it is necessary to introduce a third type of work, known as *flow work*.

To express flow work mathematically, consider a fluid of mass dm flowing into an open system through a circular duct of cross-sectional area A as shown in Fig. 4.6. As this differential mass crosses the boundary of the system, the resisting pressure P may be assumed to remain constant since the differential volume of the mass dm is very small compared to the volume of the system. The work done on the system to push the fluid of mass dm

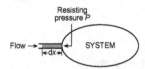

Fig. 4.6   A differential mass dm is entering an open system.

---

[6]Keep in mind that Eq. (4.2-8) holds for single component systems. In general, moles are not conserved!

against a constant force $F = PA$ is given by

$$\delta W = PA\, dx = P\, dV \qquad (4.2\text{-}11)$$

The differential volume dV of the mass dm is expressed in terms of the specific volume $\widehat{V}$ as

$$dV = d(m\widehat{V}) = \widehat{V}\, dm \qquad (4.2\text{-}12)$$

Substitution of Eq. (4.2-12) into Eq. (4.2-11) gives

$$\frac{\delta W}{dm} = P\widehat{V} \qquad (4.2\text{-}13)$$

or

$$\boxed{\widehat{W} = P\widehat{V}} \qquad (4.2\text{-}14)$$

which is known as the flow work. Note that the flow work is positive when fluid enters the system and negative when fluid leaves the system.

The term $\dot{W}$ in Eq. (4.2-10) is the rate of work done on the system by the surroundings and it is composed of the three terms

$$\dot{W} = \underbrace{- P_{sys}\frac{dV_{sys}}{dt}}_{A} + \underbrace{\dot{W}_s}_{B} + \underbrace{\left(P\widehat{V}\dot{m}\right)_{in} - \left(P\widehat{V}\dot{m}\right)_{out}}_{C} \qquad (4.2\text{-}15)$$

where terms A, B, and C represent, respectively, work associated with the expansion or compression of the system boundaries, shaft work, and flow work.

Substitution of Eq. (4.2-15) into Eq. (4.2-10) and the use of the definition of enthalpy, i.e., $\widehat{H} = \widehat{U} + P\widehat{V}$, give

$$\left[(\widehat{H} + \widehat{E}_K + \widehat{E}_P)\dot{m}\right]_{in} - \left[(\widehat{H} + \widehat{E}_K + \widehat{E}_P)\dot{m}\right]_{out} + \dot{Q} - P_{sys}\frac{dV_{sys}}{dt} + \dot{W}_s$$
$$= \frac{d}{dt}\left[(\widehat{U} + \widehat{E}_K + \widehat{E}_P)m\right]_{sys} \qquad (4.2\text{-}16)$$

which is known as the general statement of the first law of thermodynamics (or conservation of energy). In Eq. (4.2-16), the flow work term is included in the enthalpy term.

Multiplication of Eq. (4.2-16) by dt leads to

$$(\widehat{H} + \widehat{E}_K + \widehat{E}_P)_{in}\, dm_{in} - (\widehat{H} + \widehat{E}_K + \widehat{E}_P)_{out}\, dm_{out} + \delta Q - P_{sys}\, dV_{sys} + \delta W_s$$
$$= d\left[(\widehat{U} + \widehat{E}_K + \widehat{E}_P)m\right]_{sys} \qquad (4.2\text{-}17)$$

In terms of molar quantities, Eq. (4.2-17) is expressed as

$$(\tilde{H} + \tilde{E}_K + \tilde{E}_P)_{in} \, dn_{in} - (\tilde{H} + \tilde{E}_K + \tilde{E}_P)_{out} \, dn_{out} + \delta Q - P_{sys} \, dV_{sys} + \delta W_s$$
$$= d \left[(\tilde{U} + \tilde{E}_K + \tilde{E}_P)n \right]_{sys} \quad (4.2\text{-}18)$$

4.3 The First Law for Steady-state Flow Processes

Under steady-state conditions the total mass of the system does not change with respect to time, i.e., $dm_{sys}/dt = 0$. Under these circumstances Eq. (4.2-3) simplifies to

$$\boxed{\dot{m}_{in} = \dot{m}_{out}} \quad (4.3\text{-}1)$$

Under steady conditions, the total energy of the system also does not change with respect to time, i.e.,

$$\frac{d}{dt} \left[(\hat{U} + \hat{E}_K + \hat{E}_P)m \right]_{sys} = 0 \quad (4.3\text{-}2)$$

Moreover, the boundaries of the system are fixed so that $dV_{sys}/dt = 0$. Taking $\dot{m}_{in} = \dot{m}_{out} = \dot{m}$, Eq. (4.2-16) reduces to

$$(\hat{H} + \hat{E}_K + \hat{E}_P)_{in} \, \dot{m} - (\hat{H} + \hat{E}_K + \hat{E}_P)_{out} \, \dot{m} + \dot{Q} + \dot{W}_s = 0 \quad (4.3\text{-}3)$$

Noting that

$$\text{Rate} = \frac{\text{Quantity}}{\text{Time}} = \left(\frac{\text{Quantity}}{\text{Mass}} \right) \left(\frac{\text{Mass}}{\text{Time}} \right) \quad \Rightarrow \quad \boxed{\dot{\varphi} = \hat{\varphi}\,\dot{m}} \quad (4.3\text{-}4)$$

Eq. (4.3-3) takes the form

$$\boxed{\Delta\dot{H} + \Delta\dot{E}_K + \Delta\dot{E}_P = \dot{Q} + \dot{W}_s} \quad (4.3\text{-}5)$$

where the term Δ implies (out − in). Dividing each term in Eq. (4.3-5) by $\dot{m}$ leads to

$$\boxed{\Delta\hat{H} + \Delta\hat{E}_K + \Delta\hat{E}_P = \hat{Q} + \hat{W}_s} \quad (4.3\text{-}6)$$

In terms of molar quantities, Eq. (4.3-6) is expressed in the form

$$\boxed{\Delta\tilde{H} + \Delta\tilde{E}_K + \Delta\tilde{E}_P = \tilde{Q} + \tilde{W}_s} \quad (4.3\text{-}7)$$

Example 4.11 *Air is to be heated as it flows under steady conditions through a horizontal pipe of 3 cm inside diameter. Air at 2.5 bar and 25 °C enters the pipe at a velocity of 80 m/s. At the outlet of the pipe, the pressure and velocity are 1.6 bar and 450 m/s. Determine the rate of heat transfer to the air.*

Solution

The unknowns for this problem are the outlet temperature of air and the rate of heat transfer. The two equations needed to calculate these values are the conservation of mass and the conservation of energy. The schematic of the problem is given below:

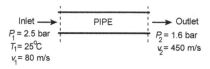

Assuming air to be an ideal gas, conservation of mass, Eq. (4.3-1), is expressed as

$$\left(\frac{P_1 M}{R T_1}\right) v_1 A_1 = \left(\frac{P_2 M}{R T_2}\right) v_2 A_2 \tag{1}$$

Since $A_1 = A_2$, Eq. (1) simplifies to

$$T_2 = T_1 \left(\frac{P_2}{P_1}\right) \left(\frac{v_2}{v_1}\right) = (25 + 273) \left(\frac{1.6}{2.5}\right) \left(\frac{450}{80}\right) = 1072.8 \, \text{K}$$

Conservation of energy, Eq. (4.3-5), simplifies to

$$\dot{Q} = \dot{m} \left[\widehat{C}_P^* (T_2 - T_1) + \frac{1}{2}(v_2^2 - v_1^2) \right] \tag{2}$$

The mass flow rate of air can be calculated from the inlet conditions as

$$\dot{m} = \left(\frac{P_1 M}{R T_1}\right) v_1 A = \left[\frac{(2.5)(29)}{(8.314 \times 10^{-2})(298)}\right] (80) \left[\frac{\pi (0.03)^2}{4}\right] = 0.165 \, \text{kg/s}$$

Substitution of the numerical values into Eq. (2) yields

$$\dot{Q} = (0.165) \left[\left(\frac{3.5 \times 8.314}{29}\right) (1072.8 - 298) + \frac{1}{2} \left(\frac{450^2 - 80^2}{1000}\right) \right] = 144.5 \, \text{kW}$$

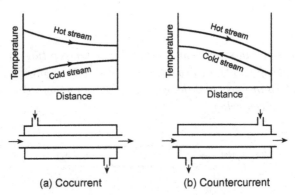

(a) Cocurrent   (b) Countercurrent

Fig. 4.7   Configurations and temperature profiles of a double-pipe heat exchanger.

4.3.1   *Heat exchanger*

A heat exchanger is a device in which two flowing streams exchange heat without mixing. In this way, transfer of energy as heat from the hot fluid to the cold one is accomplished. The simplest form of a heat exchanger is a double-pipe heat exchanger that consists of two concentric pipes of different diameters. One fluid flows in the inner pipe, and the other in the annular space between the two pipes. Heat is transferred from the hot fluid to the cold one through the pipe wall. As shown in Fig. 4.7, heat exchangers are operated in either the cocurrent (both streams flowing in the same direction) or countercurrent (streams flowing in opposite directions) configurations.

In heat exchangers no work is produced, and kinetic and potential energy changes can be assumed negligible. Considering the entire heat exchanger, i.e., the hot and the cold streams, as a system, Eq. (4.3-5) reduces to

$$\Delta \dot{H} = \dot{Q} \tag{4.3-8}$$

The outer shell of the heat exchanger is usually well insulated to prevent heat loss to the surroundings. Assuming $\dot{Q} = 0$ reduces Eq. (4.3-8) to

$$\Delta \dot{H} = 0 \quad \Rightarrow \quad \Delta \dot{H}_{\text{hot stream}} + \Delta \dot{H}_{\text{cold stream}} = 0 \tag{4.3-9}$$

or

$$\boxed{\dot{m}_{\text{hot stream}}\, \Delta \widehat{H}_{\text{hot stream}} + \dot{m}_{\text{cold stream}}\, \Delta \widehat{H}_{\text{cold stream}} = 0} \qquad (4.3\text{-}10)$$

Example 4.12 *Oil flowing at a rate of 5 kg/ min is to be cooled from 280 °C to 130 °C in a heat exchanger. Cooling water enters the heat exchanger at 20 °C and leaves it at 60 °C. Heat capacities ($\widehat{C}_P \approx \widehat{C}_V$) of oil and water are 2.5 kJ/ kg. K and 4.2 kJ/ kg. K, respectively. Calculate the flow rate of cooling water.*

Solution

Considering the heat exchanger as the system, application of Eq. (4.3-10) gives

$$\dot{m}_{oil}\widehat{C}_{P_{oil}} \left[(T_{oil})_{out} - (T_{oil})_{in} \right] + \dot{m}_{\text{H}_2\text{O}}\widehat{C}_{P_{\text{H}_2\text{O}}} \left[(T_{\text{H}_2\text{O}})_{out} - (T_{\text{H}_2\text{O}})_{in} \right] = 0 \tag{1}$$

or

$$(5)(2.5)(130 - 280) + \dot{m}_{\text{H}_2\text{O}}(4.2)(60 - 20) = 0 \tag{2}$$

Solving for $\dot{m}_{\text{H}_2\text{O}}$ gives

$$\dot{m}_{\text{H}_2\text{O}} = 11.16 \, \text{kg/ min}$$

Alternative solution: *Note that the enthalpy of a compressed liquid can be approximated as the enthalpy of saturated liquid at the given temperature. Therefore, Eq. (4.3-10) can also be expressed as*

$$\dot{m}_{oil}\widehat{C}_{P_{oil}} \left[(T_{oil})_{out} - (T_{oil})_{in} \right] + \dot{m}_{\text{H}_2\text{O}} \left(\widehat{H}^L \Big|_{60\,°\text{C}} - \widehat{H}^L \Big|_{20\,°\text{C}} \right) = 0 \tag{3}$$

From Table A.1 in Appendix A

$$\widehat{H}^L \Big|_{60\,°\text{C}} = 251.11 \, \text{kJ/ kg} \qquad \widehat{H}^L \Big|_{20\,°\text{C}} = 83.95 \, \text{kJ/ kg} \tag{4}$$

Substitution of the values into Eq. (3) gives

$$\dot{m}_{\text{H}_2\text{O}} = \frac{(5)(2.5)(280 - 130)}{251.11 - 83.95} = 11.22 \, \text{kg/ min}$$

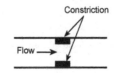

Fig. 4.8   A throttling device.

4.3.2   *Throttling device*

As shown in Fig. 4.8, a throttling device is any kind of flow-restricting device causing a significant drop in fluid pressure without involving any work or accelerating the fluid significantly. Flow through a partially opened valve, i.e., kitchen faucet, flow through a porous plug, and flow through an orifice are some examples of throttling devices. The throttling process is highly irreversible.

The throttling device produces negligible changes in kinetic and potential energies. Moreover, the residence time of a fluid particle within the device is short enough to assume $Q = 0$. Under these circumstances, Eq. (4.3-6) simplifies to

$$\boxed{\Delta \widehat{H} = 0} \quad \text{Throttling process} \qquad (4.3\text{-}11)$$

A process in which enthalpy remains constant is called an *isenthalpic process*. The variation of temperature with pressure in such a process is known as the *Joule-Thomson coefficient, μ*:

$$\mu = \left(\frac{\partial T}{\partial P} \right)_H \qquad (4.3\text{-}12)$$

The throttling process may lead to an increase or decrease in temperature depending on the value of μ, i.e.,

$$\mu \begin{cases} > 0 & \text{Temperature decreases on throttling} \\ < 0 & \text{Temperature increases on throttling} \end{cases} \qquad (4.3\text{-}13)$$

The temperature at which the Joule-Thomson coefficient changes sign is called the *inversion temperature*. At higher temperatures $\mu < 0$, and at lower temperatures $\mu > 0$. As a result, cryogenic applications require the gas temperature to be lower than the inversion temperature.

Most gases have an inversion temperature higher than the room temperature. Hydrogen, however, has an inversion temperature of $-80\,°\text{C}$. In order to liquefy hydrogen, it is first necessary to decrease its temperature below $-80\,°C$ using liquefied nitrogen and then decrease its pressure by a throttling process.

Example 4.13 *A fluid at 3.5 MPa and 350 °C enters a throttling valve and leaves it at 100 kPa. Determine the exit temperature if the fluid is (a) steam, (b) air.*

Solution

$$P_1 = 3.5\ \text{MPa}$$
$$T_1 = 350°\text{C} \qquad \bowtie \qquad P_2 = 100\ \text{kPa}$$

a) *From Table A.3 in Appendix A,* $\widehat{H}_1 = 3104.0\,\text{kJ}/\text{kg}$. *The use of Eq. (4.3-11) gives*

$$\widehat{H}_2 = \widehat{H}_1 = 3104\,\text{kJ}/\text{kg}$$

At the exit pressure of $100\,\text{kPa}$, $\widehat{H}^V = 2675.5\,\text{kJ}/\text{kg}$. *Since* $\widehat{H}_2 > 2675.5\,\text{kJ}/\text{kg}$, *then the steam is superheated at state 2. From Table A.3 in Appendix A, at* $100\,\text{kPa}$

$T\ (°C)$	$\widehat{H}\ (\text{kJ}/\text{kg})$
300	3074.3
400	3278.2

By interpolation

$$T = 300 + \left(\frac{400 - 300}{3278.2 - 3074.3} \right)(3104 - 3074.3) = 314.6\,°C$$

b) *For air (ideal gas), again* $\Delta\widehat{H} = 0$. *Since enthalpy depends only on temperature for an ideal gas, the air temperature remains constant at* $350\,°C$.

Example 4.14 *A throttling calorimeter is a device used to determine the quality of wet steam flowing in the main line. For this purpose, a portion of the steam is withdrawn from the main line through a throttling valve and admitted into a well-insulated expansion chamber in which its temperature is measured. A schematic diagram of the process is shown in the figure below. If the pressure in the main line is 7 MPa, and the temperature recorded in the throttling calorimeter is 120 °C, calculate the quality of steam in the main line.*

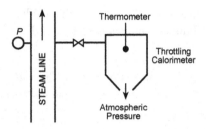

Solution

Atmospheric pressure is approximately equal to 100 kPa. *Since enthalpy does not change across the throttling valve*

$$\hat{H}_{\text{main line}} = \hat{H}(100\,\text{kPa}, 120\,^\circ\text{C})$$

From Table A.3 in Appendix A, at 100 kPa

T (°C)	$\hat{H}$ (kJ/kg)
100	2676.2
150	2776.4

By interpolation

$$\hat{H}(100\,\text{kPa}, 120\,^\circ\text{C}) = 2676.2 + \left(\frac{2776.4 - 2676.2}{150 - 100}\right)(120 - 100)$$

$$= 2716.3\,\text{kJ/kg}$$

From Table A.2 in Appendix A

$$P = 7\,\text{MPa} \left.\begin{cases} \hat{H}^L = 1267.0\,\text{kJ/kg} \\ \Delta\hat{H} = 1505.1\,\text{kJ/kg} \\ \hat{H}^V = 2772.1\,\text{kJ/kg} \end{cases}\right.$$

Therefore the quality of steam is

$$2716.3 = 1267 + x(1505.1) \qquad \Rightarrow \qquad x = 0.963$$

4.3.3 Turbine

A turbine is a device in which blades attached to a shaft (or rotor) are rotated by a gas or liquid flowing through it. In this way, the energy of a gas or liquid is converted into mechanical energy, which can then produce electricity by running a generator. The power output of a turbine may range from 0.75 kW to 1500 MW.

Steam turbines are widely used for power generation in vapor power plants. Sometimes, high reduction of steam pressure in a single-stage turbine may lead to a very high speed of the shaft, which is very difficult to control and operate. In this case, a gradual reduction in pressure is accomplished by using a multi-stage turbine.

For most steam turbines, the changes in kinetic and potential energy terms are usually quite small in comparison to the change in enthalpy, so that Eqs. (4.3-5) and (4.3-6) reduce to

$$\boxed{\Delta \dot{H} = \dot{Q} + \dot{W}_s \qquad \text{and} \qquad \Delta \widehat{H} = \widehat{Q} + \widehat{W}_s} \qquad (4.3\text{-}14)$$

The presence of moisture in a high velocity stream causes erosion of the turbine blades. Therefore, it is of utmost importance to keep the moisture content of the exit stream from the turbine as low as possible.

Example 4.15 *Steam flowing at a rate of* 0.3 kg/ s *enters an adiabatic turbine at* 3.5 MPa *and* 500 °C. *Saturated vapor leaves the turbine at* 100 kPa. *Calculate the power output of the turbine.*

Solution

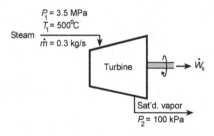

From Table A.3 in Appendix A

$$\widehat{H}_1 = 3450.9 \, \text{kJ/kg} \qquad \widehat{H}_2 = 2675.5 \, \text{kJ/kg}$$

For an adiabatic turbine $Q = 0$ *and Eq. (4.3-14) simplifies to*

$$\widehat{W}_s = \widehat{H}_2 - \widehat{H}_1 = 2675.5 - 3450.9 = -775.4 \, \text{kJ/kg}$$

The power output is

$$\dot{W}_s = \dot{m}\,\widehat{W}_s = (0.3)(-775.4) = -232.62\,\text{kW}$$

Example 4.16 *Steam at 4.5 MPa and 500 °C enters a turbine at a velocity of 65 m/s and its mass flow rate is 1.5 kg/s. The steam leaves the turbine at a point 3 m below the turbine inlet at a velocity of 300 m/s. The rate of heat loss from the turbine is 30 kW and the power output is 700 kW. A small portion of the exhaust steam from the turbine is passed through a throttling valve and discharges at an ambient pressure of 100 kPa. What is the temperature of the steam leaving the valve?*

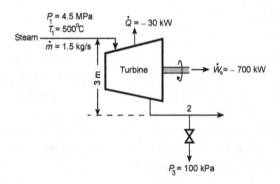

Calculation of T_3 requires one more independent intensive property besides pressure, which is specific enthalpy. Since the enthalpy change across the throttling valve is zero, i.e., $\widehat{H}_3 = \widehat{H}_2$, it is first necessary to calculate $\widehat{H}_2$ by the application of energy balance between states 1 and 2. From Eq. (4.3-5)

$$\dot{m}\left[(\widehat{H}_2 - \widehat{H}_1) + \frac{1}{2}\left(v_2^2 - v_1^2\right) + g(h_2 - h_1)\right] = \dot{Q} + \dot{W}_s \qquad (1)$$

From Table A.3 in Appendix A, $\widehat{H}_1 = 3439.6\,\text{kJ/kg}$. Substitution of the numerical values into Eq. (1) gives

$$(1.5)\left\{\widehat{H}_2 - 3439.6 + \frac{1}{2}\frac{[(300)^2 - (65)^2]}{1000} - \frac{(9.8)(3)}{1000}\right\} = -30 - 700 \qquad (2)$$

Solving Eq. (2) for $\widehat{H}_2$ *gives*

$$\widehat{H}_2 = 2910.1 \, \text{kJ/kg}$$

Thus, $\widehat{H}_3 = \widehat{H}_2 = 2910.1 \, \text{kJ/kg}$. *At* $P_3 = 100 \, \text{kPa}$, $\widehat{H}^V = 2675.5 \, \text{kJ/kg}$. *Since* $\widehat{H}_3 > 2675.5 \, \text{kJ/kg}$, *then the steam is superheated at state 3. From Table A.3 in Appendix A, at* $100 \, \text{kPa}$

T (°C)	$\widehat{H}$ (kJ/kg)
200	2875.3
250	2974.3

By interpolation

$$T = 200 + \left(\frac{250 - 200}{2974.3 - 2875.3} \right) (2910.1 - 2875.3) = 217.6 \, °\text{C}$$

Example 4.17 *Steam flowing at a rate of* 0.4 kg/s *enters an adiabatic turbine at* 2 MPa *and* 450 °C. *The power output of the turbine is* 300 kW. *The exit stream from the turbine is at* 40 kPa *and it is sent to a heat exchanger where it is condensed at constant pressure to obtain saturated liquid. As cooling medium, liquid water is used; it enters the heat exchanger at* 15 °C *and leaves at* 55 °C. *Assume no heat losses from the heat exchanger to the surroundings.*

a) *What is the condition of steam leaving the turbine?*
b) *Calculate the mass flow rate of cooling water used.*

Solution

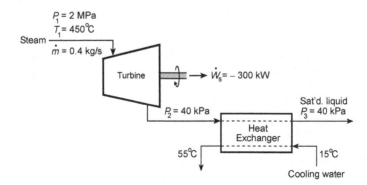

a) *The energy balance between states 1 and 2 is written as*

$$\dot{m}\left(\widehat{H}_2 - \widehat{H}_1\right) = \dot{W}_s \tag{1}$$

From Table A.3 in Appendix A, $\widehat{H}_1 = 3357.6\,\text{kJ/kg}$. *Substitution of the values into Eq. (1) gives*

$$\widehat{H}_2 = 3357.6 - \frac{300}{0.4} = 2607.6\,\text{kJ/kg}$$

From Table A.2 in Appendix A

$$\left.\begin{array}{c} P = 40\,\text{kPa} \\ (T^{sat} = 75.87\,^\circ\text{C}) \end{array}\right\} \begin{array}{l} \widehat{H}^L = 317.58\,\text{kJ/kg} \\ \Delta\widehat{H} = 2319.2\,\text{kJ/kg} \\ \widehat{H}^V = 2636.8\,\text{kJ/kg} \end{array}$$

Since $317.59 < \widehat{H}_2 < 2636.8$, *the steam at state 2 is a mixture of liquid and vapor, i.e., wet steam. Its quality is*

$$x = \frac{\widehat{H}_2 - \widehat{H}^L}{\Delta\widehat{H}} = \frac{2607.6 - 317.58}{2319.2} = 0.987$$

b) System: *Heat exchanger*

From Table A.2 in Appendix A, $\widehat{H}_3 = 317.58\,\text{kJ/kg}$. *From Eq. (4.3-10)*

$$\dot{m}_{\text{steam}}\left(\widehat{H}_3 - \widehat{H}_2\right) + \dot{m}_{\text{H}_2\text{O}}\,\widehat{C}_{P_{\text{H}_2\text{O}}}\Delta T_{\text{H}_2\text{O}} = 0 \tag{2}$$

Substitution of the numerical values into Eq. (2) gives

$$(0.4)(317.58 - 2607.6) + \dot{m}_{\text{H}_2\text{O}}(4.2)(55 - 15) = 0$$

Solving for $\dot{m}_{\text{H}_2\text{O}}$ *gives*

$$\dot{m}_{\text{H}_2\text{O}} = 5.45\,\text{kg/s}$$

Alternative solution: *Choosing the turbine and the heat exchanger as a system, Eq. (4.3-5) becomes*

$$\dot{m}_{\text{steam}}\left(\widehat{H}_3 - \widehat{H}_1\right) + \dot{m}_{\text{H}_2\text{O}}\,\Delta\widehat{H}_{\text{H}_2\text{O}} = \dot{W}_s$$

or

$$(0.4)(317.58 - 3357.6) + \dot{m}_{\text{H}_2\text{O}}(4.2)(55 - 15) = -300$$

Solving for $\dot{m}_{\mathrm{H_2O}}$ *gives*

$$\dot{m}_{\mathrm{H_2O}} = 5.45\,\mathrm{kg/s}$$

Comment: *Definition of a system plays a crucial role in the solution of a thermodynamics problem. Note that there is no need to find out the temperature and pressure at the exit of the turbine if one chooses the turbine and heat exchanger as a combined system.*

Example 4.18 *Steam flowing at a rate of* 9 kg/s *is at* 15 MPa *and* 700 °C. *It is first throttled to* 12.5 MPa *and then sent to an adiabatic two-stage turbine. Part of the steam leaving the first stage at* 3 MPa *and* 450 °C *is sent to a heat exchanger to heat* 12 kg/s *of air from* 20 °C *to* 135 °C. *The remaining part is expanded in the second stage to saturated vapor at* 100 kPa. *If the power output of the turbine is* 7 MW, *determine the condition of steam at the exit of the heat exchanger.*

Solution

From Table A.3 in Appendix A, $\widehat{H}_1 = 3840.1\,\mathrm{kJ/kg}$. *Since* $\Delta\widehat{H} = 0$ *across the throttling valve*

$$\widehat{H}_2 = \widehat{H}_1 = 3840.1\,\mathrm{kJ/kg}$$

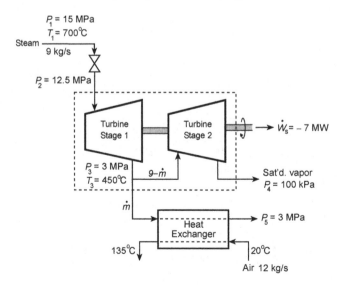

From the tables given in Appendix A

$$\widehat{H}_3 = 3344.0 \, \text{kJ/kg} \qquad \text{and} \qquad \widehat{H}_4 = 2675.5 \, \text{kJ/kg}$$

To calculate the mass flow rate of steam ($\dot{m}$) sent to the heat exchanger, the energy balance around the two-stage turbine (dotted rectangle in the figure) is needed:

$$\Delta \dot{H} = \underbrace{\dot{Q}}_{0} + \dot{W}_s$$

or

$$\left[\dot{m}\,(3344.0) + (9 - \dot{m})(2675.5) \right] - (9)(3840.1) = -7000$$

Solving for $\dot{m}$ gives

$$\dot{m} = 5.21 \, \text{kg/s}$$

The energy balance around the heat exchanger is written as

$$(5.21)(\widehat{H}_5 - 3344.0) + (12)\left(\frac{3.5 \times 8.314}{29} \right)(135 - 20) = 0$$

Solving for $\widehat{H}_5$ gives

$$\widehat{H}_5 = 3078.2 \, \text{kJ/kg}$$

At $P_5 = 3 \, \text{MPa}$, $\widehat{H}^V = 2804.2 \, \text{kJ/kg}$. Since $\widehat{H}_5 > 2804.2 \, \text{kJ/kg}$, then the steam is superheated at state 5. From Table A.3 in Appendix A, at 3 MPa

T (°C)	$\widehat{H}$ (kJ/kg)
300	2993.5
350	3115.3

By interpolation

$$T = 300 + \left(\frac{350 - 300}{3115.3 - 2993.5} \right)(3078.2 - 2993.5) = 334.8 \, ^\circ\text{C}$$

4.3.4   *Compressor and pump*

Compressors and pumps are used to increase the pressure of a fluid by doing work on the fluid through a rotating or reciprocating shaft. While compressors are used for gases, pumps are used for liquids. For most compressors and pumps it is found that the changes in kinetic and potential energy terms are usually quite small in comparison to the change in enthalpy, so that Eqs. (4.3-5) and (4.3-6) reduce to

$$\boxed{\Delta \dot{H} = \dot{Q} + \dot{W}_s \quad \text{and} \quad \Delta \widehat{H} = \widehat{Q} + \widehat{W}_s} \tag{4.3-15}$$

Example 4.19 *Air enters a compressor at* 1 bar *and* 25 °C *with a volumetric flow rate of* 10 m³/ min. *It exits the compressor at* 15 bar *and* 150 °C. *If the rate of heat transfer from the compressor is* 1.5 kW, *calculate the power input to drive the compressor under steady conditions.*

Solution

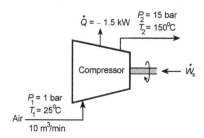

The mass flow rate of air can be calculated from the inlet conditions:

$$\dot{m} = \rho \dot{Q} = \left(\frac{P_1 M}{R T_1} \right) \dot{Q}$$

$$= \left[\frac{(1)(29)}{(8.314 \times 10^{-2})(298)} \right] \left(10 \, \frac{\text{m}^3}{\text{min}} \right) \left(\frac{1}{60} \, \frac{\text{min}}{\text{s}} \right) = 0.195 \, \text{kg/ s}$$

The use of Eq. (4.3-15) gives

$$(0.195) \left(\frac{3.5 \times 8.314}{29} \right) (150 - 25) = -1.5 + \dot{W}_s \tag{1}$$

The solution of Eq. (1) gives the power output as

$$\dot{W}_s \simeq 26 \, \text{kW}$$

Example 4.20 *Air at 1 atm and 20 °C is to be compressed in a reversible and adiabatic compressor to 5 atm. Calculate the power required to drive the compressor if the mass flow rate of air is 0.3 kg/ s.*

Solution

Since $Q = 0$, Equation (4.3-15) reduces to

$$\dot{W}_s = \Delta \dot{H} = \dot{m}\, \widehat{C}_P^*(T_2 - T_1) \tag{1}$$

The temperature at the exit of the compressor, T_2, is calculated from Eq. (4.1-77) as

$$T_2 = T_1 \left(\frac{P_2}{P_1}\right)^{(\gamma-1)/\gamma} = (293)\left(\frac{5}{1}\right)^{2/7} = 464\,\text{K}$$

Substitution of the numerical values into Eq. (1) gives

$$\dot{W}_s = (0.3)\left(\frac{3.5 \times 8.314}{29}\right)(464 - 293) = 51.5\,\text{kW}$$

Comment: *As stated in Section 4.1.3.4, adiabatic compression causes a tremendous increase in temperature. Since $\Delta H \propto \Delta T$, this leads to an increase in the power requirement of a compressor. Therefore, it is recommended to increase pressure gradually by using a multi-stage compressor with cooling of the gas between the stages.*

Example 4.21 *Air at 1 atm and 20 °C is to be compressed to 25 atm in a two-stage compressor. Assume that each stage of compression is adiabatic and reversible and that the air leaving the first stage is cooled at constant pressure to its initial temperature, i.e., 20 °C, before entering the second stage. Find the intermediate pressure, P_2, so as to minimize the power input to the compressor. The schematic diagram of the process is shown below.*

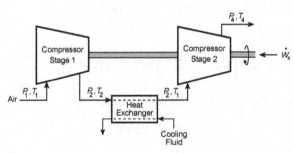

Solution

Since the compressor is adiabatic, Eq. (4.3-15) gives the power requirement of compression as

$$\dot{W}_s = \dot{m}\,\widehat{C}_P^* \left[(T_2 - T_1) + (T_4 - T_1)\right] = \dot{m}\,\widehat{C}_P^* T_1 \left(\frac{T_2}{T_1} + \frac{T_4}{T_1} - 2\right) \quad (1)$$

For stage 1

$$\frac{T_2}{T_1} = \left(\frac{P_2}{P_1}\right)^{(\gamma-1)/\gamma} \quad (2)$$

For stage 2

$$\frac{T_4}{T_1} = \left(\frac{P_4}{P_2}\right)^{(\gamma-1)/\gamma} \quad (3)$$

Substitution of Eqs. (2) and (3) into Eq. (1) gives

$$\dot{W}_s = \dot{m}\,\widehat{C}_P^* T_1 \left[\left(\frac{P_2}{P_1}\right)^{(\gamma-1)/\gamma} + \left(\frac{P_4}{P_2}\right)^{(\gamma-1)/\gamma} - 2\right] \quad (4)$$

For the minimization of power output, $d\dot{W}_s/dP_2 = 0$ or

$$\dot{m}\,\widehat{C}_P^* T_1 \left[\left(\frac{\gamma-1}{\gamma}\right)\frac{1}{P_1}\left(\frac{P_2}{P_1}\right)^{-1/\gamma} - \left(\frac{\gamma-1}{\gamma}\right)\left(\frac{P_4}{P_2^2}\right)\left(\frac{P_4}{P_2}\right)^{-1/\gamma}\right] = 0 \quad (5)$$

Simplification of Eq. (5) leads to

$$\left(\frac{P_1 P_4}{P_2^2}\right)^{1/\gamma} = \frac{P_1 P_4}{P_2^2} \quad \Longrightarrow \quad \frac{P_1 P_4}{P_2^2} = 1 \quad (6)$$

Thus

$$P_2 = \sqrt{P_1 P_4} = \sqrt{(1)(25)} = 5\,\text{atm}$$

Substitution of the numerical values into Eq. (4) gives the work per unit mass as

$$\widehat{W}_s = \frac{\dot{W}_s}{\dot{m}} = \left(\frac{3.5 \times 8.314}{29}\right)(293)\left[\left(\frac{5}{1}\right)^{2/7} + \left(\frac{25}{5}\right)^{2/7} - 2\right]$$

$$= 343.3\,\text{kJ/kg}$$

The following P-V diagram shows the paths followed during the compression process:

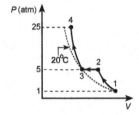

Comment: *The work required per unit mass with no intermediate cooling is*

$$\widehat{W}_s = \left(\frac{3.5 \times 8.314}{29}\right)(293)\left[\left(\frac{25}{1}\right)^{2/7} - 1\right] = 443.5\,\text{kJ/kg}$$

4.3.5 *Nozzle and diffuser*

A nozzle increases the velocity of a high-pressure fluid at the expense of its pressure and temperature. For example, a nozzle attached to the end of a garden hose results in a longer stream of water out of the hose. A diffuser is the opposite of a nozzle in which the pressure of a fluid is increased by decreasing its velocity.

Under steady conditions, the mass flow rate of a fluid remains constant, i.e.,

$$\dot{m} = \underbrace{\left(\frac{PM}{RT}\right)}_{\rho} vA = \text{constant} \qquad (4.3\text{-}16)$$

Therefore, an increase in velocity can be accomplished by decreasing cross-sectional area. On the other hand, an increase in cross-sectional area leads to a decrease in velocity. As shown in Fig. 4.9, while the cross-sectional area

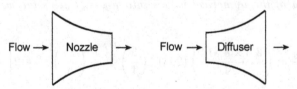

Fig. 4.9   A nozzle and a diffuser.

of a nozzle decreases in the flow direction, the cross-sectional area of a diffuser increases in the flow direction.

For nozzles and diffusers the change in potential energy is negligible. Therefore, Eqs. (4.3-5) and (4.3-6) simplify to

$$\boxed{\Delta \dot{H} + \Delta \dot{E}_K = \dot{Q} \quad \text{and} \quad \Delta \widehat{H} + \Delta \widehat{E}_K = \widehat{Q}} \tag{4.3-17}$$

Example 4.22 *Air enters a nozzle at* 450 kPa *and* 60 °C *at a velocity of* 65 m/ s, *and leaves it at* 100 kPa *and* 310 m/ s. *The heat loss is* 2 kJ/ kg *of air flowing through the nozzle.*

a) *Calculate the exit air temperature,*
b) *Calculate the exit area of the nozzle if the inlet area is* 100 cm^2.

Solution

$$\widehat{Q} = -2 \text{ kJ/kg}$$

$P_1 = 450 \text{ kPa}$
$T_1 = 60°C$
$v_1 = 65 \text{ m/s}$

Air $\longrightarrow$ 1 Nozzle

$P_2 = 100 \text{ kPa}$
$v_2 = 310 \text{ m/s}$

a) *Assuming air to be an ideal gas, Eq. (4.3-17) is expressed as*

$$\widehat{C}_P^*(T_2 - T_1) + \frac{1}{2}\left(v_2^2 - v_1^2\right) = \widehat{Q} \tag{1}$$

Solving for T_2 *gives*

$$T_2 = T_1 + \frac{\widehat{Q} - \frac{1}{2}\left(v_2^2 - v_1^2\right)}{\widehat{C}_P^*} = 60 + \frac{-2 - \dfrac{1}{2}\dfrac{\left[(310)^2 - (65)^2\right]}{1000}}{\dfrac{3.5 \times 8.314}{29}} = 12.2 °C$$

b) *Under steady conditions, the conservation of mass gives*

$$\dot{m}_1 = \dot{m}_2 \quad \Rightarrow \quad \rho_1 v_1 A_1 = \rho_2 v_2 A_2 \tag{2}$$

The density of an ideal gas is represented by

$$\rho = \frac{PM}{RT} \tag{3}$$

Substitution of Eq. (3) into Eq. (2) results in

$$\frac{P_1 v_1 A_1}{T_1} = \frac{P_2 v_2 A_2}{T_2} \tag{4}$$

Solving for A_2 gives

$$A_2 = A_1 \left(\frac{P_1}{P_2}\right)\left(\frac{v_1}{v_2}\right)\left(\frac{T_2}{T_1}\right)$$

$$= (100)\left(\frac{450}{100}\right)\left(\frac{65}{310}\right)\left(\frac{12.2 + 273}{60 + 273}\right) = 80.8\,\text{cm}^2$$

4.3.6   *Mixing chamber*

A mixing chamber is a device in which the mixing of two or more streams takes place. Consider the mixing chamber shown in Fig. 4.10[7].

The conservation of mass gives

$$\dot{m}_1 + \dot{m}_2 = \dot{m}_3 \tag{4.3-18}$$

If the changes in kinetic and potential energies are negligible, Eq. (4.3-5) reduces to[8]

$$\Delta \dot{H} = \dot{Q} \tag{4.3-19}$$

or

$$\dot{m}_3 \widehat{H}_3 - (\dot{m}_1 \widehat{H}_1 + \dot{m}_2 \widehat{H}_2) = \dot{Q} \tag{4.3-20}$$

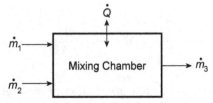

Fig. 4.10   A mixing chamber.

---

[7]The mixing chamber does not necessarily have to be a distinct chamber. For example, an ordinary T-elbow or a Y-elbow in a piping system to mix hot and cold streams also serves as a mixing chamber.

[8]Even if a stirrer is used in the mixing chamber to ensure a uniform temperature, the magnitude of the shaft work is considered negligible compared to the other terms.

The use of Eqs. (4.3-18) and (4.3-20) enables one to estimate the unknown quantities.

Example 4.23 *Liquid water flowing at a mass flow rate of 3.5 kg/s is at 0.3 MPa and 25 °C. It is mixed with superheated steam at 0.3 MPa and 250 °C in a mixing chamber to produce liquid water at 0.3 MPa and 75 °C. If the rate of heat loss from the mixing chamber is 15 kW, calculate the mass flow rate of the superheated steam.*

Solution

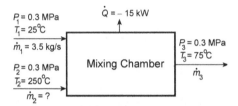

For the compressed liquid streams, from Table A.1 in Appendix A

$$\widehat{H}_1 = \widehat{H}^L|_{25\,°C} = 104.89 \, \text{kJ/kg} \qquad \text{and} \qquad \widehat{H}_3 = \widehat{H}^L|_{75\,°C} = 313.93 \, \text{kJ/kg}$$

For the superheated steam, from Table A.3 in Appendix A

$$\widehat{H}_2 = 2967.6 \, \text{kJ/kg}$$

Conservation of mass, Eq. (4.3-18), is

$$\dot{m}_1 + \dot{m}_2 = \dot{m}_3 \tag{1}$$

Conservation of energy, Eq. (4.3-20), is

$$(\dot{m}_1 + \dot{m}_2)\,\widehat{H}_3 - (\dot{m}_1\widehat{H}_1 + \dot{m}_2\widehat{H}_2) = \dot{Q} \tag{2}$$

Solving Eq. (2) for $\dot{m}_2$ gives

$$\dot{m}_2 = \frac{\dot{Q} + \dot{m}_1(\widehat{H}_1 - \widehat{H}_3)}{\widehat{H}_3 - \widehat{H}_2} \tag{3}$$

Substitution of the numerical values into Eq. (3) results in

$$\dot{m}_2 = \frac{-15 + (3.5)(104.89 - 313.93)}{313.93 - 2967.6} = 0.281 \, \text{kg/s}$$

4.3.7   *Mechanical energy balance*

For steady flow of an incompressible fluid through a pipe, Eq. (4.3-6) becomes

$$\Delta \widehat{H} + \Delta \widehat{E}_K + \Delta \widehat{E}_P = \widehat{Q} + \widehat{W}_s \tag{4.3-21}$$

The change in enthalpy is expressed as

$$\Delta \widehat{H} = \Delta \widehat{U} + \frac{\Delta P}{\rho} \tag{4.3-22}$$

On the other hand, changes in kinetic and potential energies are expressed as

$$\Delta \widehat{E}_K = \frac{1}{2} \Delta v^2 \quad \text{and} \quad \Delta \widehat{E}_P = g \, \Delta z \tag{4.3-23}$$

Substitution of Eqs. (4.3-22) and (4.3-23) into Eq. (4.3-21) gives

$$\frac{\Delta P}{\rho} + \frac{1}{2} \Delta v^2 + g \, \Delta z + \underbrace{(\Delta \widehat{U} - \widehat{Q})}_{\widehat{E}_v} = \widehat{W}_s \tag{4.3-24}$$

The term $(\Delta \widehat{U} - \widehat{Q})$ is referred to as the *friction loss per unit mass* and designated by $\widehat{E}_v$. It represents the irreversible degradation of mechanical energy into thermal energy. Thus, Eq. (4.3-24) is expressed as

$$\boxed{\frac{\Delta P}{\rho} + \frac{\Delta v^2}{2} + g \, \Delta z + \widehat{E}_v = \widehat{W}_s} \tag{4.3-25}$$

and is known as the *mechanical energy balance* (or *engineering Bernoulli equation*). Note that Eq. (4.3-25) does not address thermal energy. It is mainly used for fluid flow problems that do not involve heat transfer or chemical reaction(s). For flow in a pipe of diameter D and length L with an average velocity of v, the friction loss per unit mass is given by

$$\widehat{E}_v = \frac{2 f L \, v^2}{D} \tag{4.3-26}$$

where f is the *Fanning friction factor*.

Example 4.24 *Water is pumped from a large reservoir to an elevation of 15 m through a pipe of 5 cm internal diameter as shown in the figure below and issues from a nozzle to form a free jet. Assuming negligible frictional losses, calculate the power of the pump.*

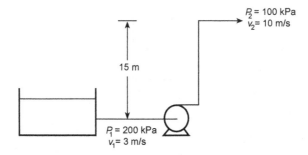

Solution

The pressure in a free jet is atmospheric, i.e., $100\,\mathrm{kPa}$. *Taking* $\widehat{E}_v = 0$,
Eq. (4.3-25) gives

$$\widehat{W}_s = \frac{P_2 - P_1}{\rho} + \frac{v_2^2 - v_1^2}{2} + g\,(z_2 - z_1)$$

$$= \frac{100 - 200}{1000} + \left[\frac{(10)^2 - (3)^2}{2}\right]\frac{1}{1000} + \frac{(9.8)(15)}{1000} = 0.0925\,\mathrm{kJ/kg}$$

The mass flow rate is calculated by using Eq. (4.2-4) as

$$\dot{m} = \rho v A = (1000)(3)\left[\frac{\pi(0.05)^2}{4}\right] = 5.89\,\mathrm{kg/s}$$

The power required is

$$\dot{W}_s = \dot{m}\,\widehat{W}_s = (5.89)(0.0925) = 0.54\,\mathrm{kW}$$

Simplification of the mechanical energy balance

▶ Static liquid

If the liquid is stagnant, i.e., $v = 0$, Eq. (4.3-25) reduces to

$$\frac{\Delta P}{\rho} + g\,\Delta z = 0 \qquad\qquad (4.3\text{-}27)$$

Consider a static liquid in a tank open to the atmosphere as shown in

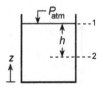

Fig. 4.11   Static liquid in a tank open to the atmosphere.

Fig. 4.11. Application of Eq. (4.3-27) between planes 1 and 2 gives

$$\frac{P_2 - P_{atm}}{\rho} + g\underbrace{(z_2 - z_1)}_{-h} = 0 \tag{4.3-28}$$

or

$$P = P_{atm} + \rho g h \tag{4.3-29}$$

indicating that the hydrostatic pressure of a liquid increases linearly with the depth of the liquid.

For water $\rho \simeq 1000\,\text{kg}/\text{m}^3$. Thus, the height of water that is equivalent to 1 atm (1.013×10^5 Pa) is

$$h = \frac{P}{\rho g} = \frac{1.013 \times 10^5}{(1000)(9.8)} = 10.3\,\text{m}$$

As a rule of thumb, keep in mind that it takes 1 atm to move water up approximately 10 m.

▶ **Bernoulli equation**

When viscous (or frictional) force is negligible compared to the other forces acting on a fluid particle, $\widehat{E}_v \simeq 0$. Moreover, if the shaft work is zero, Eq. (4.3-25) reduces to

$$\Delta\left(\frac{P}{\rho} + \frac{1}{2}v^2 + gz\right) = 0 \tag{4.3-30}$$

or

$$\boxed{\frac{P}{\rho} + \frac{1}{2}v^2 + gz = \text{constant}} \tag{4.3-31}$$

which is known as the *Bernoulli equation*. Despite its simplifying assumptions, the Bernoulli equation has a wide range of application.

The Bernoulli equation simply states that an increase in fluid velocity results in a decrease in pressure. Keeping this statement in mind, try to answer the following questions:

- How does an airplane fly? What is so special about the design of the wings?
- Why does the shower curtain always move towards water?
- In a subway station, why is it dangerous to stand near the edge of the platform as the train arrives?
- Why are the roofs of houses blown off during high winds?
- What is the reason for the rise in the oceans (surge tide) during a hurricane?

Example 4.25 *Consider a siphon to drain water from a tank as shown in the figure below.*

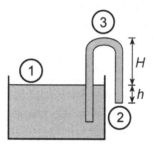

a) *Obtain an expression for the average velocity of water as it flows out of the lower end of the siphon.*

b) *Show that the siphon can never lift water more than approximately $10\,\text{m}$.*

Solution

a) *Note that the pressure at the free surface of the water in the tank is the same as that at the lower end of the siphon, i.e., $P_1 = P_2 = P_{atm}$. Application of the Bernoulli equation between points 1 and 2 leads to*

$$v_2^2 = v_1^2 + 2gh \tag{1}$$

If $A_1 \gg A_2$, then v_1 is very small compared to v_2 and can be neglected. As a result, Eq. (1) simplifies to

$$v_2 = \sqrt{2gh} \tag{2}$$

b) *Application of the Bernoulli equation between points 2 and 3 gives*

$$P_3 = P_{atm} - \rho g(h + H) \tag{3}$$

indicating that the pressure at point 3 is less than atmospheric. Thus,

$$P_{atm} > \rho g(h + H) \quad \Rightarrow \quad h + H < \frac{P_{atm}}{\rho g} = \frac{1.013 \times 10^5}{(1000)(9.8)} = 10.3\,\mathrm{m}$$

Theoretically, P_3 cannot be equal to zero since water starts to evaporate when $P_3 < P^{vap}$. The presence of vapor bubbles, known as "cavitation", interferes the flow of water, causing operating problems during the use of a siphon.

4.4 The First Law for Unsteady-State Processes

The working equations for solving unsteady-state problems are the conservation of mass and energy. The conservation of mass is given by Eq. (4.2-7), i.e.,

$$\boxed{dm_{in} - dm_{out} = dm_{sys}} \tag{4.4-1}$$

The conservation of energy is given by Eq. (4.2-17). In most practical applications, the kinetic and potential energy changes are negligible compared to changes in internal energy. Under these circumstances Eq. (4.2-17) simplifies to

$$\boxed{\widehat{H}_{in}\, dm_{in} - \widehat{H}_{out}\, dm_{out} + \delta Q + \delta W = d(m\widehat{U})_{sys}} \tag{4.4-2}$$

where the term δW includes the work associated with the displacement of system boundaries and the shaft work.

In terms of molar basis, Eqs. (4.4-1) and (4.4-2) take the form

$$\boxed{dn_{in} - dn_{out} = dn_{sys}} \tag{4.4-3}$$

and

$$\boxed{\widetilde{H}_{in}\, dn_{in} - \widetilde{H}_{out}\, dn_{out} + \delta Q + \delta W = d(n\widetilde{U})_{sys}} \tag{4.4-4}$$

While Eqs. (4.4-1) and (4.4-2) are used in problems involving steam, Eqs. (4.4-3) and (4.4-4) are preferred in problems involving an ideal gas. First let us consider how to simplify these equations for some standard type problems.

• Type I - Filling a rigid tank from a source kept at constant T & P

Choosing the contents of the tank as the system, Eqs. (4.4-1) and (4.4-2) reduce to

$$dm_{in} = dm_{sys} \tag{4.4-5}$$

$$\widehat{H}_{in}\, dm_{in} + \delta Q = d(m\widehat{U})_{sys} \tag{4.4-6}$$

Substitution of Eq. (4.4-5) into Eq. (4.4-6) results in

$$\widehat{H}_{in}\, dm_{sys} + \delta Q = d(m\widehat{U})_{sys} \tag{4.4-7}$$

Since the source is at constant T and P, in other words, $\widehat{H}_{in}$ = constant, Eq. (4.4-7) can be integrated from an initial state (state 1) to a final state (state 2)

$$\widehat{H}_{in}(m_2 - m_1) + Q = m_2\widehat{U}_2 - m_1\widehat{U}_1 \tag{4.4-8}$$

or

$$\boxed{Q = m_2(\widehat{U}_2 - \widehat{H}_{in}) - m_1(\widehat{U}_1 - \widehat{H}_{in})} \tag{4.4-9}$$

In terms of molar quantities, Eq. (4.4-9) is written as

$$\boxed{Q = n_2(\widetilde{U}_2 - \widetilde{H}_{in}) - n_1(\widetilde{U}_1 - \widetilde{H}_{in})} \tag{4.4-10}$$

For an ideal gas

$$\widetilde{H}_{in} = \widetilde{U}_{in} + (P\widetilde{V})_{in} = \widetilde{U}_{in} + RT_{in} \tag{4.4-11}$$

Substitution of Eq. (4.4-11) into Eq. (4.4-10) gives

$$Q = n_2 \left[\underbrace{(\widetilde{U}_2 - \widetilde{U}_{in})}_{\widetilde{C}_V^*(T_2 - T_{in})} - RT_{in} \right] - n_1 \left[\underbrace{(\widetilde{U}_1 - \widetilde{U}_{in})}_{\widetilde{C}_V^*(T_1 - T_{in})} - RT_{in} \right] \tag{4.4-12}$$

Simplification of Eq. (4.4-12) yields

$$Q = n_2 \left(\widetilde{C}_V^* T_2 - \widetilde{C}_P^* T_{in} \right) - n_1 \left(\widetilde{C}_V^* T_1 - \widetilde{C}_P^* T_{in} \right) \tag{4.4-13}$$

Dividing each term in Eq. (4.4-13) by $\widetilde{C}_V^*$ leads to

$$\boxed{\frac{Q}{\widetilde{C}_V^*} = n_2 \left(T_2 - \gamma T_{in}\right) - n_1 \left(T_1 - \gamma T_{in}\right)} \qquad (4.4\text{-}14)$$

Example 4.26 *An insulated rigid tank of volume $0.3\,\mathrm{m}^3$ is connected to a large pipeline carrying air at $1400\,\mathrm{kPa}$ and $300\,^\circ\mathrm{C}$ as shown in the figure below. The valve between the pipeline and the tank is opened and the tank fills with air until the pressure is $1400\,\mathrm{kPa}$ and then the valve is closed. Determine the final temperature of the air in the tank if:*

a) *The tank is initially empty,*
b) *The tank initially contains air at $350\,\mathrm{kPa}$ and $139\,^\circ\mathrm{C}$.*

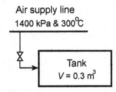

Air supply line
1400 kPa & 300°C

Tank
V = 0.3 m³

Solution

System: *Contents of the tank*

a) *Since air may be assumed to be an ideal gas, Eq. (4.4-14) is to be used for the solution. Since $Q = 0$ and $n_1 = 0$, Eq. (4.4-14) simplifies to*

$$T_2 = \gamma\, T_{in} \qquad (1)$$

Substitution of the numerical values into Eq. (1) gives

$$T_2 = \left(\frac{7}{5}\right)(300 + 273) = 802\,\mathrm{K}\ (529\,^\circ\mathrm{C})$$

b) *Since $Q = 0$, Eq. (4.4-14) simplifies to*

$$n_2 \left(T_2 - \gamma T_{in}\right) = n_1 \left(T_1 - \gamma T_{in}\right) \qquad (2)$$

Expressing n_1 and n_2 in terms of pressure and temperature yields

$$\frac{P_2}{T_2}\left(\gamma T_{in} - T_2\right) = \frac{P_1}{T_1}\left(\gamma T_{in} - T_1\right) \qquad (3)$$

Solving for T_2 leads to

$$T_2 = \frac{T_{in}}{\frac{1}{\gamma}\left(1 - \frac{P_1}{P_2}\right) + \frac{P_1}{P_2}\frac{T_{in}}{T_1}} \tag{4}$$

Substitution of the numerical values into Eq. (4) gives

$$T_2 = \frac{300 + 273}{\left(\frac{5}{7}\right)\left(1 - \frac{350}{1400}\right) + \left(\frac{350}{1400}\right)\left(\frac{300 + 273}{139 + 273}\right)} = 648.6\,\text{K}\ (375.6\,^\circ\text{C})$$

Example 4.27 *An insulated rigid tank of $0.3\,\text{m}^3$ volume is connected to a large pipeline carrying steam at $1400\,\text{kPa}$ and $300\,^\circ\text{C}$ as shown in the figure below. The valve between the pipeline and the tank is opened and steam is allowed to enter the tank until the pressure reaches $1400\,\text{kPa}$, at which point the valve is closed. Determine the final temperature of the steam for the following cases:*

a) *The tank is initially empty,*

b) *The tank initially contains saturated steam at $350\,\text{kPa}$.*

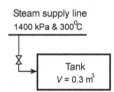

Steam supply line
1400 kPa & 300°C

Tank
$V = 0.3\,\text{m}^3$

Solution

System: *Contents of the tank*

a) *Since the properties of steam are given in terms of mass basis, it is convenient to use Eq. (4.4-9) for the solution. In this case $Q = 0$ and $m_1 = 0$. Thus, Eq. (4.4-9) simplifies to*

$$\widehat{H}_{in} = \widehat{U}_2 \tag{1}$$

From Table A.3 in Appendix A

$$\left.\begin{array}{l} P = 1.4\,\text{MPa} \\ T = 300\,^\circ\text{C} \end{array}\right\} \widehat{H}_{in} = 3040.4\,\text{kJ/kg}$$

Thus, from Eq. (1)

$$\widehat{U}_2 = 3040.4 \, \text{kJ/kg}$$

The independent intensive properties at the final state are

$$\widehat{U}_2 = 3040.4 \, \text{kJ/kg} \quad and \quad P_2 = 1400 \, \text{kPa}$$

At 1.4 MPa, $\widehat{U}^V = 2592.8 \, \text{kJ/kg}$. Since $\widehat{U}_2 > 2592.8 \, \text{kJ/kg}$, superheated steam exists at the final state. From Table A.3 in Appendix A, at 1400 kPa

$T \, (°C)$	$\widehat{U} \, (\text{kJ/kg})$
400	2952.5
500	3121.1

By interpolation

$$T = 400 + \left(\frac{500 - 400}{3121.1 - 2952.5} \right) (3040.4 - 2952.5) = 452.1 \, °C$$

Note that part of the energy represented by enthalpy is the flow energy. As the steam enters the tank, the flow stops and this flow energy is converted to internal energy. As a result, the temperature of steam increases from 300 °C to 452.1 °C.

b) *Since $Q = 0$, Eq. (4.4-9) simplifies to*

$$m_2(\widehat{U}_2 - \widehat{H}_{in}) = m_1(\widehat{U}_1 - \widehat{H}_{in}) \tag{2}$$

or

$$\widehat{H}_{in}(m_2 - m_1) = m_2 \widehat{U}_2 - m_1 \widehat{U}_1 \tag{3}$$

It is first necessary to find out the properties at the initial state. From Table A.2 in Appendix A

$$\left. \begin{array}{l} P = 0.350 \, \text{MPa} \\ (T^{sat} \simeq 139 \, °C) \end{array} \right\} \begin{array}{l} \widehat{V}_1 = 0.5243 \, \text{m}^3/\text{kg} \\ \widehat{U}_1 = 2548.9 \, \text{kJ/kg} \end{array}$$

The initial mass contained within the tank is

$$m_1 = \frac{V}{\widehat{V}_1} = \frac{0.3}{0.5243} = 0.572 \, \text{kg}$$

Substitution of the values into Eq. (3) gives

$$3040.4 \left(\frac{0.3}{\widehat{V}_2} - 0.572 \right) = \frac{0.3}{\widehat{V}_2} \widehat{U}_2 - (0.572)(2548.9) \tag{4}$$

Simplification of Eq. (4) results in

$$\widehat{U}_2 = 3040.4 - 937.1 \, \widehat{V}_2 \tag{5}$$

The pressure at the final state is 1.4 MPa *and the relationship between* $\widehat{U}_2$ *and* $\widehat{V}_2$ *is represented by Eq. (5). The temperature can be found by a trial-and-error procedure as follows:*

- *Assume a temperature* T_2,
- *At* T_2 *and* 1.4 MPa, *find out* $\widehat{U}_2$ *and* $\widehat{V}_2$ *from the steam table,*
- *Substitute the values of* $\widehat{U}_2$ *and* $\widehat{V}_2$ *into Eq. (5) and check whether it is satisfied.*

Alternatively, the temperature can be determined by a graphical procedure as follows:

- *From Table A.3 in Appendix A, plot* $\widehat{U}_2$ *versus* $\widehat{V}_2$ *at a pressure of* 1.4 MPa. *Note that the values are*

T (°C)	$\widehat{V}_2$ (m^3/ kg)	$\widehat{U}_2$ (kJ/ kg)
200	0.14302	2603.1
250	0.16350	2698.3
300	0.18228	2785.2
350	0.2003	2869.2
400	0.2178	2952.5

- *On the same graph, plot Eq. (5). The easiest way of drawing Eq. (5) is to give two values for* $\widehat{V}_2$ *and calculate* $\widehat{U}_2$ *from Eq. (5). Then join the two points with a straight line. The intersection of this straight line with the* $\widehat{U}_2$ *versus* $\widehat{V}_2$ *curve gives*

$$\widehat{U}_2 = 2855 \, \text{kJ/ kg} \quad \text{and} \quad \widehat{V}_2 = 0.197 \, \text{m}^3/ \text{kg}$$

The final temperature is found by interpolation as

$$T_2 = 300 + \left(\frac{350 - 300}{2869.2 - 2785.2} \right) (2855 - 2785.2) = 341.6 \, °\text{C}$$

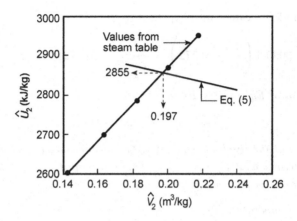

• **Type II - Emptying a rigid tank**

Choosing the contents of the tank as the system, Eqs. (4.4-1) and (4.4-2) reduce to

$$- dm_{out} = dm_{sys} \qquad (4.4\text{-}15)$$

$$- \widehat{H}_{out} \, dm_{out} + \delta Q = d(m\widehat{U})_{sys} \qquad (4.4\text{-}16)$$

Substitution of Eq. (4.4-15) into Eq. (4.4-16) results in

$$\delta Q = d(m\widehat{U})_{sys} - \widehat{H}_{out} \, dm_{sys} \qquad (4.4\text{-}17)$$

Note that $\widehat{H}_{out} = \widehat{H}_{sys}$. Since the gas properties within the tank change continuously, i.e., $\widehat{H}_{sys}$ is a function of m_{sys}, Eq. (4.4-17) cannot be integrated. To circumvent this problem, expansion of the first term on the right-hand side of Eq. (4.4-17) gives

$$\boxed{\delta Q = m_{sys} \, d\widehat{U}_{sys} + (\widehat{U}_{sys} - \widehat{H}_{sys})dm_{sys}} \qquad (4.4\text{-}18)$$

In terms of molar quantities, Eq. (4.4-18) is written as

$$\boxed{\delta Q = n_{sys} \, d\widetilde{U}_{sys} + (\widetilde{U}_{sys} - \widetilde{H}_{sys})dn_{sys}} \qquad (4.4\text{-}19)$$

For an ideal gas, Eq. (4.4-19) reduces to

$$\delta Q = n_{sys} \, \widetilde{C}_V^* \, dT_{sys} - R T_{sys} \, dn_{sys} \qquad (4.4\text{-}20)$$

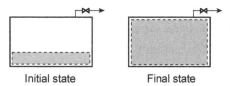

Initial state   Final state

Fig. 4.12   Choosing the gas remaining in the tank as a system during emptying of a tank.

If $\widetilde{C}_V^*$ is constant and the process is adiabatic, Eq. (4.4-20) takes the form

$$(\gamma - 1)\frac{dn_{sys}}{n_{sys}} = \frac{dT_{sys}}{T_{sys}} \tag{4.4-21}$$

Integration of Eq. (4.4-21) gives

$$\left(\frac{n_2}{n_1}\right)^{\gamma-1} = \frac{T_2}{T_1} \tag{4.4-22}$$

The use of the ideal gas equation of state to express the number of moles gives

$$\left(\frac{P_2}{P_1}\right)^{\gamma-1}\left(\frac{T_1}{T_2}\right)^{\gamma-1} = \frac{T_2}{T_1} \quad \Rightarrow \quad \frac{T_2}{T_1} = \left(\frac{P_2}{P_1}\right)^{(\gamma-1)/\gamma} \tag{4.4-23}$$

Note that Eq. (4.4-23) is identical with Eq. (4.1-77), which is developed for a closed system undergoing a reversible adiabatic process. Therefore, during the emptying process, it is possible to conclude that the "gas remaining in the tank" undergoes a reversible adiabatic expansion. Thus, one can solve the problem by choosing the system as the contents of the tank at the final state. The same amount of gas obviously occupies less volume at the initial state as shown in Fig. 4.12. This is a closed system and since the gas on one side of the imaginary boundary has precisely the same temperature as the gas on the other side, it is possible to assume that there is no heat transfer across the boundary, so that the system is adiabatic. Also, with the exception of the region near the valve (which is outside the system), the gas in the tank undergoes a uniform expansion so there will be no pressure, velocity, or temperature gradients within the tank. Therefore, it is possible to assume that the changes taking place in the system occur reversibly.

Example 4.28 *A rigid tank of $1\,\mathrm{m}^3$ volume initially contains air at $2.5\,\mathrm{MPa}$ and $20\,°\mathrm{C}$. A relief valve is opened slightly, allowing air to escape to the atmosphere. The valve is closed when the pressure in the tank reaches $350\,\mathrm{kPa}$.*

a) *Calculate the amount of heat that must be added so as to keep the tank contents at 20 °C throughout the process.*

b) *Calculate the final temperature if the process takes place adiabatically.*

Solution

a) System: *Contents of the tank*

Since $dT_{sys} = 0$, Eq. (4.4-20) reduces to

$$\delta Q = -RT_{sys} \, dn_{sys} \tag{1}$$

Integration of Eq. (1) leads to

$$Q = -RT_{sys}(n_2 - n_1) = V(P_1 - P_2) \tag{2}$$

Substitution of the numerical values into Eq. (2) gives the amount of heat transferred as

$$Q = (1)(2500 - 350) = 2150 \, \text{kJ}$$

b) System: *The gas remaining in the tank*

Since the gas remaining in the tank undergoes a reversible and adiabatic expansion, the use of Eq. (4.4-23) gives the final gas temperature in the tank as

$$T_2 = T_1 \left(\frac{P_2}{P_1}\right)^{(\gamma-1)/\gamma} = (20 + 273)\left(\frac{350}{2500}\right)^{2/7} = 167 \, \text{K} \tag{3}$$

• Type III - Filling and/or emptying a tank with fixed or variable boundaries

In this case, conservation of mass, Eq. (4.4-1), and conservation of energy, Eq. (4.4-2), should be first simplified according to the problem statement. Combination of these equations then should be integrated to obtain the unknown quantity.

Example 4.29 *A steam boiler of 80 m³ volume contains equal volumes of liquid and vapor in equilibrium with each other at 2 MPa. Over a certain time interval, saturated vapor is withdrawn from the boiler while simultaneously 2800 kg of liquid water at 50 °C is added to the boiler. Throughout the process heat is added to the boiler so as to keep pressure constant at 2 MPa. At the end of the process, the saturated liquid remaining in the boiler occupies 25% of the boiler volume, and saturated vapor fills the rest of the boiler. How much heat was added to the boiler during the process?*

Solution

System: *Steam boiler*

From Table A.2 in Appendix A, at 2 MPa

$\widehat{V}^L = 0.001177 \, \text{m}^3/\text{kg}$ $\widehat{U}^L = 906.42 \, \text{kJ}/\text{kg}$
$\widehat{V}^V = 0.09963 \, \text{m}^3/\text{kg}$ $\widehat{U}^V = 2600.3 \, \text{kJ}/\text{kg}$ $\widehat{H}^V = 2799.5 \, \text{kJ}/\text{kg}$

From Table A.1 in Appendix A

$\left. \begin{array}{l} T = 50\,^{\circ}\text{C} \\ (12.349\,\text{kPa}) \end{array} \right\} \quad \widehat{H}^L = 209.33 \, \text{kJ}/\text{kg}$

Equations (4.4-1) and (4.4-2) simplify to

$$dm_{in} - dm_{out} = dm_{sys} \tag{1}$$

$$\widehat{H}_{in}\, dm_{in} - \widehat{H}_{out}\, dm_{out} + \delta Q + \underbrace{\delta W}_{0} = d(m\widehat{U})_{sys} \tag{2}$$

Note that $\widehat{H}_{in} = 209.33 \, \text{kJ}/\text{kg}$ *and* $\widehat{H}_{out} = 2799.5 \, \text{kJ}/\text{kg}$. *Since these quantities remain constant throughout the process, integration of Eqs. (1) and (2) from the initial state (state 1) to the final state (state 2) leads to*

$$m_{in} - m_{out} = m_2 - m_1 \tag{3}$$

$$\widehat{H}_{in}\, m_{in} - \widehat{H}_{out}\, m_{out} + Q = \underbrace{m_2 \widehat{U}_2}_{U_2} - \underbrace{m_1 \widehat{U}_1}_{U_1} \tag{4}$$

• Initial state (State 1)

Since the liquid and vapor occupy equal volumes, the masses of liquid and vapor are

$$\text{Mass of liquid} = \frac{(0.5)(80)}{0.001177} = 33,985 \, \text{kg}$$

$$\text{Mass of vapor} = \frac{(0.5)(80)}{0.09963} = 401 \, \text{kg}$$

Thus,

$$m_1 = 33,985 + 401 = 34,386\,\text{kg}$$

$$U_1 = (33,985)(906.42) + (401)(2600.3) = 31,847,404\,\text{kJ}$$

• **Final state (State 2)**

Since the liquid and vapor occupy 25% and 75% of the total volume, respectively, the masses of liquid and vapor are

$$\text{Mass of liquid} = \frac{(0.25)(80)}{0.001177} = 16,992\,\text{kg}$$

$$\text{Mass of vapor} = \frac{(0.75)(80)}{0.09963} = 602\,\text{kg}$$

Thus,

$$m_2 = 16,992 + 602 = 17,594\,\text{kg}$$

$$U_2 = (16,992)(906.42) + (602)(2600.3) = 16,967,269\,\text{kJ}$$

The amount of vapor withdrawn is calculated from Eq. (3) as

$$m_{out} = m_1 - m_2 + m_{in} = 34,386 - 17,594 + 2800 = 19,592\,\text{kg}$$

Substitution of the values into Eq. (4) gives the amount of heat transferred as

$$\begin{aligned}
Q &= U_2 - U_1 - \widehat{H}_{in}\,m_{in} + \widehat{H}_{out}\,m_{out} \\
&= 16,967,269 - 31,847,404 - (209.33)(2800) + (2799.5)(19,592) \\
&= 39.38 \times 10^6\,\text{kJ}
\end{aligned}$$

Example 4.30 *An insulated rigid tank of $0.2\,\text{m}^3$ volume is connected to a large pipeline carrying nitrogen at 10 bar and 70 °C. The valve between the pipeline and the tank is opened and nitrogen is admitted to the tank at a constant molar flow rate of $0.14\,\text{mol/s}$. Simultaneously, nitrogen is withdrawn from the tank, also at a constant molar flow rate of $0.14\,\text{mol/s}$. Calculate the temperature and pressure within the tank after 1 minute if the tank initially contains nitrogen at 2 bar and 35 °C. Nitrogen may be assumed to be an ideal gas with a constant $\widetilde{C}_P^*$ of $30\,\text{J/mol.K}$.*

Solution

System: *Contents of the tank*

Contrary to the previous examples, time is involved in this problem. Therefore, the starting equations are Eqs. (4.2-3) and (4.2-16), written in terms of molar quantities:

$$\dot{n}_{in} - \dot{n}_{out} = \frac{dn_{sys}}{dt} \tag{1}$$

$$\widetilde{H}_{in}\,\dot{n}_{in} - \widetilde{H}_{out}\,\dot{n}_{out} + \dot{Q} - P_{sys}\frac{dV_{sys}}{dt} + \dot{W}_s = \frac{d(n\widetilde{U})_{sys}}{dt} \tag{2}$$

Since $\dot{n}_{in} = \dot{n}_{out} = \dot{n}$, Eq. (1) reduces to

$$\frac{dn_{sys}}{dt} = 0 \quad\Rightarrow\quad n_{sys} = \text{constant} = n_o$$

where

$$n_o = \frac{(2)(0.2)}{(8.314 \times 10^{-5})(35 + 273)} = 15.6\,\text{mol}$$

Eq. (2) simplifies to

$$\widetilde{H}_{in}\,\dot{n}_{in} - \widetilde{H}_{out}\,\dot{n}_{out} = \frac{d(n\widetilde{U})_{sys}}{dt} \tag{3}$$

or

$$\dot{n}\left(\widetilde{H}_{in} - \widetilde{H}_{out}\right) = n_o\frac{d\widetilde{U}_{sys}}{dt} \tag{4}$$

Assuming complete mixing within the tank at any instant, i.e., $\widetilde{H}_{out} = \widetilde{H}_{sys}$, and expressing enthalpy and internal energy in terms of temperature give

$$\dot{n}\,\widetilde{C}_P^*(T_{in} - T_{sys}) = n_o\widetilde{C}_V^*\frac{dT_{sys}}{dt} \tag{5}$$

Note that Eq. (5) is a separable equation. Rearrangement of Eq. (5) results in

$$\int_{T_o}^{T}\frac{dT_{sys}}{T_{in} - T_{sys}} = \frac{\dot{n}\,\gamma}{n_o}\int_0^t dt \tag{6}$$

Integration gives

$$T = T_{in} - (T_{in} - T_o)\exp\left(-\frac{\dot{n}\gamma\,t}{n_o}\right) \tag{7}$$

Substitution of the numerical values into Eq. (7) gives the temperature as

$$T = 343 - (343 - 308)\exp\left[-\frac{(0.14)(7/5)(60)}{15.6}\right] = 326.5\,\text{K}$$

The pressure is

$$P = \frac{nRT}{V} = \frac{(15.6)(8.314 \times 10^{-5})(326.5)}{0.2} = 2.12\,\text{bar}$$

4.5 Interpretation of Adages and Songs in Terms of the First Law of Thermodynamics

The following adages (Smith, 1975) express the idea of conservation:

- "You can't get something for nothing", i.e., everything costs something.
- "You can't make an omelette without breaking eggs", i.e., it is hard to achieve something important without causing unpleasant effects.
- "You can't have your cake and eat it too", i.e., once you've eaten your cake, you won't have it any more.
- "There are no gains without pains." (Benjamin Franklin)
- "There is an uphill for every downhill, and a downhill for every uphill."

The following songs express principles of conservation and cyclic processes:

- "Where Have All the Flowers Gone" by Pete Seeger
- "Spinning Wheel[9]" by Blood, Sweat and Tears

Problems

Problems related to Section 4.1

4.1 A closed thermodynamic system undergoes a cyclic process as shown in the following table:

Process	Q (kJ)	ΔU (kJ)
$1 \to 2$	100	200
$2 \to 3$	300	?
$3 \to 1$	-50	-200

---

[9]The song starts with the phrase "What goes up, must come down".

Calculate W_{23} and the net work output of the cycle.

(**Answer:** $-300\,\text{kJ}$, $-350\,\text{kJ}$)

4.2 Consider 1 mole of an ideal gas with constant heat capacities. For each of the following reversible processes, determine the trend of change, i.e., greater than zero, equal to zero, less than zero, for the quantities or changes in properties between initial and final states.

Process	ΔT	ΔP	$\Delta \widetilde{V}$	$\widetilde{Q}$	$\widetilde{W}$	$\Delta \widetilde{U}$	$\Delta \widetilde{H}$
Isothermal expansion							
Constant volume heating							

4.3 A closed thermodynamic system undergoes a cyclic process as shown in the following P-V diagram:

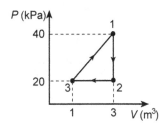

a) Calculate the net work output of the cycle.
b) Determine the trend of change for the unknown quantities or changes in properties between initial and final states in the following table:

Process	Q	W	ΔU
$1 \to 2$	< 0	?	?
$2 \to 3$	?	?	< 0
$3 \to 1$	?	?	?

(**Answer:** a) $-20\,\text{kJ}$)

4.4 Calculate $\widehat{C}_P$ for steam at:

a) 1.6 MPa and 350 °C

b) 1.6 MPa and 1100 °C

(**Answer:** a) 2.194 kJ/ kg. K, b) 2.541 kJ/ kg. K)

4.5 A cylinder fitted with a piston initially contains an ideal gas at 100 kPa and 25 °C.

a) If the gas is compressed to 500 kPa by a reversible and isothermal process, calculate the change in internal energy.

b) The gas is now compressed to 500 kPa by an isothermal but irreversible process. The actual work required is 15% greater than the reversible work for the same compression. Calculate the change in internal energy.

4.6 4.5 moles of air enclosed in a piston-cylinder arrangement undergoes a cyclic process as shown by the path $1 \rightarrow 2 \rightarrow 3 \rightarrow 1$ in the following P-V diagram:

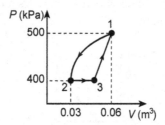

Calculate the unknown quantities in the table shown below:

Process	Nature of the Process	Q (kJ)	W (kJ)	ΔU (kJ)
$1 \rightarrow 2$	Irreversible	?	15	?
$2 \rightarrow 3$	Reversible	?	?	10
$3 \rightarrow 1$	Irreversible	?	-4	?

(**Answer:** For the entire cycle $\Delta U = 0$, $Q = -W = -7$ kJ)

4.7 A copper block of unknown mass at 90 °C is put into an insulated tank containing 0.05 m³ of water at 18 °C. A paddle mixer, driven by a 150 W motor, is started at the same time to stir the water. Thermal equilibrium is established after quarter of an hour, the final temperature being 20.5 °C. Determine the mass of the copper block.

(**Answer:** 14.8 kg)

4.8 A vertical cylinder fitted with a frictionless piston contains $2100\,\text{cm}^3$ of ideal gas ($\widetilde{C}_P^* = 21\,\text{J}/\,\text{mol}.\,\text{K}$) at $1.4\,\text{MPa}$ and $400\,°\text{C}$. Calculate the amount of heat transferred when the gas is cooled to $150\,°\text{C}$.

(**Answer:** $-2756.3\,\text{J}$)

4.9 A rigid tank of $2100\,\text{cm}^3$ volume contains ideal gas ($\widetilde{C}_P^* = 21\,\text{J}/\,\text{mol}.\,\text{K}$) at $1.4\,\text{MPa}$ and $400\,°\text{C}$. Calculate the amount of heat transferred when the gas is cooled to $150\,°\text{C}$.

(**Answer:** $-1665\,\text{J}$)

4.10 A rigid tank of $0.02\,\text{m}^3$ volume contains air at $200\,\text{kPa}$ and $30\,°\text{C}$. If $180\,\text{kJ}$ of heat is added, calculate the final pressure of the tank.

(**Answer:** $236\,\text{kPa}$)

4.11 A cylinder fitted with a frictionless piston initially contains 2 moles of nitrogen at $10\,\text{bar}$ and $330\,\text{K}$ (state 1). The nitrogen gas undergoes the following reversible processes:

- Compressed isothermally to $14\,\text{bar}$ (state 2),
- Heated at constant pressure to $660\,\text{K}$ (state 3),
- Cooled at constant volume until the pressure drops to $6\,\text{bar}$ (state 4),
- Compressed adiabatically to $10\,\text{bar}$ (state 5),
- Heated at constant pressure to state 1.

a) Show each process in a single P-V diagram.
b) Calculate Q, W, ΔU, and ΔH for each process and for the entire cycle.
(**Answer:** For the entire cycle $\Delta U = \Delta H = 0$, $Q = -W = 1862\,\text{J}$)

4.12 $0.5\,\text{kmol}$ of ideal gas ($\widetilde{C}_V^* = 1.5R$) undergoes the following reversible changes in a series of nonflow processes:

- From an initial state of $1\,\text{bar}$ and $12\,\text{m}^3$ (state 1), it is compressed adiabatically to $3\,\text{bar}$ (state 2),
- It is then cooled at constant pressure to initial temperature (state 3),
- Finally, it is compressed isothermally to $8\,\text{bar}$ (state 4).

a) Show each process in a single P-V diagram.
b) Calculate Q, W, and ΔU for each process and for the overall process.
(**Answer:** For the overall process $\Delta U = 0$, $Q = -W = -2832.6\,\text{kJ}$)

4.13 Five moles of nitrogen are expanded from an initial state of $3\,\text{bar}$ and $88\,°\text{C}$ to a final state of $1\,\text{bar}$ and $88\,°\text{C}$ by the following reversible processes:

a) Isothermal,

b) Heating at constant pressure followed by cooling at constant volume.

Calculate Q, W, ΔU, and ΔH for each process.

(**Answer:** a) $\Delta U = \Delta H = 0$, $Q = -W = 16,487$ J b) $\Delta U = \Delta H = 0$, $Q = -W = 30,013$ J)

4.14   Air contained in a cylinder fitted with a frictionless piston undergoes a reversible process in which its volume doubles. Which type of process leads to a larger work output: an isothermal process or an adiabatic process?

(**Answer:** Isothermal)

4.15   Two cubic meters of air at 1500 kPa and 327 °C expands to six times its initial volume. Calculate the final temperature, pressure, and the work done by the gas for the following processes:

a) Reversible isothermal process.
b) Reversible adiabatic process.
c) Irreversible adiabatic process in which expansion is against an external pressure of 200 kPa.

(**Answer:** a) 600 K, 250 kPa, -5375 kJ, b) 293 K, 122.1 kPa, -3838 kJ, c) 440 K, 183.3 kPa, -2000 kJ)

4.16   Air at 500 kPa and 200 °C expands by a reversible and isothermal process to such a pressure that when cooled at constant volume to 10 °C its final pressure is 100 kPa. Calculate $\widetilde{Q}$, $\widetilde{W}$, and $\Delta \widetilde{U}$ for the overall process.

(**Answer:** $\widetilde{Q} = 360.8$ J/ mol, $\widetilde{W} = -4310.1$ J/ mol, $\Delta \widetilde{U} = -3949.2$ J/ mol)

4.17   Helium undergoes the following reversible changes in a series of nonflow processes:

• From an initial state of 100 kPa and 27 °C (state 1), it is compressed isothermally to 700 kPa (state 2).
• It is then expanded adiabatically back to 100 kPa (state 3).

a) Show each process in a single P-V diagram.
b) Calculate the net work required per mole of helium.

(**Answer:** b) 2829.5 J/ mol)

4.18   Air undergoes the following reversible changes in a series of nonflow processes:

• From an initial state of 300 kPa and 40 °C (state 1), it is compressed isothermally to 500 kPa (state 2),
• It is then heated at constant pressure to 554 °C (state 3),
• Finally, it is cooled at constant volume to state 1 with the removal of 2500 J as heat.

a) Show each process on a single *P-V* diagram.
b) Calculate Q, W, and ΔU for each process.

(**Answer:** b) For the entire cycle $\Delta U = 0$, $Q = -W = 688.94$ J)

4.19 A cylinder fitted with a frictionless piston initially contains 3 kmol of air at 3 bar and 20 °C (state 1). The air undergoes the following reversible processes:

• Compressed isothermally to 5 bar (state 2),
• Heated at constant pressure to 250 °C (state 3),
• Compressed adiabatically to state 4 such that when cooled at constant volume it returns to state 1.

a) Show each process in a single *P-V* diagram.
b) Calculate Q, W, ΔU, and ΔH for each process and for the entire cycle.

(**Answer:** b) For the entire cycle $\Delta U = \Delta H = 0$, $Q = -W = 1098.8$ kJ)

4.20 An insulated vertical cylinder fitted with a frictionless piston contains ideal gas ($\widetilde{C}_P^* = 29$ J/ mol. K) at 100 bar and 21 °C. The piston is initially held in place by a pin as shown in the figure below. The initial volume of the gas is 0.006 m³. The piston has a mass of 250 kg and a cross-sectional area of 0.02 m². The ambient pressure is 100 kPa. The pin is removed and the piston moves upward to a locked position so that the gas volume is doubled.

a) Determine the final temperature and pressure of the gas.
b) Determine the final temperature and pressure of the gas if the cylinder and piston were rotated 90 ° before removing the pin.

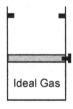

Ideal Gas

(**Answer:** a) 291.4 K, 49.56 bar, b) 292.8 K, 49.8 bar)

4.21   A rigid tank A is connected to a spherical elastic balloon B as shown in the figure below. Both contain air at the ambient temperature of 25 °C. The volume of tank A is 0.05 m³ and the initial pressure is 250 kPa. The initial diameter of the balloon is 0.3 m and the pressure inside is 110 kPa. The valve connecting A and B is now opened and remains open. It may be assumed that the pressure inside the balloon is directly proportional to its diameter, and also that the final temperature of air is uniform throughout at 25 °C.

a) Determine the final diameter of the balloon and the final pressure in the system.

b) Calculate Q if the atmospheric pressure is $100\,kPa$.

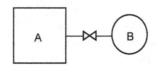

(**Answer:** a) 0.425 m, 155.8 kPa, b) 2.61 kJ)

4.22   A rigid tank A of 0.1 m³ volume initially contains methane at 1500 kPa and 25 °C. Tank A is connected through a valve to a cylinder B containing a frictionless piston of a mass such that a pressure of 115 kPa inside the cylinder is required to raise the piston. Initially the piston is at the bottom of the cylinder. The valve is opened until the pressure in tank A drops to 750 kPa. Assume methane to be an ideal gas with $\gamma = 1.31$.

a) Calculate the moles of methane in cylinder B if the process is slow enough to keep the temperature in tank A constant.

b) Calculate the moles of methane in cylinder B and its temperature if the process occurs very rapidly.

(**Answer:** a) 30.27 mol, b) 24.87 mol, 276.9 K)

4.23   A rigid tank contains wet steam at 200 kPa and 90% quality. The mass of the water vapor is 5 kg. Calculate the heat that must be added to produce a saturated vapor.

(**Answer:** 1148 kJ)

4.24   A rigid tank of 0.5 m³ volume contains steam at 800 kPa and 350 °C. How much heat must be transferred from the tank to bring the steam's temperature to 145 °C?

(**Answer:** − 1020.7 kJ)

4.25 A rigid tank of $0.035\,\mathrm{m}^3$ volume contains $0.021\,\mathrm{m}^3$ liquid water at $100\,^\circ$C in equilibrium with its vapor, which fills the rest of the tank. Heat is added to the tank until just one phase remains.

a) What is the final temperature? Is the resultant final phase liquid or vapor?
b) Calculate the amount of heat transferred.

(**Answer:** a) $350\,^\circ$C, liquid, b) $24,589\,\mathrm{kJ}$)

4.26 A rigid tank contains $2\,\mathrm{kg}$ of saturated steam at $5\,\mathrm{bar}$. Calculate the amount of heat that must be removed to reduce the quality to 50%.

(**Answer:** $-2057.7\,\mathrm{kJ}$)

4.27 A cylinder fitted with a piston contains wet steam at $100\,\mathrm{kPa}$ and 65% quality. Heat is added at constant pressure until the temperature reaches $200\,^\circ$C. The work done during the expansion process is $300\,\mathrm{kJ}$.

a) Calculate the mass of water in the cylinder.
b) Determine the amount of heat transferred.

(**Answer:** a) $2.8\,\mathrm{kg}$, b) $2772.7\,\mathrm{kJ}$)

4.28 A vertical cylinder fitted with a frictionless piston contains $2\,\mathrm{kg}$ of wet steam at $125\,\mathrm{kPa}$ and 15% quality. Heat is added at constant pressure while the piston moves upward. This continues until the quality reaches 40%, at which point the piston becomes stuck and cannot move any further. Heat continues to be added until the pressure reaches $250\,\mathrm{kPa}$.

a) Calculate the total heat added to the system.
b) Show the process in a *P-V* diagram.

(**Answer:** a) $2714\,\mathrm{kJ}$)

4.29 An insulated cylindrical tank of $0.1\,\mathrm{m}^3$ volume is divided into two equal parts by a frictionless piston. The piston conducts heat and is initially held in place by a pin. While one side of the cylinder contains helium ($\widetilde{C}_P^* = 2.5R$) at $0.3\,\mathrm{MPa}$ and $20\,^\circ$C, the other side contains nitrogen ($\widetilde{C}_P^* = 3.5R$) at $0.1\,\mathrm{MPa}$ and $20\,^\circ$C. The cross-sectional area of the piston is $150\,\mathrm{cm}^2$. Calculate the final temperature and pressure when the pin is removed.

(**Answer:** $20\,^\circ$C, $0.2\,\mathrm{MPa}$)

4.30 An insulated cylindrical tank of $0.028\,\mathrm{m}^3$ volume is divided into two equal parts by a frictionless piston as shown in the figure below. The piston does not conduct heat. Initially both gases are at $1.4\,\mathrm{bar}$ and $40\,^\circ$C. Heat is supplied to the air side by electrical resistance coils until the pressure of both gases reaches 2.8 bar. Calculate the final temperature of air.

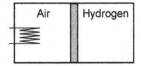

(**Answer:** 870.4 K)

4.31   An insulated cylindrical tank of 0.14 m^3 volume is divided into two equal parts by a frictionless piston as shown in the figure below. The piston does not conduct heat. Initially air is at 40 °C and the water is at 90 °C and 15% quality. Heat is supplied to the water side by electrical resistance coils until it exists as saturated vapor.

a) Calculate the final pressure.
b) Calculate the amount of heat transferred.

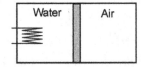

(**Answer:** a) 310 kPa, b) 371 kJ)

4.32   An insulated tank A of 0.6 m^3 volume is connected to an insulated tank B of 0.3 m^3 volume. Initially, tank A contains steam at 0.2 MPa and 200 °C, and tank B contains wet steam at 0.5 MPa and 85% quality. The valve is opened and flow occurs until equilibrium (both mechanical and thermal) is attained. Estimate the final pressure.

(**Answer:** 281 kPa)

4.33   An insulated rigid tank is divided into two chambers by a metal partition, each having a volume of 0.03 m^3. Initially, one chamber contains saturated steam at 700 kPa, and the other contains steam at 700 kPa and 450 °C. The two chambers come to equilibrium by heat transfer through the metal partition. Estimate the final temperature.

(**Answer:** 251 °C)

Problem related to Section 4.2

4.34   A fluid at 8 bar and 200 °C flows into a rigid tank of 1 m^3 volume through a 10 cm internal diameter pipe at an average velocity of 3 m/ s. The tank is initially empty. After 30 s, the temperature in the tank is recorded as 200 °C.

a) Calculate the pressure in the tank if the fluid is steam.
b) Calculate the pressure in the tank if the fluid is air.

(**Answer:** a) 5.77 bar, b) 5.65 bar)

Problems related to Section 4.3

4.35 Water at 15 MPa and 20 °C flows under steady conditions through a 2.5 cm internal diameter pipe at an average velocity of 1.5 m/s. If the rate of heat loss from the pipe is 11 kJ/s, calculate the enthalpy of water at the exit of the pipe. Kinetic and potential energy changes are considered negligible.

(**Answer:** 83.13 kJ/kg)

4.36 Air flowing at a mass flow rate of 1.2 kg/s is to be heated from 10 °C to 40 °C in a heat exchanger. Steam enters the heat exchanger at 300 kPa and 250 °C, and leaves at 300 kPa and 90% quality. Calculate the mass flow rate of steam.

(**Answer:** 7.87×10^{-2} kg/s)

4.37 Steam generators are heat exchangers used to convert liquid water into steam using hot combustion gases as a heating medium. Consider a steam generator in which liquid water at 1.6 MPa and 20 °C is converted into saturated steam at 1.6 MPa. Hot combustion gases, flowing at a molar flow rate of 130 mol/s, enter the steam generator at 1 bar and 425 °C and exit at 1 bar and 220 °C. The heat capacity of the combustion gases is given by

$$\widetilde{C}_P = 27.8 + 9.3 \times 10^{-3}T$$

where $\widetilde{C}_P$ is in J/mol.K and T is in K. Calculate the mass flow rate of steam produced in the steam generator.

(**Answer:** 0.328 kg/s)

4.38 A fluid at 50 MPa and 600 °C enters a throttling valve and leaves at 100 kPa. Determine the exit temperature if the fluid is

a) Steam,
b) Air.

(**Answer:** a) 385 °C, b) 600 °C)

4.39   Wet steam of unknown quality at 2 MPa enters a throttling valve and leaves at 100 kPa and 100 °C. Calculate the quality of steam.

(**Answer:** 0.935)

4.40   Steam at 4 MPa and 600 °C expands in an adiabatic turbine to 100 kPa and 200 °C. Calculate the mass flow rate of steam if the power output of the turbine is 650 kW?

(**Answer:** 0.813 kg/s)

4.41   Steam flowing at a mass flow rate of 5 kg/s enters an adiabatic turbine at 6 MPa. It exits at 10 kPa and 90% quality. Calculate the inlet temperature of steam if the power output of the turbine is 5 MW.

(**Answer:** 468.1 °C)

4.42   Steam at 5 MPa and 400 °C enters a turbine at a velocity of 60 m/s and its mass flow rate is 11 kg/s. The steam leaves the turbine as a saturated vapor at 0.1 MPa and at a point 2 m below the turbine inlet at a velocity of 100 m/s. The power output of the turbine is 4.5 MW.

a) Determine the inlet area,
b) Calculate the rate of heat exchange between the turbine and its surroundings.

(**Answer:** a) 0.0106 m², b) − 1187 kW)

4.43   Air enters an adiabatic turbine at negligible velocity at 3.5 MPa and 250 °C. The air expands to an exit pressure of 100 kPa. The exit velocity and temperature are 100 m/s and 80 °C, respectively. The diameter of the exit pipe is 0.5 m. Determine the power output of the turbine.

(**Answer:** − 3213 kW)

4.44   A chemical plant requires process steam at a rate of 0.3 kg/s at 500 kPa with not less than 90% quality and not more than 20 °C superheat. The available steam supply is at 2 MPa and 400 °C. The following two schemes have been suggested for this purpose:

Scheme I - Once the steam supply is throttled to 500 kPa, it is cooled at constant pressure in a heat exchanger to the required condition.
Scheme II - Steam supply is expanded in an adiabatic turbine to 500 kPa.

a) Calculate the minimum rate of heat removal in the heat exchanger for Scheme I.
b) Calculate the maximum power output of the turbine for Scheme II.

(**Answer:** a) − 136.4 kW, b) − 212.9 kW)

4.45 Air at 6 atm and 25 °C flows through a main line. While checking the pipeline for maintenance purposes, you discover a leak. Is it possible to obtain work from this leak? If yes, estimate the maximum value of this work.

(**Answer:** − 3474.4 J/ mol)

4.46 Air enters an adiabatic and reversible compressor at 100 kPa and 20 °C. If the discharge pressure is 800 kPa, calculate the work input per kg of air.

(**Answer:** 238.6 kJ/ kg)

4.47 Air flowing at 30 kmol/ h is compressed from 100 kPa and 25 °C to 700 kPa and 260 °C. If the power input is 90 kW, calculate the rate of heat transfer from the compressor.

(**Answer:** 147 kW)

4.48 Air at 1 bar and 20 °C is first compressed in an adiabatic and reversible compressor to 5 bar and then heated in a heat exchanger to 400 °C. Assume that the pressure drop in the heat exchanger is negligible.

a) Determine the amount of work that must be supplied to the compressor per mole of air passing through it.
b) Determine the heat load on the heat exchanger per mole of air passing through it.

(**Answer:** a) 4978.8 J/ mol, b) 6078.8 J/ mol)

4.49 A desuperheater is a device in which saturated steam is produced by spraying water into superheated steam. If superheated steam at 8 MPa and 600 °C enters the desuperheater at a rate of 750 kg/ h, at what rate should liquid water at 5 MPa and 20 °C be added to the desuperheater to produce saturated steam at 4 MPa?

(**Answer:** 232.4 kg/ h)

4.50 It is desired to produce a water stream at 35 °C by mixing a hot water stream at 60 °C with a stream of cold water at 20 °C. If the mass flow rate of the hot water stream is 0.5 kg/ s, determine the mass flow rate of the cold water stream. Assume all the streams are at a pressure of 300 kPa.

(**Answer:** 0.83 kg/ s)

4.51 Wet steam at 200 °C and 85% quality enters a chamber with a mass flow rate of 1 kg/ s. The chamber has two exits. Saturated liquid at 100 kPa leaves one exit at a rate of 0.15 kg/ s. Saturated vapor at 500 kPa leaves the other exit. Calculate the rate of heat flow into or out of the chamber.

(**Answer:** -103 kW)

4.52 Water flows steadily through an inclined circular pipe as shown in the figure below. What is the direction of flow?

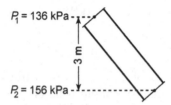

4.53 A flat-plate solar collector with an area of $3 \, m^2$ is placed on the roof of a house for water heating. Incoming energy flux from the sun is $750 \, W/m^2$ and 35% of it is lost to the surroundings. If the flow rate of water is 50 L/ h, determine the increase in water temperature under steady conditions.

(**Answer:** 25 °C)

4.54 Consider the process shown in the figure below. The work output of an adiabatic turbine is used to drive an adiabatic compressor. Calculate the mass flow rate of cold air used in the heat exchanger.

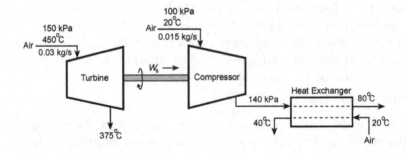

(**Answer:** 0.0675 kg/ s)

4.55 Consider the process shown in the figure below. The work output of an adiabatic and reversible turbine is used to drive an adiabatic and reversible compressor.

a) Calculate the molar flow rate of air through the system.

b) Determine the temperature of air at the exit of the combustor.

c) If the cross-sectional area of the nozzle at the exit is $40 \, \text{cm}^2$, calculate the exhaust velocity of air.

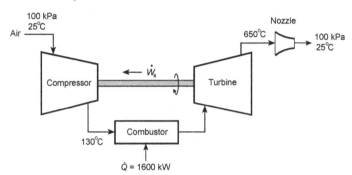

(**Answer:** a) $0.088 \, \text{kmol/s}$, b) $754.8 \, ^\circ\text{C}$, c) $545 \, \text{m/s}$)

Problems related to Section 4.4

4.56 A cylinder fitted with a frictionless piston initially contains $0.5 \, \text{kg}$ of water with a volume of $0.1 \, \text{m}^3$. The mass of the piston is such that it maintains a constant pressure of $0.6 \, \text{MPa}$ inside the cylinder. The cylinder is connected to a steam line at $5 \, \text{MPa}$ and $600 \, ^\circ\text{C}$. The valve between the steam line and the cylinder is opened and steam is allowed to flow slowly into the cylinder until the volume of the cylinder doubles and the temperature reaches $200 \, ^\circ\text{C}$, at which point the valve is closed.

a) Determine the mass of the steam that has entered the cylinder.

b) Determine the amount of heat transferred to the surroundings.

(**Answer:** a) $0.068 \, \text{kg}$, b) $376.9 \, \text{kJ}$)

4.57 An insulated vertical cylinder fitted with a frictionless piston initially contains $8 \, \text{kg}$ of wet steam at $325 \, \text{kPa}$ and 62.5% quality. The cylinder is connected to a steam line at $1.2 \, \text{MPa}$ and $400 \, ^\circ\text{C}$. The valve between the steam line and the cylinder is opened and steam is allowed to enter the cylinder until all the liquid in the cylinder has vaporized, at which point the valve is closed.

a) Determine the final temperature in the cylinder.

b) Determine the mass of the steam that has entered.

(**Answer:** a) $136.3 \, ^\circ\text{C}$, b) $12.16 \, \text{kg}$)

4.58 A cylinder fitted with a piston initially contains 0.5 kg of steam at 700 kPa and 200 °C. The cylinder is connected to a steam line at 700 kPa and 200 °C. The valve between the pipeline and the cylinder is opened and 0.8 kg of steam slowly enters the cylinder. The final condition of the 1.3 kg of steam in the cylinder is 100 kPa and 125 °C. Calculate the work done during this process if 100 kJ of heat is transferred from the tank to the surroundings.

(**Answer:** − 185 kJ)

4.59 A rigid tank of 4 m³ volume initially contains 200 kg of liquid water at 300 °C in equilibrium with its vapor, which fills the rest of the tank. A quantity of 75 kg of water at 60 °C is pumped into the tank. How much heat must be added during the process if the temperature in the tank is not to change?

(**Answer:** 74, 879 kJ)

4.60 A tank open to the atmosphere initially contains 60 kg of liquid water at 20 °C. Calculate the amount of steam at 0.2 MPa and 400 °C that should be added from the supply line so as to increase the final temperature of the water in the tank to 85 °C? Assume no heat loss from the water.

(**Answer:** 5.59 kg)

4.61 An adiabatic turbine receives air from a supply line at 800 kPa and 50 °C. The exit stream from the turbine is sent to an insulated, initially evacuated tank of 10 m³ volume. If the final pressure and temperature of the air in the tank are 800 kPa and 20 °C, respectively, calculate the work output of the turbine during the filling process.

Hint: Consider the turbine and the tank together as a system.

(**Answer:** − 10, 867 kJ)

4.62 An adiabatic turbine receives steam from a supply line at 1 MPa and 450 °C. The exit stream from the turbine is sent to an insulated, initially evacuated tank of 10 m³ volume. If the final pressure and temperature of the steam in the tank are 1 MPa and 400 °C, respectively, calculate the work output of the turbine during the filling process.

Hint: Consider the turbine and the tank together as a system.

(**Answer:** − 13, 500 kJ)

4.63 An insulated vertical cylinder fitted with a frictionless piston initially contains 10 kg of wet steam at 1.3 MPa and 5% quality. The cylinder is connected to a steam line at 2 MPa and unknown temperature. The valve between the steam line and the cylinder is opened and 2.5 kg of steam enters the cylinder. The valve is then closed and the contents of the cylinder are allowed to come to equilibrium. If the final volume of the cylinder is three times the initial volume, determine the temperature in the steam line.

(**Answer:** 212.42 °C)

4.64 A rigid tank of 20 m³ volume initially contains 0.015 m³ of liquid water at 60 °C in equilibrium with its vapor, which fills the rest of the tank. Wet steam at 100 kPa and 95% quality enters the tank until the pressure in the tank reaches 100 kPa. Determine the mass of the wet steam that has entered.

(**Answer:** 10.144 kg)

4.65 An evacuated rigid tank of 0.03 m³ volume is connected to a steam line at 1.8 MPa and 300 °C. The valve between the steam line and the tank is opened and steam is allowed to enter the tank. If the temperature of the tank contents remains constant at 300 °C, calculate the amount of heat transferred when the flow of steam into the tank finally stops.

(**Answer:** − 54 kJ)

4.66 An insulated cylindrical tank is attached to a steam line at 8 MPa and 400 °C as shown in the figure below. Initially the piston is at the extreme left-hand side of the tank and the spring force is zero. The valve between the steam line and the tank is opened and steam is slowly admitted into the tank. Find the temperature at which the pressure in the tank reaches 8 MPa.

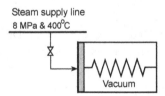

(**Answer:** 458.5 °C)

4.67 A cylinder fitted with a piston initially contains 1 kg of steam at 800 kPa and 250 °C. The cylinder is connected to a steam line at 800 kPa and 250 °C. The valve between the steam line and the cylinder is opened and 3 kg of steam slowly enters the cylinder. During this process 300 kJ of heat is transferred from the contents of the cylinder to the surroundings.

The final condition of 4 kg of steam in the cylinder is 200 kPa and 150 °C. Calculate the work done by the steam during this process.

(**Answer:** − 957.9 kJ)

4.68 A rigid tank of $0.3 \, \text{m}^3$ volume is initially evacuated. Atmospheric air at 100 kPa and 20 °C leaks into the tank through a tiny hole. Since the process is slow, it is possible to assume that the tank contents may undergo an isothermal process as a result of heat transfer between the tank contents and ambient air. Calculate the amount of heat transferred when the tank pressure reaches 100 kPa.

(**Answer:** − 30 kJ)

4.69 A tank containing $1000 \, \text{cm}^3$ of air at 0.5 MPa and 30 °C is connected to a large pipeline carrying air at 1 MPa and 30 °C. The valve between the pipeline and the tank is opened and air flows into the tank until the pressure is 1 MPa and then the valve is closed. If this process occurs adiabatically, what is the final temperature of the air in the tank?

(**Answer:** 353.5 K)

4.70 A rigid tank of $3 \, \text{m}^3$ volume initially contains wet steam at 3.5 MPa and 10% quality. Saturated liquid is slowly withdrawn from the bottom of the tank until the final total mass in the tank is one-half of the initial total mass. Calculate the amount of heat transferred to keep the temperature of the tank contents constant during this process.

(**Answer:** 8485 kJ)

4.71 A rigid tank of $0.1 \, \text{m}^3$ volume initially contains wet steam at 2 MPa and 35% quality. Saturated vapor is withdrawn from the top of the tank until the quality of wet steam becomes 60%. Determine the amount of heat transferred to keep the pressure constant during withdrawal of the saturated vapor.

(**Answer:** 2192.9 kJ)

4.72 A rigid tank of $3 \, \text{m}^3$ volume contains air at 3 MPa and 25 °C. A relief valve is opened slightly, allowing air to escape to the atmosphere. The valve is closed when the pressure in the tank drops to 0.6 MPa. Heat is added during the process so as to keep the temperature of the tank contents constant. Calculate the amount of heat transferred.

(**Answer:** 7200 kJ)

4.73 An insulated cylindrical storage tank of $80\,m^3$ volume initially contains $75\,m^3$ of nonvolatile liquid and $5\,m^3$ of air at $100\,kPa$ and $300\,K$. Solubility of air in the liquid is negligible. It is proposed to transfer this liquid to another storage tank by pressurizing the tank with air from a supply line at $3\,MPa$ and $300\,K$. For this purpose, first air is admitted into the tank, while the exit valve is kept closed, until the pressure in the tank reaches $3\,MPa$. Then the exit valve is opened slightly to transfer the liquid while keeping the pressure constant at $3\,MPa$. Calculate the temperature in the tank when it is completely filled with air.

(**Answer:** $305.3\,K$)

4.74 In one of your weekly homework assignments, you are faced with the following problem:

A rigid tank A of known volume is connected to an insulated tank B of known volume. Initially, tank A contains an ideal gas at T^* and P_A, and tank B contains the same ideal gas at T^* and P_B, with $P_A > P_B$. The valve connecting the tanks is now opened and flow occurs until equilibrium (both mechanical and thermal) is attained. Throughout the process, the gas temperature in tank A is maintained at T^* by adding or removing heat, Q_A. Is Q_A positive, negative, or zero?

While discussing the problem's solution with two of your friends, Henry and George, they proposed two different solutions:

Considering tank A as the system, Henry simplified the conservation statements as

$$- dn_{out} = dn_{sys} \tag{1}$$

$$- \widetilde{H}_{out}\, dn_{out} + \delta Q_A = d(n\widetilde{U})_{sys} \tag{2}$$

Since $\widetilde{H}_{out} = \widetilde{H}_{sys}$ and $\widetilde{U}_{sys} = \text{constant}$ (T^* remains constant), combination of Eqs. (1) and (2) leads to

$$\delta Q_A = (\widetilde{U}_{sys} - \widetilde{H}_{sys})\, dn_{sys} = - RT^* dn_{sys} \tag{3}$$

Integration of Eq. (3) gives

$$Q_A = RT^*(n_{A_{initial}} - n_{A_{final}}) > 0 \tag{4}$$

George, on the other hand, considered both tanks as the system and applied the first law as

$$\Delta U = Q + W \tag{5}$$

Since tank B is insulated, $Q = Q_A$. The system produces no work, i.e., $W = 0$. Once equilibrium is attained, the gas in both A and B will be at T^*; thus $\Delta U = 0$. Under these conditions, Eq. (5) gives

$$Q_A = 0$$

Who do you think proposed the correct solution? For more details on this problem see Müller (2000).

References

Müller, E.A., 2000, *Chem. Eng. Ed.*, **34** (4), 366-368.

Smith, W.L., 1975, *J. Chem. Ed.*, **52** (2), 97-98.

Chapter 5

The Second Law of Thermodynamics

The first law of thermodynamics provides the means for energy bookkeeping. Conservation of energy, however, cannot predict the preferred direction of processes. For example, is it possible for a solid block to slide up an inclined plane with a rough surface and become cooler? At this point, the second law of thermodynamics comes into the picture by defining a new thermodynamic property called *entropy* such that the entropy of the universe can only increase. The second law of thermodynamics predicts whether the given process is feasible or not. Instead of competing with each other, the first and second laws of thermodynamics complement each other. The purpose of this chapter is to show how to use the concept of entropy in energy related engineering problems.

5.1 Carnot Cycle

How to convert heat continuously into mechanical energy and work? One way of answering this practical question is to convert the thermal energy of a flowing fluid into mechanical energy by using steady-state flow devices, such as turbines. It is also possible to accomplish this task by a *thermodynamic cycle*, i.e., a process in which a working fluid (air, steam, etc.) undergoes a series of state changes and finally returns to its initial state.

The most efficient thermodynamic cycle is known as the *Carnot*[1] *cycle* in which an ideal gas enclosed in a frictionless piston-cylinder device undergoes the following four reversible successive processes:

- Isothermal and reversible expansion of an ideal gas by the addition of heat from a hot reservoir at T_H (heat source),

---

[1]Nicolas Leonard Sadi Carnot (1796-1832), French military engineer and physicist.

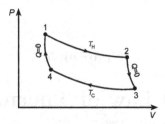

Fig. 5.1   The Carnot cycle.

- Adiabatic and reversible expansion of an ideal gas,
- Isothermal and reversible compression of an ideal gas by discarding heat to a cold reservoir at T_C (heat sink),
- Adiabatic and reversible compression of an ideal gas to its initial state.

These four processes are shown on a P-V diagram in Fig. 5.1. Considering the ideal gas in a frictionless piston-cylinder device as a system, the analysis of the Carnot cycle is as follows:

- **Process $1 \to 2$ (Reversible isothermal expansion at T_H)**

The first law of thermodynamics for a closed system is

$$\underbrace{\Delta \widetilde{U}_{12}}_{0} = \widetilde{Q}_{12} + \widetilde{W}_{12} \tag{5.1-1}$$

The work done by the ideal gas is

$$\widetilde{W}_{12} = -\int_{\widetilde{V}_1}^{\widetilde{V}_2} P \, d\widetilde{V} = -RT_H \int_{\widetilde{V}_1}^{\widetilde{V}_2} \frac{d\widetilde{V}}{\widetilde{V}}$$

$$= -RT_H \ln\left(\frac{\widetilde{V}_2}{\widetilde{V}_1}\right) = -RT_H \ln\left(\frac{P_1}{P_2}\right) \tag{5.1-2}$$

Thus, the amount of heat addition is

$$\widetilde{Q}_{12} = RT_H \ln\left(\frac{P_1}{P_2}\right) \tag{5.1-3}$$

Since $P_1 > P_2$, $\widetilde{W}_{12} < 0$ and $\widetilde{Q}_{12} > 0$ as expected.

- **Process $2 \to 3$ (Reversible adiabatic expansion)**

The first law of thermodynamics for a closed system is

$$\Delta \widetilde{U}_{23} = \underbrace{\widetilde{Q}_{23}}_{0} + \widetilde{W}_{23} \tag{5.1-4}$$

Therefore, the work done by the ideal gas is

$$\widetilde{W}_{23} = \Delta \widetilde{U}_{23} = \widetilde{C}_V^* (T_C - T_H) \tag{5.1-5}$$

Note that since $T_C < T_H$, $\widetilde{W}_{23} < 0$ as expected.

• **Process $3 \to 4$ (Reversible isothermal compression)**

The first law of thermodynamics for a closed system is

$$\underbrace{\Delta \widetilde{U}_{34}}_{0} = \widetilde{Q}_{34} + \widetilde{W}_{34} \tag{5.1-6}$$

The work done on the ideal gas is

$$\widetilde{W}_{34} = - \int_{\widetilde{V}_3}^{\widetilde{V}_4} P \, d\widetilde{V} = - RT_C \int_{\widetilde{V}_3}^{\widetilde{V}_4} \frac{d\widetilde{V}}{\widetilde{V}}$$

$$= - RT_C \ln \left(\frac{\widetilde{V}_4}{\widetilde{V}_3} \right) = - RT_C \ln \left(\frac{P_3}{P_4} \right) \tag{5.1-7}$$

Thus, the amount of heat removed from the system is

$$\widetilde{Q}_{34} = RT_C \ln \left(\frac{P_3}{P_4} \right) \tag{5.1-8}$$

Since $P_3 < P_4$, $\widetilde{W}_{34} > 0$ and $\widetilde{Q}_{34} < 0$ as expected.

• **Process $4 \to 1$ (Reversible adiabatic compression)**

The first law of thermodynamics for a closed system is

$$\Delta \widetilde{U}_{41} = \underbrace{\widetilde{Q}_{41}}_{0} + \widetilde{W}_{41} \tag{5.1-9}$$

Therefore, the work done on the ideal gas is

$$\widetilde{W}_{41} = \Delta \widetilde{U}_{41} = \widetilde{C}_V^* (T_H - T_C) \tag{5.1-10}$$

Note that since $T_H > T_C$, $\widetilde{W}_{41} > 0$ as expected.

The net work output of the cycle is

$$-\widetilde{W}_{cycle} = -\widetilde{W}_{12} - \widetilde{W}_{23} - \widetilde{W}_{34} - \widetilde{W}_{41}$$

$$= RT_H \ln\left(\frac{P_1}{P_2}\right) - \widetilde{C}_V^* (T_C - T_H) + RT_C \ln\left(\frac{P_3}{P_4}\right) - \widetilde{C}_V^* (T_H - T_C)$$

$$= RT_H \ln\left(\frac{P_1}{P_2}\right) + RT_C \ln\left(\frac{P_3}{P_4}\right) \qquad (5.1\text{-}11)$$

The application of the first law of thermodynamics for the cycle leads to

$$\underbrace{\Delta\widetilde{U}_{cycle}}_{0} = \widetilde{Q}_{cycle} + \widetilde{W}_{cycle} \qquad (5.1\text{-}12)$$

or

$$-\widetilde{W}_{cycle} = \widetilde{Q}_{12} + \widetilde{Q}_{34} \qquad (5.1\text{-}13)$$

Since $\widetilde{Q}_{34} < 0$, Eq. (5.1-13) indicates that the work output of the cycle is always less than the heat added, i.e.,

$$\boxed{-\widetilde{W}_{cycle} < \widetilde{Q}_{12}} \qquad (5.1\text{-}14)$$

The efficiency of the cycle, η, is defined as

$$\eta = \frac{\text{Net work output}}{\text{Heat added}} = \frac{-\widetilde{W}_{cycle}}{\widetilde{Q}_{12}} < 1 \qquad (5.1\text{-}15)$$

Substitution of Eqs. (5.1-3) and (5.1-11) into Eq. (5.1-15) gives

$$\eta = \frac{RT_H \ln\left(\frac{P_1}{P_2}\right) + RT_C \ln\left(\frac{P_3}{P_4}\right)}{RT_H \ln\left(\frac{P_1}{P_2}\right)} = 1 + \left(\frac{T_C}{T_H}\right)\frac{\ln(P_3/P_4)}{\ln(P_1/P_2)}$$

$$= 1 - \left(\frac{T_C}{T_H}\right)\frac{\ln(P_3/P_4)}{\ln(P_2/P_1)} \qquad (5.1\text{-}16)$$

Since the processes $2 \to 3$ and $4 \to 1$ are reversible and adiabatic, then

$$\frac{T_C}{T_H} = \left(\frac{P_3}{P_2}\right)^{(\gamma-1)/\gamma} = \left(\frac{P_4}{P_1}\right)^{(\gamma-1)/\gamma} \qquad \Rightarrow \qquad \frac{P_3}{P_4} = \frac{P_2}{P_1} \qquad (5.1\text{-}17)$$

The use of Eq. (5.1-17) in Eq. (5.1-16) leads to

$$\boxed{\eta = 1 - \frac{T_C}{T_H}} \qquad (5.1\text{-}18)$$

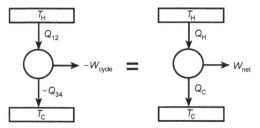

Fig. 5.2   Alternative representation for the Carnot heat engine.

Substitution of Eq. (5.1-13) into Eq. (5.1-15) also gives the efficiency as

$$\eta = 1 + \frac{\widetilde{Q}_{34}}{\widetilde{Q}_{12}} \tag{5.1-19}$$

Combination of Eqs. (5.1-18) and (5.1-19) results in

$$\boxed{\frac{\widetilde{Q}_{12}}{T_H} + \frac{\widetilde{Q}_{34}}{T_C} = 0} \tag{5.1-20}$$

Any mechanism, operating cyclically, whose primary purpose is to partially convert heat into work is called a *heat engine*. The Carnot cycle is the most efficient heat engine operating between two constant heat reservoirs T_H and T_C; its efficiency indicates the upper limit. It is also possible to schematically represent the Carnot heat engine as shown in Fig. 5.2. The circle represents the heat engine, $Q_{12} = Q_H$, $|Q_{34}| = Q_C$ and $|W_{cycle}| = W_{net}$. Note that

$$Q_H = W_{net} + Q_C \tag{5.1-21}$$

Using the new notation, Eqs. (5.1-15) and (5.1-18) are combined as

$$\boxed{\eta = \frac{W_{net}}{Q_H} = 1 - \frac{T_C}{T_H}} \tag{5.1-22}$$

The use of Eq. (5.1-21) in Eq. (5.1-22) leads to

$$\boxed{\frac{Q_C}{Q_H} = \frac{T_C}{T_H}} \tag{5.1-23}$$

Example 5.1 *A Carnot heat engine receives* 1000 kJ *of heat at* 800 °C *and rejects heat at* 300 °C*. Calculate the work output.*

Solution

From Eq. (5.1-22)

$$\eta = \frac{W_{net}}{Q_H} = 1 - \frac{T_C}{T_H} \tag{1}$$

Substitution of the values into Eq. (1) gives

$$\frac{W_{net}}{1000} = 1 - \frac{300 + 273}{800 + 273}$$

Solving for W_{net} results in

$$W_{net} = 466 \, \text{kJ}$$

Comment: *Application of Eq. (5.1-22) gives the upper limit of the work that can be theoretically obtained. In reality, the actual work output of a heat engine would be less than 466 kJ.*

5.2 Heat Engines and Heat Pumps

As stated in Section 5.1, a heat engine partially converts heat into work. The working fluid in the Carnot heat engine is an ideal gas. In practical applications, however, the working fluid is steam. In a typical heat engine as shown in Fig. 5.3-a, high pressure steam produced in the boiler is expanded in the turbine to obtain work. The low pressure steam at the exit of the turbine is sent to a condenser in which heat is discarded to the surroundings. Before returning the resulting liquid to the boiler, its pressure is increased by the pump. A small portion of the work obtained from the turbine is used for the pumping process. A schematic representation of a heat engine is shown in Fig. 5.3-b in which the circle represents the boiler-turbine-condenser-pump assembly.

The performance of a heat engine is measured in terms of its *efficiency*, η, defined by

$$\boxed{\eta = \frac{W_{net}}{Q_H}} \tag{5.2-1}$$

The Carnot heat engine gives the maximum efficiency (or maximum work

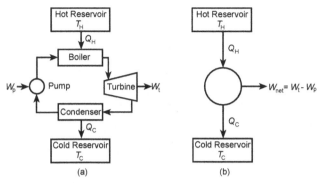

Fig. 5.3   A heat engine.

output) as

$$\eta_{carnot} = \frac{W_{net,rev}}{Q_H} = 1 - \frac{T_C}{T_H} \qquad (5.2\text{-}2)$$

Heat pumps are considered reversed heat engines, i.e., they are used to transfer heat from a low temperature reservoir to a high temperature reservoir by the help of an external work[2]. A schematic representation of a heat pump is shown in Fig. 5.4. The performance of a heat pump is measured in terms of its *coefficient of performance*, COP, defined by

$$\boxed{\text{COP} = \frac{Q_C}{W_{net}}} \qquad (5.2\text{-}3)$$

The Carnot heat pump gives the maximum COP (or minimum work requirement) as

$$\text{COP}_{carnot} = \frac{Q_C}{W_{net,rev}} = \frac{Q_C}{Q_H - Q_C} = \frac{1}{\dfrac{Q_H}{Q_C} - 1} = \frac{1}{\dfrac{T_H}{T_C} - 1} = \frac{T_C}{T_H - T_C} \qquad (5.2\text{-}4)$$

Example 5.2 *It is required to keep the interior temperature of the house constant at 23 °C all year long. The rate of heat loss from the house is estimated as 250 J per second per degree temperature difference between the inside and outside.*

---

[2]A domestic refrigerator is a heat pump.

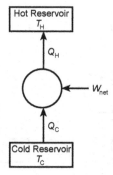

Fig. 5.4   Schematic representation of a heat pump.

a) *A heat pump is used to heat the house in the winter, the atmosphere being the source of energy. If the outside temperature is* − 5 °C *in the winter time, what is the minimum power required to drive the heat pump?*

b) *A heat pump is used to cool the house in the summer, the atmosphere being the heat sink. If the outside temperature is* 30 °C *in the summer time, what is the minimum power required to drive the heat pump?*

Solution

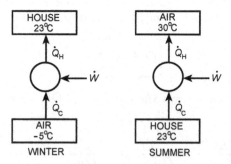

a) *The rate of heat that must be supplied to the house is*

$$\dot{Q}_H = (250)(23 + 5) = 7000\,\text{W}$$

The Carnot heat pump requires the minimum power input. From Eq. (5.2-4)

$$\frac{\dot{Q}_C}{\dot{W}_{net,rev}} = \frac{T_C}{T_H - T_C} \qquad \Rightarrow \qquad \frac{\dot{Q}_H - \dot{W}_{net,rev}}{\dot{W}_{net,rev}} = \frac{T_C}{T_H - T_C}$$

or

$$\dot{W}_{net,rev} = \dot{Q}_H \left(\frac{T_H - T_C}{T_H} \right) = (7000) \left(\frac{23 + 5}{23 + 273} \right) = 662.2 \, \text{W}$$

b) *The rate of heat that must be removed from the house is*

$$\dot{Q}_C = (250)(30 - 23) = 1750 \, \text{W}$$

The Carnot heat pump requires the minimum power input. From Eq. (5.2-4)

$$\frac{\dot{Q}_C}{\dot{W}_{net,rev}} = \frac{T_C}{T_H - T_C}$$

or

$$\frac{1750}{\dot{W}_{net,rev}} = \frac{23 + 273}{30 - 23} \qquad \Rightarrow \qquad \dot{W}_{net,rev} = 41.4 \, \text{W}$$

5.2.1 Statements of the second law

The second law of thermodynamics can be expressed in terms of the Clausius[3] and Kelvin-Planck[4] statements:

• **Clausius statement:** As shown in Fig. 5.5-a, a heat pump cannot transfer heat from a cooler body to a hotter body without receiving external work. In other words, spontaneous flow of heat is only possible in the direction of decreasing temperature, i.e., from a hotter body to a colder one.

• **Kelvin-Planck statement:** As shown in Fig. 5.5-b, a heat engine cannot convert all the heat it receives from an energy source into work. In other words, part of the heat is converted into work and the remainder is discarded as waste.

---

[3]Rudolf Julius Emanuel Clausius (1822-1888), German physicist and mathematician.
[4]William Thomson (Lord Kelvin) (1824-1907), British mathematical physicist and engineer; Max Karl Ernst Ludwig Planck (1858-1947), German theoretical physicist.

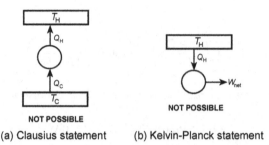

(a) Clausius statement   (b) Kelvin-Planck statement

Fig. 5.5   Schematic representations of the Clausius and Kelvin-Planck statements.

In the first instance, the Clausius and Kelvin-Planck statements seem to be unrelated. However, these two statements are equivalent to each other and one necessarily implies the other. For example, contrary to the Clausius statement, suppose that it is possible to construct a heat pump A that transfers 40 kJ of heat from a low temperature reservoir to a high temperature reservoir without using external work. If this heat pump is combined with a heat engine B discarding 40 kJ of heat to a low temperature reservoir as shown in Fig. 5.6, the combined system violates the Kelvin-Planck statement.

On the other hand, contrary to the Kelvin-Planck statement, suppose that a heat engine A receives 40 kJ of heat and converts it to work. If this heat engine is combined with a heat pump B receiving 40 kJ of work as shown in Fig. 5.7, the combined system violates the Clausius statement.

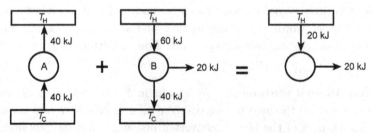

Fig. 5.6   Violation of the Clausius statement leads to violation of the Kelvin-Planck statement.

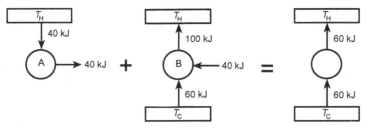

Fig. 5.7 Violation of the Kelvin-Planck statement leads to violation of the Clausius statement.

Example 5.3 *Show that all reversible heat engines operating between the same temperature limits must have identical efficiencies.*

Proof

Consider two reversible heat engines A *and* B *operating between the same temperature limits as shown in the figure below.*

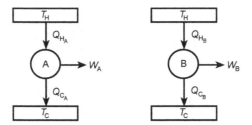

Assume that the heat engine A *has an efficiency greater than that of* B *and set* $Q_{HA} = Q_{HB} = Q_H$. *Then*

$$\frac{W_A}{Q_H} > \frac{W_B}{Q_H} \tag{1}$$

implying that

$$W_A > W_B \quad \text{and} \quad Q_{CA} < Q_{CB} \tag{2}$$

Since the heat engines are reversible, the heat engine B *could be reversed and combined with* A. *The result shown in the figure below obviously violates the Kelvin-Planck statement. Therefore, the assumption that the heat engine* A *has an efficiency greater than that of the heat engine* B *must be wrong.*

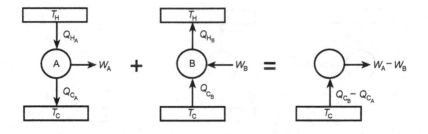

Example 5.4 *Show that an irreversible heat engine cannot have an efficiency greater than a reversible one if both operate between the same temperature limits.*

Proof

Consider an irreversible heat engine A and a reversible heat engine B, both operating between the same temperature limits as shown in the figure below.

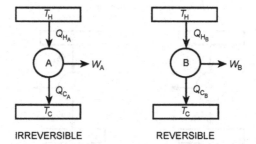

Assume that the heat engine A has an efficiency greater than that of B and set $W_A = W_B = W$. Then

$$\frac{W}{Q_{HA}} > \frac{W}{Q_{HB}} \tag{1}$$

implying that

$$Q_{HA} < Q_{HB} \quad \text{and} \quad Q_{CA} < Q_{CB} \tag{2}$$

Since the heat engine B is reversible, it could be reversed and combined with A. The result shown in the figure below obviously violates the Clausius statement. Therefore, the assumption that the irreversible heat engine A has an efficiency greater than that of the reversible heat engine B must be wrong.

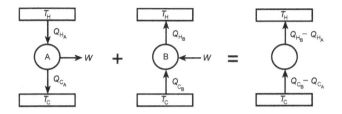

5.3 Entropy

Consider a closed system undergoing a reversible process from the initial state A to the final state B by following a process path as shown in Fig. 5.8-a. It is always possible to draw the adiabatic lines passing from the initial and final states as shown in Fig. 5.8-b. The work done by the system in going from state A to B is the area under the process path on a P-V diagram. Between the two adiabatic lines, as shown in Fig. 5.8-c, it is possible to find an isothermal path $\overline{CD}$ such that the area under the path $\overline{ACDB}$ is equal to the actual work, i.e.,

$$W_{AB} = W_{ACDB} \tag{5.3-1}$$

From the first law of thermodynamics

$$\Delta U_{AB} = Q_{AB} + W_{AB} \tag{5.3-2}$$

and

$$\Delta U_{ACDB} = Q_{ACDB} + W_{ACDB} \tag{5.3-3}$$

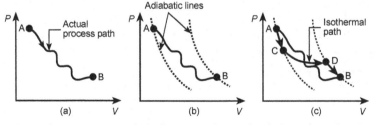

Fig. 5.8   A zigzag route (adiabatic-isothermal-adiabatic) replacing an actual process path (Tester and Modell, 1997).

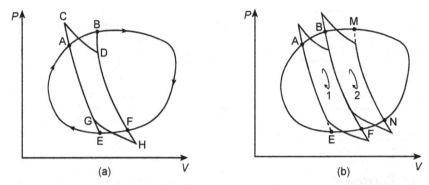

Fig. 5.9   Representation of a reversible cyclic process by series of Carnot cycles.

Since internal energy is a state function, then $\Delta U_{AB} = \Delta U_{ACDB}$ and Eqs. (5.3-1)-(5.3-3) lead to

$$Q_{AB} = Q_{ACDB} = \underbrace{Q_{AC}}_{0} + Q_{CD} + \underbrace{Q_{DB}}_{0} \qquad (5.3\text{-}4)$$

Therefore, the process path of any given reversible process can be replaced by a reversible zigzag route consisting of adiabatic-isothermal-adiabatic paths such that the heat interaction in the isothermal path is equal to the heat interaction in the actual process. In the literature, this is known as the *theorem of Clausius*.

Consider a reversible cyclic process as shown in Fig. 5.9-a and take any two states, A and B, on the process path. The adiabatic lines passing through the states A and B intersect the process path also at the states E and F, respectively. Using the theorem of Clausius, it is possible to replace the paths $\overline{AB}$ and $\overline{FE}$ by adiabatic-isothermal-adiabatic paths as $\overline{ACDB}$ and $\overline{FHGE}$, respectively. Note that $Q_{AB} = Q_{CD}$ and $Q_{FE} = Q_{HG}$.

Suppose that Q_{AB} amount of heat is received from a heat reservoir at T_H in going from state A to state B, and Q_{FE} amount of heat is rejected to a heat reservoir at T_C in going from state F to state E. The cycle A-C-D-F-H-G-A is nothing but a Carnot cycle. Thus, the use of Eq. (5.1-20) leads to

$$\frac{Q_{CD}}{T_H} + \frac{Q_{HG}}{T_C} = 0 \qquad (5.3\text{-}5)$$

or

$$\frac{Q_{AB}}{T_H} + \frac{Q_{FE}}{T_C} = 0 \tag{5.3-6}$$

Now consider two more states, M and N, on the process path as shown in Fig. 5.9-b. For the resulting Carnot cycles, 1 and 2, Eq. (5.3-6) is expressed as

$$\frac{Q_{AB}}{T_H} + \frac{Q_{FE}}{T_C} + \frac{Q_{BM}}{T_H^*} + \frac{Q_{NF}}{T_C^*} = 0 \tag{5.3-7}$$

where T_H^* and T_C^* are the temperatures of the heat reservoirs during heat addition and rejection in Carnot cycle 2, respectively. As the number of Carnot cycles increases, Eq. (5.3-7) is written as

$$\sum_{i=1}^{N} \frac{Q_{i,rev}}{T_i} = 0 \tag{5.3-8}$$

As the number of Carnot cycles is increased, the zigzag route can be made to approach the original cyclic process. For infinitesimally small Carnot cycles, i.e., $N \rightarrow \infty$, Eq. (5.3-8) takes the following form:

$$\oint \frac{\delta Q_{rev}}{T} = 0 \tag{5.3-9}$$

The property $\delta Q_{rev}/T$ is a state function[5] and is called the *entropy, dS*[6], i.e.,

$$\boxed{dS = \frac{\delta Q_{rev}}{T}} \tag{5.3-10}$$

The term entropy comes from a Greek word meaning "transformation." Integration of Eq. (5.3-10) gives

$$\boxed{\Delta S = \int \frac{\delta Q_{rev}}{T}} \tag{5.3-11}$$

To evaluate ΔS for an irreversible process, the real process path is replaced by a convenient hypothetical reversible path(s) between the same initial and final equilibrium states and then the term $\delta Q_{rev}/T$ is integrated over

---

[5] Although δQ_{rev} is a path function, $\delta Q_{rev}/T$ is a state function. Therefore, $1/T$ can be interpreted as the integrating factor.

[6] "Where does S for entropy come from?" Interested readers may refer to Battino *et al.* (1997) and Howard (2001) for the answer.

the reversible path(s). Keep in mind that entropy change is zero for a cyclic process.

Specific entropy, $\widehat{S}$, is an intensive property and can be used to specify the state of a system. Besides specific volume, specific internal energy, and specific enthalpy values, the Steam Tables also provide specific entropy values as a function of temperature and pressure.

5.3.1   *Reversible and isothermal process*

For a reversible and isothermal process, integration of Eq. (5.3-11) yields

$$\boxed{Q = T\,\Delta S}\qquad \text{Reversible and isothermal process}\qquad (5.3\text{-}12)$$

Consider a closed system containing a real gas, like steam. If the real gas undergoes reversible expansion or compression, calculation of work by using

$$W = -\int P\,dV \qquad (5.3\text{-}13)$$

is not straightforward. Since P changes as a function of V, numerical integration is necessary to evaluate the integral. Instead of using this rather tedious procedure, once Q is calculated from Eq. (5.3-12) work can be calculated indirectly from the application of the first law as

$$W = \Delta U - Q = \Delta U - T\,\Delta S \qquad (5.3\text{-}14)$$

Example 5.5 *A cylinder fitted with a piston initially contains* 5 kg *of steam at* 100 kPa *and* 150 °C. *The steam is then compressed reversibly and isothermally until it becomes saturated liquid. Determine the work required for such compression.*

Solution

• **Initial state (State 1)**

From Table A.3 in Appendix A

$$\left.\begin{array}{l} P_1 = 100\,\text{kPa} \\ T_1 = 150\,°\text{C} \end{array}\right\} \begin{array}{l} \widehat{U}_1 = 2582.8\,\text{kJ/kg} \\ \widehat{S}_1 = 7.6134\,\text{kJ/kg. K} \end{array}$$

• **Final state (State 2)**

For saturated liquid at 150 °C, *from Table A.1 in Appendix A*

$$\widehat{U}_2 = 631.68\,\text{kJ/kg} \qquad \widehat{S}_2 = 1.8418\,\text{kJ/kg. K}$$

The use of Eq. (5.3-12) gives the amount of heat transferred as

$$\widehat{Q} = T\,\Delta\widehat{S} = (150 + 273)(1.8418 - 7.6134) = -2441.4\,\text{kJ/kg}$$

The change in internal energy is

$$\Delta\widehat{U} = 631.68 - 2582.80 = -1951.12\,\text{kJ/kg}$$

Therefore, the work required for compression is

$$W = m\,(\Delta\widehat{U} - \widehat{Q}) = (5)(-1951.12 + 2441.4) = 2451.4\,\text{kJ}$$

5.3.2 *Reversible and adiabatic (isentropic) process*

For a reversible and adiabatic process $\delta Q_{rev} = 0$ and Eq. (5.3-11) gives

$$\boxed{\Delta S = 0} \qquad \text{Reversible and adiabatic process} \qquad (5.3\text{-}15)$$

A process in which entropy remains constant is called an *isentropic* process. As stated in Section 1.2.2, $\mathcal{F} = 2$ for single-phase, single-component systems. In other words, two independent intensive properties are needed to specify the state of such systems. During isentropic expansion and/or compression, one of the specific properties is then specific entropy.

Example 5.6 *Steam flowing at a rate of* 40 kg/s *enters an adiabatic and reversible turbine at* 4 MPa *and* 800 °C. *The exit stream from the turbine is sent to a heat exchanger, where it is condensed at constant pressure to obtain saturated liquid at* 100 kPa.

a) *Calculate the power output of the turbine.*
b) *Calculate the rate of heat load of the heat exchanger.*

Solution

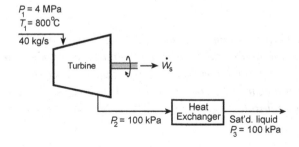

a) *The power output can be calculated from Eq. (4.3-14) as*

$$\dot{W}_s = \dot{m}\,(\hat{H}_2 - \hat{H}_1) \tag{1}$$

From Table A.3 in Appendix A

$$\left.\begin{array}{l} P_1 = 4\,\text{MPa} \\ T_1 = 800\,°\text{C} \end{array}\right\} \begin{array}{l} \hat{H}_1 = 4141.5\,\text{kJ/kg} \\ \hat{S}_1 = 7.8502\,\text{kJ/kg. K} \end{array}$$

Since the turbine operates reversibly and adiabatically, $\hat{S}_2 = \hat{S}_1 = 7.8502\,\text{kJ/kg. K}$. *From Table A.3 in Appendix A, at 100 kPa*

$T\,(°\text{C})$	$\hat{S}\,(\text{kJ/kg. K})$	$\hat{H}\,(\text{kJ/kg})$
200	7.8343	2875.3
250	8.0333	2974.3

By interpolation

$$\hat{H}_2 = 2875.3 + \left(\frac{2974.3 - 2875.3}{8.0333 - 7.8343}\right)(7.8502 - 7.8343) = 2883.2\,\text{kJ/kg}$$

Substitution of the numerical values into Eq. (1) gives

$$\dot{W}_s = (40)(2883.2 - 4141.5) = -50,332\,\text{kW}$$

b) *The rate of heat load is*

$$\dot{Q} = \dot{m}\,(\hat{H}_3 - \hat{H}_2)$$

From Table A.2 in Appendix A, $\hat{H}_3 = 417.46\,\text{kJ/kg}$. *Thus, the rate of heat removal is*

$$\dot{Q} = (40)(417.46 - 2883.2) = -98,630\,\text{kW}$$

Example 5.7 *Air flowing at a rate of 0.2 kg/s is at 0.1 MPa and 20 °C. It is compressed in an adiabatic and reversible compressor to 0.8 MPa. The power required for the compressor is supplied by an isentropic turbine through which steam is flowing at a rate of 0.75 kg/s. If the steam exits the turbine as saturated vapor at 0.5 MPa, determine its inlet pressure and temperature.*

Solution

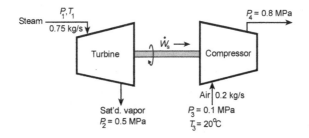

From the problem statement

$$-\dot{W}_{s,turbine} = \dot{W}_{s,comp} \tag{1}$$

Power required for the compression of air is

$$\dot{W}_{s,comp} = \dot{m}_{air}\widehat{C}_P^*(T_4 - T_3) = \dot{m}_{air}\widehat{C}_P^*T_3\left(\frac{T_4}{T_3} - 1\right)$$

$$= \dot{m}_{air}\widehat{C}_P^*T_3\left[\left(\frac{P_4}{P_3}\right)^{(\gamma-1)/\gamma} - 1\right]$$

$$= (0.2)\left(\frac{3.5 \times 8.314}{29}\right)(293)\left[\left(\frac{0.8}{0.1}\right)^{2/7} - 1\right] = 47.7\,\text{kW} \tag{2}$$

Power output of the turbine is

$$\dot{W}_{s,turbine} = \dot{m}_{steam}(\widehat{H}_2 - \widehat{H}_1) \tag{3}$$

From Table A.2 in Appendix A

$$\begin{array}{l} P = 0.5\,\text{MPa} \\ (T^{sat} = 151.86\,^\circ\text{C}) \end{array} \left\} \begin{array}{l} \widehat{H}_2 = 2748.7\,\text{kJ/kg} \\ \widehat{S}_2 = 6.8213\,\text{kJ/kg. K} \end{array} \right.$$

Substitution of Eq. (3) into Eq. (1) and rearrangement give

$$\widehat{H}_1 = \widehat{H}_2 + \frac{\dot{W}_{s,comp}}{\dot{m}_{steam}} = 2748.7 + \frac{47.7}{0.75} = 2812.3\,\text{kJ/kg}$$

Since the turbine is isentropic, then $\widehat{S}_1 = \widehat{S}_2 = 6.8213\,\text{kJ/kg. K.}$ *Using the procedure given in Example 3.9*

$$P_1 = 0.7\,\text{MPa} \qquad and \qquad T_1 = 186\,^\circ\text{C}$$

Example 5.8 *A cylinder fitted with a frictionless piston contains* 2 kg *of steam at* 800 kPa *and* 600 °C. *Calculate the work done by steam for the following cases:*

a) *The steam expands to* 300 kPa *by a reversible and isothermal process,*
b) *The steam expands to* 300 kPa *by a reversible and adiabatic process.*

Solution

System: *Contents of the tank*

a) *The first law for a closed system is*

$$W = \Delta U - Q \tag{1}$$

From Table A.3 in Appendix A

$$\left. \begin{array}{l} P_1 = 800\,\text{kPa} \\ T_1 = 600\,°\text{C} \end{array} \right\} \begin{array}{l} \widehat{U}_1 = 3297.9\,\text{kJ/kg} \\ \widehat{S}_1 = 8.1332\,\text{kJ/kg. K} \end{array}$$

$$\left. \begin{array}{l} P_2 = 300\,\text{kPa} \\ T_2 = 600\,°\text{C} \end{array} \right\} \begin{array}{l} \widehat{U}_2 = 3300.8\,\text{kJ/kg} \\ \widehat{S}_2 = 8.5892\,\text{kJ/kg. K} \end{array}$$

The amount of heat transfer is calculated from

$$Q = T\,\Delta S = mT\,\Delta\widehat{S} = (2)(600 + 273)(8.5892 - 8.1332) = 796\,\text{kJ}$$

The change in internal energy is

$$\Delta U = m\,\Delta\widehat{U} = (2)(3300.8 - 3297.9) = 5.8\,\text{kJ}$$

Substitution of the values into Eq. (1) leads to

$$W = 5.8 - 796 = -790.2\,\text{kJ}$$

b) *In this case the first law simplifies to*

$$W = \Delta U - \underbrace{Q}_{0} = m\,\Delta\widehat{U} \tag{2}$$

Since the process is reversible and adiabatic (or isentropic), $\widehat{S}_2 = \widehat{S}_1 = 8.1332\,\text{kJ/kg. K}$. *From Table A.3 in Appendix A, at* 300 kPa

T (°C)	$\widehat{S}$ (kJ/kg. K)	$\widehat{U}$ (kJ/kg)
400	8.0330	2965.6
500	8.3251	3130.0

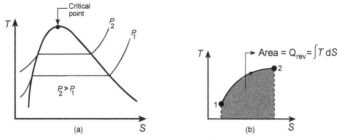

Fig. 5.10   (a) T-S diagram for water, (b) The area under the reversible
process path is equal to heat.

By interpolation

$$\widehat{U}_2 = 2965.6 + \left(\frac{3130.0 - 2965.6}{8.3251 - 8.0330}\right)(8.1332 - 8.0330) = 3022\,\text{kJ}/\text{kg}$$

Substitution of the numerical values into Eq. (2) gives

$$W = (2)(3022 - 3297.9) = -551.8\,\text{kJ}$$

5.3.3   *Thermodynamic diagrams involving entropy*

Thermodynamic diagrams are used to display the relationship between T,
P, V, H, and S. So far we have considered P-V, T-V, and P-T diagrams.
The thermodynamic diagrams involving entropy include T-S and H-S diagrams.

T-S diagram (Carnot cycle - revisited)

T-S diagrams are especially useful in the analysis of power and refrigeration
cycles, which will be covered in Chapter 6. A typical T-S diagram for water
is shown in Fig. 5.10-a. Note that it is very similar to a P-V diagram.
According to Eq. (5.3-10), the area under the reversible process path on a
T-S diagram is equal to the heat transferred between the system and its
surroundings, as shown in Fig. 5.10-b.

The ideal Carnot cycle, shown on a P-V diagram in Fig. 5.1, is represented on a T-S diagram as shown in Fig. 5.11. Since process $1 \rightarrow 2$ is
reversible and isothermal, then from Eq. (5.3-12)

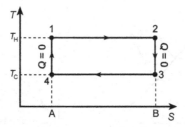

Fig. 5.11 T-S diagram of a Carnot cycle.

$$Q_H = T_H(S_2 - S_1) \tag{5.3-16}$$

In other words, the area under the process path $1 \to 2$ (Area 1-2-B-A-1) is the heat added to the cycle. On the other hand, process $3 \to 4$ is also reversible and isothermal. Therefore, the heat rejected from the cycle is simply the area under the process path $3 \to 4$ (Area 3-4-A-B-3) given by

$$|Q_C| = T_C(S_3 - S_4) = T_C(S_2 - S_1) \tag{5.3-17}$$

The efficiency of the cycle is given by

$$\eta = \frac{W_{net}}{Q_H} = \frac{Q_H - |Q_C|}{Q_H} = \frac{(\text{Area 1-2-B-A-1}) - (\text{Area 3-4-A-B-3})}{\text{Area 1-2-B-A-1}}$$

$$= \frac{\text{Area 1-2-3-4-1}}{\text{Area 1-2-B-A-1}} \tag{5.3-18}$$

Substitution of Eqs. (5.3-16) and (5.3-17) into Eq. (5.3-18) leads to

$$\eta = 1 - \frac{T_C}{T_H} \tag{5.3-19}$$

which is identical with Eq. (5.1-22).

H-S (Mollier) diagram

A plot of enthalpy versus entropy is known as the *Mollier*[7] *diagram*. A typical Mollier diagram is shown in Fig. 5.12. In the Mollier diagram, constant pressure and constant temperature lines coincide and form straight lines within the two-phase region. To prove this, note that the slope of a constant pressure line is given by

[7]Richard Mollier (1863-1935), German professor of applied physics and mechanics at Göttingen and Dresden universities.

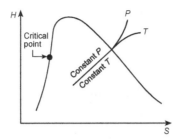

Fig. 5.12 H-S (Mollier) diagram.

$$\text{Slope} = \left(\frac{\partial H}{\partial S}\right)_P \tag{5.3-20}$$

For an isobaric and reversible process $dH = \delta Q = T\,dS$. Thus, Eq. (5.3-20) becomes

$$\text{Slope} = \left(\frac{\partial H}{\partial S}\right)_P = \frac{T\,dS}{dS} = T \tag{5.3-21}$$

5.4 Entropy Generation

Entropy, like mass and energy, is an extensive property that may be used to characterize the state of thermodynamic systems at equilibrium. Like mass and energy, the entropy of a system changes as a result of a process. However, entropy is not a conserved quantity. Generation of entropy, S_{gen}, is equal to the entropy change of the universe. The entropy change of the universe, on the other hand, is the sum of the entropy change of the system and the entropy change of the surroundings:

$$\boxed{S_{gen} = \Delta S_{universe} = \Delta S_{sys} + \Delta S_{surr}} \tag{5.4-1}$$

To determine the entropy generation (or the entropy change of the universe), let us consider two mechanical and two thermal processes[8]:

Case 1: Adiabatic expansion

A cylinder fitted with a piston contains a gas at P_1 and T_1. The gas (system) undergoes an adiabatic and reversible expansion to P_2 and T_2. Since work is done by the gas during expansion, $T_2 < T_1$. The process is shown in a P-T diagram in Fig. 5.13. As a result of the adiabatic and reversible

---

[8]These examples are taken from Van Ness (1983).

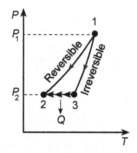

Fig. 5.13   Reversible and irreversible adiabatic expansion of a gas from P_1 to P_2.

process $S_1 = S_2$ and $\Delta S_{sys} = 0$. Note that the process $1 \rightarrow 2$ causes no change in the entropy of the surroundings because there is no exchange of heat between the system and its surroundings. Thus,

$$\boxed{S_{gen,rev} = \Delta S_{universe,rev} = 0} \qquad (5.4\text{-}2)$$

What will be the entropy generation if the expansion from P_1 to P_2 takes place irreversibly, the final state being state 3? To answer this question, it is necessary to have an idea about the final state temperature, T_3, of the gas.

For the processes $1 \rightarrow 2$ and $1 \rightarrow 3$, the initial and final state pressures are the same, and $(-W)$ is a positive quantity. Part of the work done during the irreversible expansion is used to overcome friction. As a result

$$-W_{rev} > -W_{irrev} \qquad (5.4\text{-}3)$$

The process is adiabatic $(Q = 0)$ and the first law of thermodynamics reduces to

$$\Delta U = W \qquad (5.4\text{-}4)$$

The use of Eq. (5.4-4) in Eq. (5.4-3) gives

$$-\Delta U_{rev} > -\Delta U_{irrev} \;\Rightarrow\; -(U_2 - U_1) > -(U_3 - U_1) \;\Rightarrow\; U_3 > U_2 \qquad (5.4\text{-}5)$$

indicating $T_3 > T_2$ and state 3 must be placed on the right-hand side of state 2.

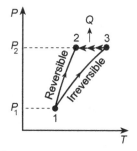

Fig. 5.14   Reversible and irreversible adiabatic compression of a gas from P_1 to P_2.

Consider a reversible process in going from state 3 to state 2 by removing heat. The change in entropy is

$$S_2 - S_3 = \int_3^2 \frac{\delta Q_{rev}}{T} \qquad (5.4\text{-}6)$$

Since heat is removed from the system, i.e., $\delta Q < 0$, then $S_3 > S_2$ or $S_3 > S_1$, implying $\Delta S_{sys} > 0$. The process $1 \to 3$ causes no change in the entropy of the surroundings because there is no exchange of heat between the system and its surroundings. Thus,

$$\boxed{S_{gen,irrev} = \Delta S_{universe,irrev} > 0} \qquad (5.4\text{-}7)$$

Case 2: Adiabatic compression

A cylinder fitted with a piston contains a gas at P_1 and T_1. The gas (system) undergoes adiabatic and reversible compression to P_2 and T_2. Since work is done on the gas during compression, $T_2 > T_1$. The process is shown in a *P-T* diagram in Fig. 5.14. As a result of the adiabatic and reversible process $S_1 = S_2$ and $\Delta S_{sys} = 0$. Note that the process $1 \to 2$ causes no change in the entropy of the surroundings because there is no exchange of heat between the system and its surroundings. Thus,

$$\boxed{S_{gen,rev} = \Delta S_{universe,rev} = 0} \qquad (5.4\text{-}8)$$

What will be the entropy generation if the compression from P_1 to P_2 takes place irreversibly, the final state being state 3? To answer this question, it is necessary to have an idea about the final state temperature, T_3, of the gas.

For the processes $1 \rightarrow 2$ and $1 \rightarrow 3$, the initial and final state pressures are the same, and W is a positive quantity. The irreversible compression requires extra work to overcome friction and as a result

$$W_{irrev} > W_{rev} \tag{5.4-9}$$

The process is adiabatic ($Q = 0$) and the first law of thermodynamics reduces to

$$\Delta U = W \tag{5.4-10}$$

The use of Eq. (5.4-10) in Eq. (5.4-9) gives

$$\Delta U_{irrev} > \Delta U_{rev} \quad \Rightarrow \quad (U_3 - U_1) > (U_2 - U_1) \quad \Rightarrow \quad U_3 > U_2 \tag{5.4-11}$$

indicating $T_3 > T_2$ and state 3 must be placed on the right-hand side of state 2.

Consider a reversible process in going from state 3 to state 2 by removing heat. The change in entropy is

$$S_2 - S_3 = \int_3^2 \frac{\delta Q_{rev}}{T} \tag{5.4-12}$$

Since heat is removed from the system, i.e., $\delta Q < 0$, then $S_3 > S_2$ or $S_3 > S_1$, implying $\Delta S_{sys} > 0$. The process $1 \rightarrow 3$ causes no change in the entropy of the surroundings because there is no exchange of heat between the system and its surroundings. Thus,

$$\boxed{S_{gen,irrev} = \Delta S_{universe,irrev} > 0} \tag{5.4-13}$$

Case 3: Heat engine

Consider a reversible heat engine and an irreversible heat engine operating between the same constant temperature heat reservoirs T_H and T_C as shown in Fig. 5.15. Both engines receive the same amount of heat, Q_H, from the high temperature reservoir.

The entropy generation is given by

$$S_{gen} = \Delta S_{universe} = \Delta S_H + \Delta S_{heat\,engine} + \Delta S_C \tag{5.4-14}$$

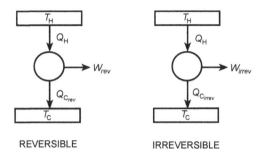

REVERSIBLE   IRREVERSIBLE

Fig. 5.15   Reversible and irreversible heat engines operating between the same temperature limits.

Since the heat engine operates in cycles $\Delta S_{heat\,engine} = 0$ and Eq. (5.4-14) simplifies to

$$S_{gen} = \Delta S_{universe} = \Delta S_H + \Delta S_C \qquad (5.4\text{-}15)$$

For a reversible heat engine, Eq. (5.4-15) is expressed as

$$S_{gen,rev} = \Delta S_{universe,rev} = -\frac{Q_H}{T_H} + \frac{Q_{C_{rev}}}{T_C} \qquad (5.4\text{-}16)$$

Moreover, when the heat engine is reversible, from Eq. (5.1-23)

$$\frac{Q_H}{T_H} = \frac{Q_{C_{rev}}}{T_C} \qquad (5.4\text{-}17)$$

Therefore, the use of Eq. (5.4-17) in Eq. (5.4-16) leads to

$$\boxed{S_{gen,rev} = \Delta S_{universe,rev} = 0} \qquad (5.4\text{-}18)$$

For an irreversible heat engine, Eq. (5.4-14) is expressed as

$$S_{gen,irrev} = \Delta S_{universe,irrev} = -\frac{Q_H}{T_H} + \frac{Q_{C_{irrev}}}{T_C}$$

$$= \underbrace{-\frac{Q_H}{T_H} + \frac{Q_{C_{rev}}}{T_C}}_{0} + \frac{Q_{C_{irrev}} - Q_{C_{rev}}}{T_C} \qquad (5.4\text{-}19)$$

For a heat engine $\Delta U = 0$ and $-W = Q$. Thus,

$$Q_H - Q_{C_{rev}} = -W_{rev} \quad \text{and} \quad Q_H - Q_{C_{irrev}} = -W_{irrev} \qquad (5.4\text{-}20)$$

Since both heat engines operate between the same temperature limits, the work output of the reversible heat engine is always greater than that of the irreversible one, i.e.,

$$- W_{rev} > - W_{irrev} \tag{5.4-21}$$

The use of Eq. (5.4-20) in Eq. (5.4-21) leads to

$$Q_{C_{irrev}} > Q_{C_{rev}} \tag{5.4-22}$$

Thus, Eq. (5.4-19) indicates that

$$\boxed{S_{gen,irrev} = \Delta S_{universe,irrev} > 0} \tag{5.4-23}$$

Case 4: Heat pump

Consider a reversible heat pump and an irreversible heat pump (refrigerators) operating between the same constant temperature heat reservoirs T_H and T_C as shown in Fig. 5.16. Both heat pumps extract the same amount of heat, Q_C, from the low temperature reservoir.

The entropy generation is given by

$$S_{gen} = \Delta S_{universe} = \Delta S_H + \Delta S_{heat\,pump} + \Delta S_C \tag{5.4-24}$$

Since the heat pump operates in cycles $\Delta S_{heat\,pump} = 0$ and Eq. (5.4-24) simplifies to

$$S_{gen} = \Delta S_{universe} = \Delta S_H + \Delta S_C \tag{5.4-25}$$

For a reversible heat pump, Eq. (5.4-25) is expressed as

$$S_{gen,rev} = \Delta S_{universe,rev} = \frac{Q_{H_{rev}}}{T_H} - \frac{Q_C}{T_C} \tag{5.4-26}$$

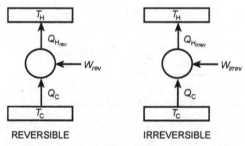

REVERSIBLE   IRREVERSIBLE

Fig. 5.16   Reversible and irreversible heat pumps operating between the same temperature limits.

The use of Eq. (5.1-23) in Eq. (5.4-26) gives

$$\boxed{S_{gen,rev} = \Delta S_{universe,rev} = 0} \qquad (5.4\text{-}27)$$

For an irreversible heat pump, Eq. (5.4-25) is expressed as

$$S_{gen,irrev} = \Delta S_{universe,irrev} = \frac{Q_{H_{irrev}}}{T_H} - \frac{Q_C}{T_C}$$

$$= \frac{Q_{H_{irrev}} - Q_{H_{rev}}}{T_H} - \underbrace{\left(\frac{Q_C}{T_C} - \frac{Q_{H_{rev}}}{T_H} \right)}_{0} \qquad (5.4\text{-}28)$$

For a heat pump $\Delta U = 0$ and $-W = Q$. Thus,

$$Q_C - Q_{H_{rev}} = -W_{rev} \qquad \text{and} \qquad Q_C - Q_{H_{irrev}} = -W_{irrev} \qquad (5.4\text{-}29)$$

Since both heat pumps operate between the same temperature limits, the work received by the reversible heat pump is always less than that of the irreversible one, i.e.,

$$W_{rev} < W_{irrev} \qquad (5.4\text{-}30)$$

The use of Eq. (5.4-29) in Eq. (5.4-30) leads to

$$Q_{H_{irrev}} > Q_{H_{rev}} \qquad (5.4\text{-}31)$$

Thus, Eq. (5.4-28) indicates that

$$\boxed{S_{gen,irrev} = \Delta S_{universe,irrev} > 0} \qquad (5.4\text{-}32)$$

The mechanical and thermal examples covered above indicate that the entropy generation is either zero for a reversible process or greater than zero for an irreversible one, i.e.,

$$\boxed{S_{gen} = \Delta S_{universe} = \Delta S_{sys} + \Delta S_{surr} \geq 0} \qquad (5.4\text{-}33)$$

Equation (5.4-33) was originally stated by Clausius in 1865 as *"Die Energie der Welt ist Konstant; die Entropie der Welt strebt einem Maximum zu"*, translating into English as "The energy of the world is constant; the entropy of the world tends toward a maximum."

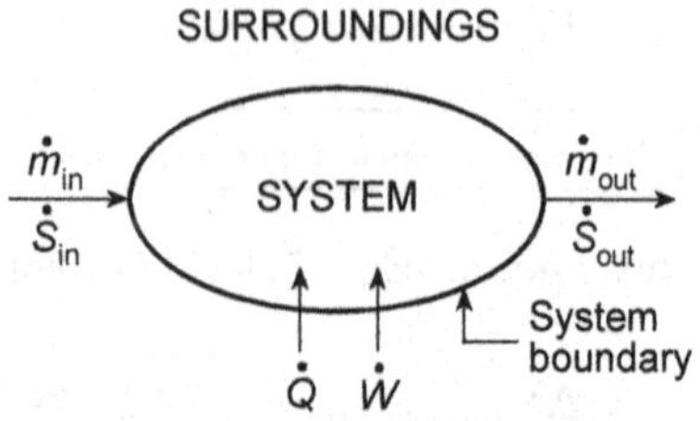

SURROUNDINGS

Fig. 5.17   An open system exchanging mass and energy with its surroundings.

5.4.1   *Entropy balance*

Equation (5.4-33) indicates that entropy generation (or entropy change of the universe) is either equal to zero for reversible processes or greater than zero for irreversible ones. In other words, real processes always generate entropy with a concomitant increase in the entropy of the universe. Therefore, there is no such thing as the conservation of entropy.

For the systematic evaluation of entropy generation, consider an open system as shown in Fig. 5.17. The rate of entropy generation is expressed as

$$\dot{S}_{gen} = \dot{S}_{system} + \dot{S}_{surroundings} \geq 0 \qquad (5.4\text{-}34)$$

The rate of entropy change of the system is expressed as

$$\dot{S}_{system} = \frac{d}{dt}\left(m\widehat{S}\right)_{sys} \qquad (5.4\text{-}35)$$

The entropy of the surroundings may change by two means: (*i*) mass flow, (*ii*) heat transfer. Note that entropy transfer has nothing to do with work. Thus,

$$\dot{S}_{surroundings} = \dot{m}_{out}\,\widehat{S}_{out} - \dot{m}_{in}\,\widehat{S}_{in} - \frac{\dot{Q}}{T_{surr}} \qquad (5.4\text{-}36)$$

Substitution of Eqs. (5.4-35) and (5.4-36) into Eq. (5.4-34) leads to

$$\dot{S}_{gen} = \frac{d}{dt}\left(m\widehat{S}\right)_{sys} + \dot{m}_{out}\,\widehat{S}_{out} - \dot{m}_{in}\,\widehat{S}_{in} - \frac{\dot{Q}}{T_{surr}} \geq 0 \qquad (5.4\text{-}37)$$

Multiplication of Eq. (5.4-37) by dt leads to

$$\boxed{\delta S_{gen} = d(m\widehat{S})_{sys} + \widehat{S}_{out}\,dm_{out} - \widehat{S}_{in}\,dm_{in} - \frac{\delta Q}{T_{surr}} \geq 0} \qquad (5.4\text{-}38)$$

which is also known as the *entropy balance*. The term Q in Eq. (5.4-38) is associated with the system. Therefore, it is positive when heat is added to the system, and negative when heat is removed from the system. Simplification of Eq. (5.4-38) for different types of systems is given below:

▶ **Isolated system**

When there is no exchange of mass and energy between the system and its surroundings, Eq. (5.4-38) simplifies to

$$\delta S_{gen} = d(m\widehat{S})_{sys} \geq 0 \qquad (5.4\text{-}39)$$

or

$$\boxed{S_{gen} = \Delta S_{sys} \geq 0} \qquad (5.4\text{-}40)$$

Since the universe is an isolated system, Eq. (5.4-40) simply tells us that the entropy of the universe is always increasing as a result of irreversibilities.

▶ **Closed system**

Since there is no exchange of mass between the system and its surroundings, i.e., $dm_{in} = dm_{out} = 0$, Eq. (5.4-38) simplifies to

$$\delta S_{gen} = m\,d\widehat{S}_{sys} - \frac{\delta Q}{T_{surr}} \geq 0 \qquad (5.4\text{-}41)$$

or

$$\boxed{\widehat{S}_{gen} = \Delta\widehat{S}_{sys} - \frac{\widehat{Q}}{T_{surr}} \geq 0} \qquad (5.4\text{-}42)$$

▶ **Steady-state flow system**

In this case

$$dm_{in} = dm_{out} = dm \qquad \text{and} \qquad d(m\widehat{S})_{sys} = 0 \qquad (5.4\text{-}43)$$

Thus, Eq. (5.4-38) reduces to

$$\delta S_{gen} = (\widehat{S}_{out} - \widehat{S}_{in})\,dm - \frac{\delta Q}{T_{surr}} \geq 0 \qquad (5.4\text{-}44)$$

Dividing each term by dm results in

$$\boxed{\widehat{S}_{gen} = \Delta\widehat{S} - \frac{\widehat{Q}}{T_{surr}} \geq 0}$$

(5.4-45)

Multiplication of Eq. (5.4-45) by the mass flow rate leads to

$$\boxed{\dot{S}_{gen} = \Delta\dot{S} - \frac{\dot{Q}}{T_{surr}} \geq 0}$$

(5.4-46)

5.4.2 *Feasibility of a process*

As stated in Section 1.1, any feasible process in nature should satisfy both the first and the second laws of thermodynamics. To satisfy the first law, one should carry out an energy balance and show that the total energy does not change. To satisfy the second law, one should calculate the entropy change of the system, ΔS_{sys}, and the entropy change of its surroundings, ΔS_{surr}, and show that the summation of these two quantities, i.e., entropy generation, is either equal to or greater than zero.

It should be kept in mind that while the entropy generation is either equal to or greater than zero, the entropy change of a system may be greater than, less than, or equal to zero. Therefore, if $\Delta S_{sys} < 0$, this does not necessarily imply violation of the second law.

Example 5.9 *Steam flowing at a rate of* 450 kg/h *is at* 3.5 MPa *and* 700 °C. *It is expanded in a turbine to produce work. Two exit streams leave the turbine. One of the streams is at* 1.4 MPa *and* 200 °C *and has a flow rate equal to one-third of that of the inlet stream. The other one is at* 0.1 MPa *and* 150 °C. *The heat loss from the turbine is estimated to be* 19.43 kW. *The temperature of the surroundings is* 20 °C. *The work to be obtained from the turbine is expected to be* 121 kW. *Determine if this process is possible.*

Solution

System: *Turbine*

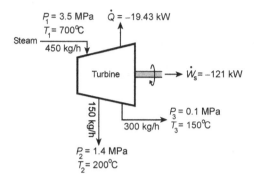

From Table A.3 in Appendix A

$$P_1 = 3.5\,\text{MPa} \atop T_1 = 700\,°\text{C} \left.\right\} \begin{array}{l} \widehat{H}_1 = 3908.8\,\text{kJ/kg} \\ \widehat{S}_1 = 7.6837\,\text{kJ/kg.K} \end{array}$$

$$P_2 = 1.4\,\text{MPa} \atop T_2 = 200\,°\text{C} \left.\right\} \begin{array}{l} \widehat{H}_2 = 2803.3\,\text{kJ/kg} \\ \widehat{S}_2 = 6.4975\,\text{kJ/kg.K} \end{array}$$

$$P_3 = 0.1\,\text{MPa} \atop T_3 = 150\,°\text{C} \left.\right\} \begin{array}{l} \widehat{H}_3 = 2776.4\,\text{kJ/kg} \\ \widehat{S}_3 = 7.6134\,\text{kJ/kg.K} \end{array}$$

First, it is necessary to check whether the first law is satisfied. The rate of enthalpy change is

$$\Delta \dot{H} = (150)(2803.3 - 3908.8) + (300)(2776.4 - 3908.8) = -505,545\,\text{kJ/h}$$

From the problem statement

$$\dot{Q} + \dot{W}_s = (-19.43 - 121)(3600) = -505,548\,\text{kJ/h}$$

Since

$$\Delta \dot{H} = \dot{Q} + \dot{W}_s$$

the first law is satisfied. To check whether the second law is satisfied, the rate of entropy generation must be calculated from Eq. (5.4-46), i.e.,

$$\dot{S}_{gen} = \Delta \dot{S} - \frac{\dot{Q}}{T_{surr}} \tag{1}$$

Since heat is removed from the system, $\dot{Q} = -19.43\,\text{kW}$. *Substitution of the numerical values into Eq. (1) gives*

$$\dot{S}_{gen} = (150)(6.4975 - 7.6837) + (300)(7.6134 - 7.6837) + \frac{(19.43)(3600)}{20 + 273}$$

$$= 39.71\,\text{kJ}/\,\text{h}.\,\text{K} > 0$$

Therefore, the process is possible.

Example 5.10 *A cylinder fitted with a piston contains 2 kg of steam at 400 kPa and 250 °C. The steam is then compressed isothermally until it becomes saturated vapor, with a work input of 430 kJ. During compression, heat transfer takes place with the surroundings at 25 °C through the cylinder wall. Is this process reversible, irreversible, or impossible?*

Solution

System: *Contents of the cylinder*

To determine whether this process is reversible, irreversible, or impossible, entropy generation must be calculated from Eq. (5.4-42), i.e.,

$$S_{gen} = \Delta S_{sys} - \frac{Q}{T_{surr}} \tag{1}$$

From Table A.3 in Appendix A

$$\left.\begin{array}{l} P_1 = 0.4\,\text{MPa} \\ T_1 = 250\,^\circ\text{C} \end{array}\right\} \begin{array}{l} \widehat{U}_1 = 2726.1\,\text{kJ}/\,\text{kg} \\ \widehat{S}_1 = 7.3789\,\text{kJ}/\,\text{kg}.\,\text{K} \end{array}$$

From Table A.1 in Appendix A

$$T_2 = 250\,^\circ\text{C} \left.\right\} \begin{array}{l} \widehat{U}_2 = 2602.4\,\text{kJ}/\,\text{kg} \\ \widehat{S}_2 = 6.0730\,\text{kJ}/\,\text{kg}.\,\text{K} \end{array}$$

The change in internal energy is

$$\Delta U = (2)(2602.4 - 2726.1) = -247.4\,\text{kJ}$$

The amount of heat transferred is

$$Q = \Delta U - W = -247.4\,\text{kJ} - 430 = -677.4\,\text{kJ}$$

Substitution of the numerical values into Eq. (1) gives

$$S_{gen} = (2)(6.0730 - 7.3789) + \frac{677.4}{25 + 273} = -0.34\,\text{kJ/K} < 0$$

Therefore, the process is impossible.

Alternative solution: *If compression takes place reversibly, the amount of heat transferred is*

$$Q_{rev} = T\,\Delta S_{sys} = (250 + 273)(-2.6118) = -1366\,\text{kJ}$$

The work associated with this reversible compression is

$$W_{rev} = \Delta U - Q_{rev} = -247.4 + 1366 = 1118.6\,\text{kJ}$$

If work is done on the system, W_{irrev} should always be greater than W_{rev}. According to the calculations, however, while $W_{rev} = 1118.6\,\text{kJ}$, $W_{irrev} = 430\,\text{kJ}$. Therefore, the process is not possible.

5.5 Calculation of Entropy Change

The first law of thermodynamics for a closed system is given by

$$d\widetilde{U} = \delta\widetilde{Q} + \delta\widetilde{W} \tag{5.5-1}$$

For a reversible process in which the shaft work is zero, Eq. (5.5-1) is expressed in the form

$$d\widetilde{U} = T\,d\widetilde{S} - P\,d\widetilde{V} \tag{5.5-2}$$

Solving for $d\widetilde{S}$ yields

$$\boxed{d\widetilde{S} = \frac{d\widetilde{U}}{T} + \frac{P}{T}\,d\widetilde{V}} \tag{5.5-3}$$

Since entropy is a state function, i.e., it depends only on the initial and final states, Eq. (5.5-3) holds for any kind of process. Note that Eq. (5.5-3) is valid for all substances.

The enthalpy is expressed as

$$d\widetilde{H} = d\widetilde{U} + d(P\widetilde{V}) \tag{5.5-4}$$

Substitution of Eq. (5.5-2) into Eq. (5.5-4) yields

$$d\widetilde{H} = T\,d\widetilde{S} + \widetilde{V}\,dP \qquad (5.5\text{-}5)$$

Solving for $d\widetilde{S}$ gives

$$\boxed{d\widetilde{S} = \frac{d\widetilde{H}}{T} - \frac{\widetilde{V}}{T}\,dP} \qquad (5.5\text{-}6)$$

Again, Eq. (5.5-6) holds for any substance and for any kind of process.

5.5.1 *Entropy change for liquids and solids*

For solids and liquids, density is almost independent of pressure and temperature. As a result, Eq. (5.5-3) reduces to

$$d\widetilde{S} = \frac{d\widetilde{U}}{T} \qquad (5.5\text{-}7)$$

Furthermore, for a solid or a liquid $\widetilde{C}_V \simeq \widetilde{C}_P$ and we have

$$d\widetilde{U} = \widetilde{C}_V\,dT \simeq \widetilde{C}_P\,dT \qquad (5.5\text{-}8)$$

Substitution of Eq. (5.5-8) into Eq. (5.5-7) and integration give

$$\boxed{\Delta\widetilde{S} = \int_{T_1}^{T_2} \widetilde{C}_P\,\frac{dT}{T}} \qquad (5.5\text{-}9)$$

If $\widetilde{C}_P$ is constant, Eq. (5.5-9) reduces to

$$\boxed{\Delta\widetilde{S} = \widetilde{C}_P \ln\left(\frac{T_2}{T_1}\right)} \quad \widetilde{C}_P = \text{constant} \qquad (5.5\text{-}10)$$

Example 5.11 *Calculate the change in entropy when saturated liquid at 25 °C is compressed to 50 bar and 60 °C in a piston-cylinder assembly.*

Solution

From Table A.4 in Appendix A

$$\left.\begin{array}{l} P = 5\,\text{MPa} \\ T = 60\,^\circ\text{C} \end{array}\right\} \widehat{S} = 0.8285\,\text{kJ/kg.\,K}$$

At $25\,^\circ$C, $\widehat{S}^L = 0.3674\,\text{kJ}/\text{kg}.\,\text{K}$. *Thus, the change in entropy is*

$$\Delta\widehat{S} = 0.8285 - 0.3674 = 0.4611\,\text{kJ}/\text{kg}.\,\text{K}$$

Alternative solution: *The entropy change can also be calculated from Eq. (5.5-10) as*

$$\Delta\widehat{S} = \widehat{C}_P \ln\left(\frac{T_2}{T_1}\right) = (4.2)\ln\left(\frac{60+273}{25+273}\right) = 0.4664\,\text{kJ}/\text{kg}.\,\text{K}$$

Since the heat capacity is considered constant, the result is slightly different from $0.4611\,\text{kJ}/\text{kg}.\,\text{K}$.

Comment: *If the value of the entropy for a compressed (or subcooled) liquid is not available, then the entropy of a compressed liquid can be approximated as the entropy of a saturated liquid at the given temperature. In this specific case,* $\widehat{S}^L = 0.8312\,\text{kJ}/\text{kg}.\,\text{K}$ *at* $60\,^\circ$C. *Therefore, the change in entropy is*

$$\Delta\widehat{S} = 0.8312 - 0.3674 = 0.4638\,\text{kJ}/\text{kg}.\,\text{K}$$

5.5.2 *Entropy change for a phase change*

The phase diagram (or *P-T* diagram) given in Figure 3.4 shows that phase transitions take place at constant temperature and pressure. Under constant pressure, $\delta\widehat{Q}_{rev} = d\widehat{H}$ and the change in entropy is

$$\Delta\widehat{S} = \int \frac{\delta\widehat{Q}_{rev}}{T} = \frac{1}{T}\int d\widehat{H} = \frac{\Delta\widehat{H}}{T} \tag{5.5-11}$$

Consider the phase change from liquid to vapor. In this case, Eq. (5.5-11) becomes

$$\widehat{S}^V - \widehat{S}^L = \frac{\widehat{H}^V - \widehat{H}^L}{T_b} \quad\Rightarrow\quad \boxed{\Delta\widehat{S}^{vap} = \frac{\Delta\widehat{H}^{vap}}{T_b}} \tag{5.5-12}$$

where $\Delta\widehat{S}^{vap}$ is the entropy change on vaporization, $\Delta\widehat{H}^{vap}$ is the enthalpy change on vaporization (or heat of vaporization), and T_b is the boiling point temperature. In the literature an empirical relation, called *Trouton's rule*, states that the entropy change on vaporization of liquids at their normal boiling point[9], T_{nbp}, is constant:

$$\Delta \widetilde{S}^{vap} = \frac{\Delta \widetilde{H}^{vap}}{T_{nbp}} = 88 \, \text{J}/\text{mol. K} \tag{5.5-13}$$

Keep in mind that Trouton's rule is not valid for hydrogen-bonded liquids, i.e., water and alcohols.

When the phase change is from solid to liquid, Eq. (5.5-11) takes the form

$$\widehat{S}^L - \widehat{S}^S = \frac{\widehat{H}^L - \widehat{H}^S}{T_m} \qquad \Rightarrow \qquad \boxed{\Delta \widehat{S}^{fus} = \frac{\Delta \widehat{H}^{fus}}{T_m}} \tag{5.5-14}$$

where $\Delta \widehat{S}^{fus}$ is the entropy change on fusion, $\Delta \widehat{H}^{fus}$ is the enthalpy change on fusion (or heat of fusion), and T_m is the melting (or freezing) temperature.

When the phase change is from solid to vapor, Eq. (5.5-11) takes the form

$$\widehat{S}^V - \widehat{S}^S = \frac{\widehat{H}^V - \widehat{H}^S}{T_s} \qquad \Rightarrow \qquad \boxed{\Delta \widehat{S}^{sub} = \frac{\Delta \widehat{H}^{sub}}{T_s}} \tag{5.5-15}$$

where $\Delta \widehat{S}^{sub}$ is the entropy change on sublimation, $\Delta \widehat{H}^{sub}$ is the enthalpy change on sublimation (or heat of sublimation), and T_s is the sublimation temperature.

5.5.3 *Entropy change of an ideal gas*

For an ideal gas

$$d\widetilde{U}^{IG} = \widetilde{C}_V^* \, dT \tag{5.5-16}$$

and

$$\frac{P}{T} = \frac{R}{\widetilde{V}} \tag{5.5-17}$$

---

[9]The normal boiling point of a liquid is the temperature at which the vapor pressure equals 1 atm.

Substitution of Eqs. (5.5-16) and (5.5-17) into Eq. (5.5-3) and integration give

$$\Delta \widetilde{S}^{IG} = \widetilde{S}_2 - \widetilde{S}_1 = \int_{T_1}^{T_2} \widetilde{C}_V^* \frac{dT}{T} + R \ln \left(\frac{\widetilde{V}_2}{\widetilde{V}_1} \right) \tag{5.5-18}$$

If $\widetilde{C}_V^*$ is constant, Eq. (5.5-18) reduces to

$$\Delta \widetilde{S}^{IG} = \widetilde{C}_V^* \ln \left(\frac{T_2}{T_1} \right) + R \ln \left(\frac{\widetilde{V}_2}{\widetilde{V}_1} \right) \qquad \widetilde{C}_V^* = \text{constant} \tag{5.5-19}$$

For an ideal gas, the following relationships also hold:

$$d\widetilde{H}^{IG} = \widetilde{C}_P^* \, dT \tag{5.5-20}$$

and

$$\frac{\widetilde{V}}{T} = \frac{R}{P} \tag{5.5-21}$$

Substitution of Eqs. (5.5-20) and (5.5-21) into Eq. (5.5-6) and integration give

$$\Delta \widetilde{S}^{IG} = \widetilde{S}_2 - \widetilde{S}_1 = \int_{T_1}^{T_2} \widetilde{C}_P^* \frac{dT}{T} - R \ln \left(\frac{P_2}{P_1} \right) \tag{5.5-22}$$

If $\widetilde{C}_P^*$ is constant, Eq. (5.5-22) reduces to

$$\Delta \widetilde{S}^{IG} = \widetilde{C}_P^* \ln \left(\frac{T_2}{T_1} \right) - R \ln \left(\frac{P_2}{P_1} \right) \qquad \widetilde{C}_P^* = \text{constant} \tag{5.5-23}$$

Example 5.12 *Determine the change in entropy when* 2 kmol *of* N_2 *(ideal gas) at* 10 bar *and* 380 K *(state 1) undergoes the following reversible changes:*

a) *Compressed isothermally to* 15 bar *(state 2),*
b) *Heated at constant pressure to* 660 K *(state 3),*
c) *Cooled at constant volume until the pressure drops to* 6 bar *(state 4),*
d) *Compressed adiabatically to* 10 bar *(state 5),*
e) *Heated at constant pressure to state 1.*

Solution

The process paths in a P-V diagram are shown below.

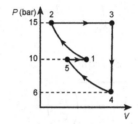

• Process 1 → 2 (Reversible isothermal)

Since temperature remains constant, Eq. (5.5-23) simplifies to

$$\Delta S_{12} = -nR \ln \left(\frac{P_2}{P_1}\right) = -(2)(8.314) \ln \left(\frac{15}{10}\right) = -6.74 \, \text{kJ/K}$$

• Process 2 → 3 (Reversible isobaric)

Since pressure remains constant, Eq. (5.5-23) simplifies to

$$\Delta S_{23} = n\widetilde{C}_P^* \ln \left(\frac{T_3}{T_2}\right) = (2)(3.5 \times 8.314) \ln \left(\frac{660}{380}\right) = 32.13 \, \text{kJ/K}$$

• Process 3 → 4 (Reversible isometric)

For a constant volume process, Eq. (5.5-19) reduces to

$$\Delta S_{34} = n\widetilde{C}_V^* \ln \left(\frac{T_4}{T_3}\right) \tag{1}$$

From the ideal gas law

$$\frac{T_4}{T_3} = \frac{P_4}{P_3} \tag{2}$$

Substitution of Eq. (2) into Eq. (1) yields

$$\Delta S_{34} = n\widetilde{C}_V^* \ln \left(\frac{P_4}{P_3}\right) = (2)(2.5 \times 8.314) \ln \left(\frac{6}{15}\right) = -38.09 \, \text{kJ/K}$$

• Process 4 → 5 (Reversible adiabatic)

In this case entropy remains constant, i.e.,

$$\Delta S_{45} = 0$$

• **Process $5 \to 1$ (Reversible isobaric)**

For a constant pressure process, Eq. (5.5-23) reduces to

$$\Delta S_{51} = n\widetilde{C}_P^* \ln\left(\frac{T_1}{T_5}\right) \tag{3}$$

in which T_5 is an unknown. Since the process $4 \to 5$ is adiabatic and reversible, then

$$\frac{T_5}{T_4} = \left(\frac{P_5}{P_4}\right)^{(\gamma-1)/\gamma} \tag{4}$$

To calculate T_5, it is first necessary to determine T_4. From Eq. (2)

$$T_4 = T_3\left(\frac{P_4}{P_3}\right) = (660)\left(\frac{6}{15}\right) = 264\,\text{K}$$

The use of this value in Eq. (4) gives

$$T_5 = (264)\left(\frac{10}{6}\right)^{2/7} = 305.5\,\text{K}$$

Therefore, the entropy change is

$$\Delta S_{51} = (2)(3.5 \times 8.314)\ln\left(\frac{380}{305.5}\right) = 12.7\,\text{kJ/K}$$

The total entropy change for the overall process is

$$\Delta S_{\text{cycle}} = \Delta S_{12} + \Delta S_{23} + \Delta S_{34} + \Delta S_{45} + \Delta S_{51}$$
$$= -6.74 + 32.13 - 38.09 + 0 + 12.7 = 0$$

Since entropy is a state function, this is an expected result.

Example 5.13 *A cylinder fitted with a frictionless piston contains $1.5\,\text{kg}$ of steam at $1\,\text{MPa}$ and 40% quality (state 1). The steam undergoes the following reversible processes:*

- *It is heated at constant volume to $4\,\text{MPa}$ (state 2),*
- *It is then expanded isothermally to $1\,\text{MPa}$ (state 3),*
- *Finally, it is cooled at constant pressure to state 1.*

a) *Calculate Q, W, and ΔU for each process and for the entire cycle,*
b) *Show each process in a P-V diagram.*

Solution

a) State 1

From Table A.2 in Appendix A

$$P = 1\,\text{MPa} \left\{ \begin{array}{ll} \widehat{V}^L = 0.001127\,\text{m}^3/\text{kg} & \widehat{U}^L = 761.68\,\text{kJ}/\text{kg} \\ \widehat{V}^V = 0.19444\,\text{m}^3/\text{kg} & \Delta\widehat{U} = 1822.0\,\text{kJ}/\text{kg} \end{array} \right.$$

$$\widehat{V}_1 = 0.001127 + (0.4)(0.19444 - 0.001127) = 0.07845\,\text{m}^3/\text{kg}$$

$$\widehat{U}_1 = 761.68 + (0.4)(1822) = 1490.7\,\text{kJ}/\text{kg}$$

State 2

Note that $P_2 = 4\,\text{MPa}$ *and* $\widehat{V}_2 = \widehat{V}_1 = 0.07845\,\text{m}^3/\text{kg}$. *From Table A.3 in Appendix A, at* $4\,\text{MPa}$

$T\,(^\circ\text{C})$	$\widehat{V}\,(\text{m}^3/\text{kg})$	$\widehat{U}\,(\text{kJ}/\text{kg})$	$\widehat{S}\,(\text{kJ}/\text{kg. K})$
400	0.07341	2919.9	6.7690
450	0.08002	3010.2	6.9363

By interpolation

$$\widehat{U}_2 = 2919.9 + \left(\frac{3010.2 - 2919.9}{0.08002 - 0.07341}\right)(0.07845 - 0.07341) = 2988.8\,\text{kJ}/\text{kg}$$

$$\widehat{S}_2 = 6.7690 + \left(\frac{6.9363 - 6.7690}{0.08002 - 0.07341}\right)(0.07845 - 0.07341) = 6.8966\,\text{kJ}/\text{kg. K}$$

$$T_2 = 400 + \left(\frac{450 - 400}{0.08002 - 0.07341}\right)(0.07845 - 0.07341) = 438.1\,^\circ\text{C}$$

State 3

Note that $P_3 = P_1 = 1\,\text{MPa}$ *and* $T_3 = T_2 = 438.1\,^\circ\text{C}$. *From Table A.3 in Appendix A, at* $1\,\text{MPa}$

$T\,(^\circ\text{C})$	$\widehat{V}\,(\text{m}^3/\text{kg})$	$\widehat{U}\,(\text{kJ}/\text{kg})$	$\widehat{S}\,(\text{kJ}/\text{kg. K})$
400	0.3066	2957.3	7.4651
500	0.3541	3124.4	7.7622

By interpolation

$$\widehat{V}_3 = 0.3066 + \left(\frac{0.3541 - 0.3066}{500 - 400} \right) (438.1 - 400) = 0.3247 \, \text{m}^3/\text{kg}$$

$$\widehat{U}_3 = 2957.3 + \left(\frac{3124.4 - 2957.3}{500 - 400} \right) (438.1 - 400) = 3021.0 \, \text{kJ}/\text{kg}$$

$$\widehat{S}_3 = 7.4651 + \left(\frac{7.7622 - 7.4651}{500 - 400} \right) (438.1 - 400) = 7.5783 \, \text{kJ}/\text{kg. K}$$

Process $1 \rightarrow 2$ (Reversible constant volume)

$$W_{12} = 0$$

$$Q_{12} = \Delta U_{12} = (1.5)(2988.8 - 1490.7) = 2247.2 \, \text{kJ}$$

Process $2 \rightarrow 3$ (Reversible isothermal)

$$Q_{23} = T_2 \, \Delta S_{23} = (438.1 + 273)(1.5)(7.5783 - 6.8966) = 727.14 \, \text{kJ}$$

$$\Delta U_{23} = (1.5)(3021 - 2988.8) = 48.3 \, \text{kJ}$$

$$W_{23} = \Delta U_{23} - Q_{23} = 48.3 - 727.14 = -678.84 \, \text{kJ}$$

Process $3 \rightarrow 1$ (Reversible isobaric)

$$W_{31} = -P_3 \Delta V = -(1000)(1.5)(0.07845 - 0.3247) = 369.38 \, \text{kJ}$$

$$\Delta U_{31} = (1.5)(1490.7 - 3021) = -2295.5 \, \text{kJ}$$

$$Q_{31} = \Delta U_{31} - W_{31} = -2295.5 - 369.38 = -2664.88 \, \text{kJ}$$

The following table summarizes the values of Q, W, and ΔU for each process as well as for the entire process:

Process	Q (kJ)	W (kJ)	ΔU (kJ)
$1 \rightarrow 2$	2247.2	0	2247.2
$2 \rightarrow 3$	727.14	-678.84	48.3
$3 \rightarrow 1$	-2664.88	369.38	-2295.5
$\sum$	309.46	-309.46	0

Since internal energy is a state function, for the overall cyclic process $\Delta U = 0$. When $\Delta U_{cycle} = 0$, from the first law $Q_{cycle} = -W_{cycle}$.

b)

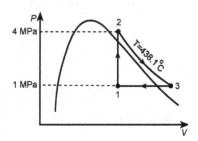

Example 5.14 *A rigid tank A and cylinder B are connected as shown in the figure below. Tank A initially contains 5 kg of steam at 1 MPa and 500 °C. Cylinder B contains a frictionless piston of a mass such that a pressure of 300 kPa inside the cylinder is required to raise the piston. Initially the piston rests on stops; the volume of cylinder B is 0.1 m³ and it is evacuated. The connecting valve is opened until the pressure in tank A drops to 300 kPa. Calculate the final temperature of the steam in cylinder B and the work done during the process. Assume the entire process to be adiabatic.*

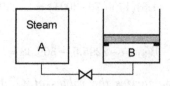

Solution

System: *Contents of the tank* A *and cylinder* B

The final temperatures in tank A, T_{A_2}, *and in cylinder* B, T_{B_2}, *together with the final volume of cylinder* B, V_{B_2}, *are unknown quantities. Therefore, besides mass and energy balances, one more equation is needed. For this purpose, we assume that the steam remaining in tank* A *undergoes an adiabatic and reversible expansion, i.e.,* $\widehat{S}_{A_1} = \widehat{S}_{A_2}$.

From Table A.3 in Appendix A

$$\left. \begin{array}{l} P_{A_1} = 1\,\mathrm{MPa} \\ T_{A_1} = 500\,^\circ\mathrm{C} \end{array} \right\} \begin{array}{l} \widehat{V}_{A_1} = 0.3541\,\mathrm{m^3/kg} \\ \widehat{U}_{A_1} = 3124.4\,\mathrm{kJ/kg} \\ \widehat{S}_{A_1} = 7.7622\,\mathrm{kJ/kg.\,K} \end{array}$$

The pressure and specific entropy of steam in tank A *at the final state are* $P_{A_2} = 300\,\mathrm{kPa}$ *and* $\widehat{S}_{A_2} = 7.7622\,\mathrm{kJ/kg.\,K}$. *From Table A.3 in Appendix A, at* $300\,\mathrm{kPa}$

$T\ (^\circ\mathrm{C})$	$\widehat{V}\ (\mathrm{m^3/kg})$	$\widehat{U}\ (\mathrm{kJ/kg})$	$\widehat{S}\ (\mathrm{kJ/kg.\,K})$
300	0.8753	2806.7	7.7022
400	1.0315	2965.6	8.0330

By interpolation

$$\widehat{V}_{A_2} = 0.8753 + \left(\frac{1.0315 - 0.8753}{8.0330 - 7.7022} \right) (7.7622 - 7.7022) = 0.9036\,\mathrm{m^3/kg}$$

$$\widehat{U}_{A_2} = 2806.7 + \left(\frac{2965.6 - 2806.7}{8.0330 - 7.7022} \right) (7.7622 - 7.7022) = 2835.5\,\mathrm{kJ/kg}$$

The volume of tank A, V_A, *is*

$$V_A = (5)(0.3541) = 1.77\,\mathrm{m^3}$$

The final amount of steam in tank A, m_{A_2}, *is*

$$m_{A_2} = \frac{V_A}{\widehat{V}_{A_2}} = \frac{1.77}{0.9036} = 1.96\,\mathrm{kg}$$

The final amount of steam in cylinder B, m_{B_2}, *is*

$$m_{B_2} = 5 - 1.96 = 3.04\,\mathrm{kg}$$

Application of the first law of thermodynamics gives

$$\Delta U = \underbrace{Q}_{0} + W \qquad \Rightarrow \qquad \Delta U_A + \Delta U_B = W \tag{1}$$

The term W in Eq. (1) can be obtained from

$$W = -\int_{V_{B_1}}^{V_{B_2}} P_{ex}\, dV = -300\,(V_{B_2} - 0.1) \tag{2}$$

Thus, Eq. (1) can be represented as

$$\left[(1.96)(2835.5) - (5)(3124.4)\right] + (3.04)\,\widehat{U}_{B_2} = -300\,(V_{B_2} - 0.1) \tag{3}$$

Note that

$$\widehat{U}_{B_2} = \widehat{H}_{B_2} - (300)\,\widehat{V}_{B_2} \tag{4}$$

$$V_{B_2} = (3.04)\,\widehat{V}_{B_2} \tag{5}$$

Substitution of Eqs. (4) and (5) into Eq. (3) gives

$$\widehat{H}_{B_2} = 3320.5\,\mathrm{kJ/kg}$$

From Table A.3 in Appendix A, at 300 kPa

T (°C)	$\widehat{V}$ (m³/ kg)	$\widehat{H}$ (kJ/ kg)
400	1.0315	3275.0
500	1.1867	3486.0

By interpolation

$$T_{B_2} = 400 + \left(\frac{500 - 400}{3486 - 3275}\right)(3320.5 - 3275) = 421.6\,^{\circ}\mathrm{C}$$

$$\widehat{V}_{B_2} = 1.0315 + \left(\frac{1.1867 - 1.0315}{3486 - 3275}\right)(3320.5 - 3275) = 1.065\,\mathrm{m^3/kg}$$

Thus, the work done can be calculated from Eq. (2) as

$$W = -300\left[(3.04)(1.065) - 0.1\right] = -941.3\,\mathrm{kJ}$$

Example 5.15 *Consider the system shown in the figure below. The cylindrical tank is insulated and the frictionless piston does not conduct heat.*

Initially, the piston is at the extreme left-hand side of the tank and the tank contains 1.5 kg of steam at 100 kPa and 200 °C. The tank is connected to a large pipeline carrying air at 0.5 MPa and 250 °C. The valve between the pipeline and the tank is opened slightly and air is allowed to enter the tank slowly until the pressure reaches 0.5 MPa, at which point the valve is closed.

a) *Calculate the final temperature of the air.*
b) *Calculate the final temperature of the air if the steam temperature is kept constant at 200 °C by a cooling coil.*

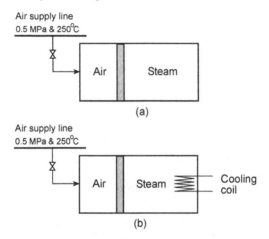

(a)

(b)

Solution

Let the subscripts A and S stand for air and steam, respectively. Since steam occupies the whole volume of the tank initially, the volume of the tank is calculated from the initial conditions. From Table A.3 in Appendix A

$$\left. \begin{array}{l} P_{S_1} = 100\,\text{kPa} \\ T_{S_1} = 200\,°\text{C} \end{array} \right\} \begin{array}{l} \widehat{V}_{S_1} = 2.172\,\text{m}^3/\text{kg} \\ \widehat{U}_{S_1} = 2658.1\,\text{kJ}/\text{kg} \\ \widehat{S}_{S_1} = 7.8343\,\text{kJ}/\text{kg. K} \end{array}$$

Thus

$$V_{S_1} = (1.5)(2.172) = 3.258\,\text{m}^3$$

a) *Note that the steam undergoes a reversible and adiabatic compression, i.e., $\widehat{S}_{S_1} = \widehat{S}_{S_2} = 7.8343\,\text{kJ}/\text{kg. K}$. At the final state, the steam properties are $P_{S_2} = 0.5\,\text{MPa}$ and $\widehat{S}_{S_2} = 7.8343\,\text{kJ}/\text{kg. K}$. From Table A.3 in Appendix A, at 0.5 MPa*

$T\,(°C)$	$\widehat{V}\,(m^3/kg)$	$\widehat{U}\,(kJ/kg)$	$\widehat{S}\,(kJ/kg.\,K)$
400	0.6173	2963.2	7.7938
500	0.7109	3128.4	8.0873

$$\widehat{V}_{S_2} = 0.6173 + \left(\frac{0.7109 - 0.6173}{8.0873 - 7.7938}\right)(7.8343 - 7.7938) = 0.6302\,m^3/kg$$

$$\widehat{U}_{S_2} = 2963.2 + \left(\frac{3128.4 - 2963.2}{8.0873 - 7.7938}\right)(7.8343 - 7.7938) = 2986\,kJ/kg$$

The final volume occupied by the steam is

$$V_{S_2} = (1.5)(0.6302) = 0.9453\,m^3$$

and the final volume of the air in the tank is

$$V_{A_2} = 3.258 - 0.9453 = 2.3127\,m^3$$

Considering the steam in the tank as the system, the first law of thermodynamics gives

$$W_S = \Delta U_S = (1.5)(2986 - 2658.1) = 491.9\,kJ$$

Considering the air in the tank as the system, the unsteady-state mass and energy balances take the form

$$(dn_A)_{in} = (dn_A)_{sys} \tag{1}$$

$$(\widetilde{H}_A)_{in}\,(dn_A)_{in} + \delta W_A = d(n_A\widetilde{U}_A)_{sys} \tag{2}$$

Combination of Eqs. (1) and (2) and integration lead to

$$(\widetilde{H}_A)_{in}(n_{A_2} - n_{A_1}) + W_A = n_{A_2}\widetilde{U}_{A_2} - n_{A_1}\widetilde{U}_{A_1} \tag{3}$$

where subscripts 1 and 2 refer to the initial and final states, respectively. Since $n_{A_1} = 0$, Eq. (3) simplifies to

$$n_{A_2}(\widetilde{C}_P^* T_{A_{in}} - \widetilde{C}_V^* T_{A_2}) + W_A = 0 \tag{4}$$

Noting that

$$n_{A_2} = \frac{P_{A_2} V_{A_2}}{R T_{A_2}} \qquad \widetilde{C}_P^* = \frac{7}{2} R \qquad \widetilde{C}_V^* = \frac{5}{2} R \qquad W_A = -W_S \qquad (5)$$

Eq. (4) simplifies to

$$T_{A_2} = \frac{7 T_{A_{in}}}{5 + 2 \left(\dfrac{W_S}{P_{A_2} V_{A_2}} \right)} = \frac{(7)(250 + 273)}{5 + 2 \left[\dfrac{491.9}{(500)(2.3127)} \right]} = 625.7 \, \text{K}$$

b) *In this case, the steam undergoes a reversible and isothermal compression. Thus, the properties of the steam at the final state are*

$$\left. \begin{array}{l} P_{S_2} = 0.5 \, \text{MPa} \\ T_{S_2} = 200 \, °C \end{array} \right\} \begin{array}{l} \widehat{V}_{S_2} = 0.4249 \, \text{m}^3 / \text{kg} \\ \widehat{U}_{S_2} = 2642.9 \, \text{kJ} / \text{kg} \\ \widehat{S}_{S_2} = 7.0592 \, \text{kJ} / \text{kg. K} \end{array}$$

The final volume occupied by the steam is

$$V_{S_2} = (1.5)(0.4249) = 0.6374 \, \text{m}^3$$

and the final volume of the air in the tank is

$$V_{A_2} = 3.258 - 0.6374 = 2.621 \, \text{m}^3$$

Considering the steam in the tank as the system, the first law of thermodynamics gives

$$\begin{aligned} W_S &= \Delta U_S - Q_S = \Delta U_S - T \, \Delta S_S \\ &= \left[(1.5)(2642.9 - 2658.1) \right] - (200 + 273)(1.5)(7.0592 - 7.8342) \\ &= 527.1 \, \text{kJ} \end{aligned}$$

Considering the air in the tank as the system, the unsteady-state mass and energy balances again lead to

$$T_{A_2} = \frac{7 T_{A_{in}}}{5 + 2 \left(\dfrac{W_S}{P_{A_2} V_{A_2}} \right)} = \frac{(7)(250 + 273)}{5 + 2 \left[\dfrac{527.1}{(500)(2.621)} \right]} = 630.7 \, \text{K}$$

Example 5.16 *Consider the system shown in the figure below. The cylindrical tank is insulated and the frictionless piston does not conduct heat. Initially, the piston is at the extreme left-hand side of the tank and the tank contains 5 moles of air at 100 kPa and 30 °C. The tank is connected to a* *large pipeline carrying steam at 0.5 MPa and 250 °C. The valve between the pipeline and the tank is opened slightly and steam is allowed to enter the tank slowly until the pressure reaches 0.5 MPa, at which point the valve is closed.*

a) *Calculate the final temperature of the steam.*
b) *Calculate the final temperature of the steam if the air temperature is kept constant at 30 °C by a cooling coil.*

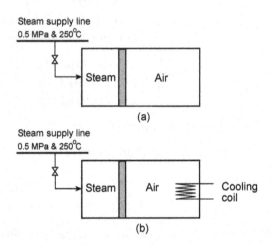

Solution

Let the subscripts A and S stand for air and steam, respectively. Initially the air occupies the whole tank volume. Therefore,

$$V_{A_1} = \frac{n_A R T_{A_1}}{P_{A_1}} = \frac{(5)(8.314 \times 10^{-3})(30 + 273)}{100} = 0.126\,\mathrm{m}^3$$

a) Note that the air undergoes a reversible and adiabatic compression. Thus, the final values of volume and temperature are

$$V_{A_2} = V_{A_1} \left(\frac{P_{A_1}}{P_{A_2}}\right)^{1/\gamma} = (0.126)\left(\frac{100}{500}\right)^{5/7} = 0.04\,\mathrm{m}^3$$

$$T_{A_2} = T_{A_1} \left(\frac{P_{A_2}}{P_{A_1}}\right)^{(\gamma-1)/\gamma} = (303)\left(\frac{500}{100}\right)^{2/7} = 480\,\mathrm{K}$$

Considering the air in the tank as the system, the first law of thermodynamics gives

$$W_A = \Delta U_A = n_A \widetilde{C}_V^* (T_{A_2} - T_{A_1})$$
$$= (5)(2.5 \times 8.314)(480 - 303) = 18,395 \, \text{J} \simeq 18.4 \, \text{kJ}$$

Considering the steam in the tank as the system, the unsteady-state mass and energy balances take the form

$$(dm_S)_{in} = (dm_S)_{sys} \tag{1}$$

$$(\widehat{H}_S)_{in} \, (dm_S)_{in} + \delta W_S = d(m_S \widehat{U}_S)_{sys} \tag{2}$$

Combination of Eqs. (1) and (2) and integration lead to

$$(\widehat{H}_S)_{in}(m_{S_2} - m_{S_1}) + W_S = m_{S_2}\widehat{U}_{S_2} - m_{S_1}\widehat{U}_{S_1} \tag{3}$$

where subscripts 1 and 2 refer to the initial and final states, respectively. Since $m_{S_1} = 0$, Eq. (3) simplifies to

$$m_{S_2}(\widehat{H}_S)_{in} + W_S = m_{S_2}\widehat{U}_{S_2} \tag{4}$$

From Table A.3 in Appendix A

$$\left. \begin{array}{l} P = 0.5 \, \text{MPa} \\ T = 250 \, °\text{C} \end{array} \right\} (\widehat{H}_S)_{in} = 2960.7 \, \text{kJ/kg}$$

The volume of the steam at the final state is

$$V_{S_2} = 0.126 - 0.04 = 0.086 \, \text{m}^3$$

and $W_S = -W_A = -18.4 \, \text{kJ}$. Thus, Eq. (4) can be expressed as

$$\frac{0.086}{\widehat{V}_{S_2}} (2960.7) - 18.4 = \frac{0.086}{\widehat{V}_{S_2}} \widehat{U}_{S_2} \tag{5}$$

Simplification of Eq. (5) results in

$$\widehat{U}_{S_2} = 2960.7 - 214 \, \widehat{V}_{S_2} \tag{6}$$

The temperature can be determined by a graphical procedure as follows:

• *From Table A.3 in Appendix A, plot $\widehat{U}_2$ versus $\widehat{V}_2$ at a pressure of 0.5 MPa. Note that the values are*

T (°C)	$\widehat{U}_{S_2}$ (kJ/ kg)	$\widehat{V}_{S_2}$ (m³/ kg)
200	2642.9	0.4249
250	2723.5	0.4744
300	2802.9	0.5226
350	2882.6	0.5701
400	2963.2	0.6173

• *On the same graph, plot also Eq. (6). The intersection of this straight line with the* $\widehat{U}_{S_2}$ *versus* $\widehat{V}_{S_2}$ *curve gives*

$$\widehat{U}_{S_2} = 2844\,\text{kJ/ kg} \qquad and \qquad \widehat{V}_{S_2} = 0.547\,\text{m}^3/\text{kg}$$

The final temperature of the steam is found by interpolation as

$$T_{S_2} = 300 + \left(\frac{350 - 300}{2882.6 - 2802.9} \right) (2844 - 2802.9) = 325.8\,°\text{C}$$

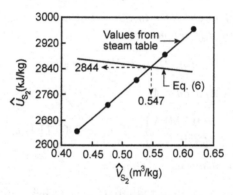

b) *In this case, the air undergoes a reversible and isothermal compression. Thus, the final volume of the air can be calculated from the ideal gas equation of state as*

$$V_{A_2} = V_{A_1} \left(\frac{P_{A_1}}{P_{A_2}} \right) = (0.126) \left(\frac{100}{500} \right) = 0.0252\,\text{m}^3$$

The work done on the air is given by

$$W_A = - \int_{V_{A_1}}^{V_{A_2}} P_A \, dV_A = - n_A R T_A \ln \left(\frac{V_{A_2}}{V_{A_1}} \right) = - n_A R T_A \ln \left(\frac{P_{A_1}}{P_{A_2}} \right)$$

$$= - (5)(8.314)(303) \ln \left(\frac{100}{500} \right) = 20,272\,\text{J} = 20.3\,\text{kJ}$$

Considering the steam in the tank as the system, the unsteady-state mass and energy balances give

$$m_{S_2}(\widehat{H}_S)_{in} + W_S = m_{S_2}\widehat{U}_{S_2} \tag{7}$$

The volume of the steam at the final state is

$$V_{S_2} = 0.126 - 0.0252 = 0.1008\,\mathrm{m}^3$$

Thus, substitution of values into Eq. (7) results in

$$\widehat{U}_{S_2} = 2960.7 - 201.4\,\widehat{V}_{S_2} \tag{8}$$

Application of the graphical procedure yields

$$\widehat{U}_{S_2} = 2850\,\mathrm{kJ/kg} \qquad and \qquad \widehat{V}_{S_2} = 0.551\,\mathrm{m}^3/\mathrm{kg}$$

The final temperature of the steam is found by interpolation as

$$T_{S_2} = 300 + \left(\frac{350 - 300}{2882.6 - 2802.9}\right)(2850 - 2802.9) = 329.6\,^{\circ}\mathrm{C}$$

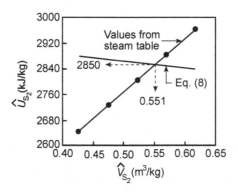

Example 5.17 *A cylinder fitted with a piston contains 2 moles of ideal gas at 200 kPa and 150 °C. The gas is compressed isothermally but irreversibly to 800 kPa. The actual work required for compression is 30% greater than the reversible work for the same compression. If the heat removed from the gas is discarded to a heat reservoir at 25 °C, calculate the entropy generation.*

Solution

System: *Contents of the cylinder*

The entropy change of the system is calculated from Eq. (5.5-23) as

$$\Delta S_{sys} = -nR \ln \left(\frac{P_2}{P_1} \right) = -(2)(8.314) \ln \left(\frac{800}{200} \right) = -23.05 \text{ J/K}$$

To calculate the entropy change of the surroundings, it is necessary to calculate the amount of heat removed from the system. The first law of thermodynamics gives

$$\underbrace{\Delta U}_{0} = Q + W \qquad \Rightarrow \qquad Q = -W$$

The reversible work is given by

$$W_{rev} = -\int P dV = -nRT \ln \left(\frac{V_2}{V_1} \right) = -nRT \ln \left(\frac{P_1}{P_2} \right)$$

$$= -(2)(8.314)(150 + 273) \ln \left(\frac{200}{800} \right) = 9750.7 \text{ J}$$

Thus, the actual work done on the system is

$$W = 1.3 \, W_{rev} = (1.3)(9750.7) = 12,676 \text{ J}$$

Therefore

$$Q = -W = -12,676 \text{ J}$$

The entropy change of the surroundings, i.e., heat reservoir at 25 °C, is

$$\Delta S_{surr} = \frac{12,676}{25 + 273} = 42.5 \text{ J/K}$$

Therefore, the entropy generation is

$$S_{gen} = \Delta S_{sys} + \Delta S_{surr} = -23.05 + 42.5 = 19.45 \text{ J/K}$$

Comment: *Note that it is also possible to calculate the entropy change of the system as*

$$\Delta S_{sys} = \frac{Q_{rev}}{T} = \frac{-9750.7}{150 + 273} = -23.05 \text{ J/K}$$

Example 5.18 *One of your friends claims to have come up with a process that will satisfy the heating and cooling requirements of a plant simultaneously. The following is a schematic of his/her proposal:*

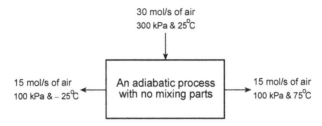

30 mol/s of air
300 kPa & 25°C

15 mol/s of air
100 kPa & – 25°C

An adiabatic process
with no mixing parts

15 mol/s of air
100 kPa & 75°C

Is this process possible?

Solution

If the proposed process works, it has to satisfy both the first and second laws of thermodynamics. The first law reduces to

$$\Delta \dot{H} = \underbrace{\dot{Q}}_{0} + \underbrace{\dot{W}_s}_{0} \quad \Rightarrow \quad \Delta \dot{H}_{\text{hot air}} + \Delta \dot{H}_{\text{cold air}} = 0 \qquad (1)$$

Substitution of the numerical values into Eq. (1) gives

$$(15)(3.5 \times 8.314)(75 - 25) + (15)(3.5 \times 8.314)(-25 - 25) = 0$$

Therefore, the first law is satisfied. Now it is necessary to check whether $\dot{S}_{gen} \geq 0$. Since there is no heat interaction between the system and its surroundings, the rate of entropy generation is equal to

$$\dot{S}_{gen} = \Delta \dot{S}_{\text{hot air}} + \Delta \dot{S}_{\text{cold air}}$$

Rate of entropy changes of hot and cold air are calculated with the help of Eq. (5.5-23) as

$$\dot{S}_{gen} = (15) \left[(3.5 \times 8.314) \ln \left(\frac{75 + 273}{25 + 273} \right) - (8.314) \ln \left(\frac{100}{300} \right) \right]$$
$$+(15) \left[(3.5 \times 8.314) \ln \left(\frac{-25 + 273}{25 + 273} \right) - (8.314) \ln \left(\frac{100}{300} \right) \right] = 261.6 \, \text{J/K.s}$$

Therefore, the proposed process is possible.

Example 5.19 *150 mol/s of an air stream at a pressure of 150 kPa is to be heated from 20 °C to 80 °C in a heat exchanger with steam. Superheated steam at 800 kPa and 600 °C enters the heat exchanger and leaves it as saturated liquid at 800 kPa.*

a) *How much steam would be required if a conventional heat exchanger were used?*

b) *How much steam would be required if heating of the air were carried out by using the heat transferred from the steam to drive a reversible heat engine and by using the air stream as the heat sink? What would be the power output of the engine?*

Solution

From Appendix A

$$\left. \begin{array}{l} P = 800\,\text{kPa} \\ T = 600\,^{\circ}\text{C} \end{array} \right\} \quad \begin{array}{l} \widehat{H} = 3699.4\,\text{kJ/kg} \\ \widehat{S} = 8.1333\,\text{kJ/kg. K} \end{array}$$

$$\left. \begin{array}{l} \\ P = 800\,\text{kPa} \\ \\ \end{array} \right\} \quad \begin{array}{l} \widehat{H}^L = 721.11\,\text{kJ/kg} \\ \widehat{S}^L = 2.0462\,\text{kJ/kg. K} \end{array}$$

a) System: *Heat exchanger*

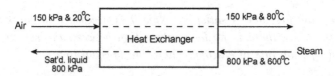

From the first law of thermodynamics

$$\Delta \dot{H} = \dot{n}_{air}\, \Delta \widetilde{H}_{air} + \dot{m}_{steam}\, \Delta \widehat{H}_{steam} = 0 \tag{1}$$

Substitution of the numerical values into Eq. (1) gives

$$\left(\frac{150}{1000} \right)(3.5 \times 8.314)(80 - 20) + \dot{m}_{steam}\,(721.11 - 3699.4) = 0$$

Solving for $\dot{m}_{steam}$ *results in*

$$\dot{m}_{steam} = 0.088\,\text{kg/s}$$

b) *A schematic of the proposed arrangement is shown below.*

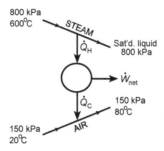

Since the Carnot cycle operates reversibly, then

$$\dot{S}_{gen} = \Delta\dot{S}_{steam} + \underbrace{\Delta\dot{S}_{heat\ engine}}_{0} + \Delta\dot{S}_{air} = 0 \qquad (2)$$

or

$$\dot{m}_{steam}\,\Delta\widehat{S}_{steam} + \dot{n}_{air}\,\Delta\widetilde{S}_{air} = 0 \qquad (3)$$

Substitution of the numerical values into Eq. (3) gives

$$\dot{m}_{steam}(2.0462 - 8.1333) + \left(\frac{150}{1000}\right)(3.5 \times 8.314)\ln\left(\frac{80+273}{20+273}\right) = 0$$

Solving for $\dot{m}_{steam}$ results in

$$\dot{m}_{steam} = 0.134\,\text{kg/s}$$

The rate of heat received by the heat engine is

$$\dot{Q}_H = \dot{m}_{steam}\left|\Delta\widehat{H}_{steam}\right| = (0.134)(3699.4 - 721.11) = 399.09\,\text{kW}$$

The rate of heat rejected by the heat engine (or the rate of heat received by air) is

$$\dot{Q}_C = \dot{n}_{air}\,\Delta\widetilde{H}_{air} = \left(\frac{150}{1000}\right)(3.5 \times 8.314)(80 - 20) = 261.89\,\text{kW}$$

Thus, the power output of the heat engine is

$$\dot{W}_{net} = \dot{Q}_H - \dot{Q}_C = 399.09 - 261.89 = 137.2\,\text{kW}$$

5.6   Shaft Work and Isentropic Efficiency

5.6.1   *Shaft work in reversible steady-state flow processes*

In differential form, the first law of thermodynamics for a closed system is given by

$$dU = \delta Q + \delta W \qquad (5.6\text{-}1)$$

If the process is reversible, Eq. (5.6-1) takes the form

$$\boxed{dU = T\,dS - P\,dV} \qquad (5.6\text{-}2)$$

The definition of enthalpy is given by

$$H = U + PV \qquad (5.6\text{-}3)$$

The differential form of Eq. (5.6-3) becomes

$$dH = dU + P\,dV + V\,dP \qquad (5.6\text{-}4)$$

Substitution of Eq. (5.6-2) into Eq. (5.6-4) gives

$$\boxed{dH = T\,dS + V\,dP} \qquad (5.6\text{-}5)$$

Note that Eqs. (5.6-2) and (5.6-5) consist of only properties and their differential changes. These properties and their changes are state functions and are not dependent on the path or process involved. Therefore, both equations hold for all processes (reversible or irreversible) and for a change of state in either a closed system or a steady-state flow system.

In differential form, the first law of thermodynamics for a steady-state flow system is given by

$$dH = \delta Q + \delta W_s \qquad (5.6\text{-}6)$$

For a reversible process, Eq. (5.6-6) becomes

$$dH = T\,dS + \delta W_{s,rev} \qquad (5.6\text{-}7)$$

The use of Eq. (5.6-5) in Eq. (5.6-7) leads to

$$\delta W_{s,rev} = V\,dP \qquad (5.6\text{-}8)$$

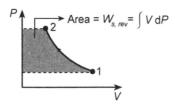

Fig. 5.18   Reversible shaft work on a P-V diagram.

Thus, the reversible shaft work in a steady-state flow process with negligible changes in kinetic and potential energies is given by

$$\boxed{W_{s,rev} = \int_{P_1}^{P_2} V\, dP}$$

(5.6-9)

The area representing the reversible shaft work is shown in a P-V diagram in Fig. 5.18.

Equation (5.6-9) is valid for turbines, compressors, and pumps. For a turbine in which work is produced, it is necessary to make V (or $\widehat{V}$) as large as possible. On the other hand, for a compressor or a pump in which work is supplied, it is necessary to make V (or $\widehat{V}$) as small as possible. Since $\widehat{V}_{vapor} \gg \widehat{V}_{liquid}$, the work requirement for a compressor is much larger than that for a pump. Now, let us evaluate the integral in Eq. (5.6-9) for two types of fluids.

▶ **Ideal gas**

The use of the ideal gas equation of state in Eq. (5.6-9) gives the reversible shaft work for isothermal and adiabatic processes as

$$\widetilde{W}_{s,rev}^{isothermal} = RT \ln\left(\frac{P_2}{P_1}\right)$$

(5.6-10)

$$\widetilde{W}_{s,rev}^{adiabatic} = \frac{\gamma RT_1}{\gamma - 1}\left[\left(\frac{P_2}{P_1}\right)^{(\gamma-1)/\gamma} - 1\right]$$

(5.6-11)

▶ **Incompressible fluid**

For an incompressible fluid

$$\widehat{V} = \frac{1}{\rho} = \text{constant}$$

(5.6-12)

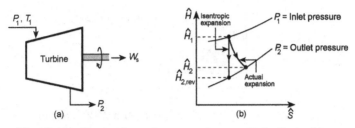

Fig. 5.19   Expansion of a vapor in an adiabatic turbine.

and the pump work per unit mass is given by

$$\widehat{W}_s = \frac{P_2 - P_1}{\rho} = \frac{|\Delta P|}{\rho} \tag{5.6-13}$$

The power required can be obtained by multiplying Eq. (5.6-13) by the mass flow rate, $\dot{m} = \dot{Q}\rho$. The result is

$$\boxed{\dot{W}_s = \dot{Q}\,|\Delta P|} \tag{5.6-14}$$

indicating that the power required to transport a liquid from one location to another is simply the volumetric flow rate times pressure drop.

5.6.2   Isentropic efficiency of a turbine

Consider the adiabatic turbine shown in Fig. 5.19-a in which vapor expands from P_1 to P_2. In turbines, as stated in Section 4.3.3, the changes in kinetic and potential energy terms are usually quite small in comparison to the change in enthalpy. Thus, the first law reduces to

$$-\widehat{W}_s = \widehat{H}_1 - \widehat{H}_2 \tag{5.6-15}$$

Since the value of $\widehat{H}_1$ is fixed by the inlet conditions, $\widehat{H}_2$ should be kept as small as possible to maximize the work output. Referring to Fig. 5.19-b, the smallest value of $\widehat{H}_2$ corresponds to a reversible process in which $\widehat{S}_2 = \widehat{S}_1$. Thus,

$$-\widehat{W}_{s,rev} = \widehat{H}_1 - \widehat{H}_{2,rev} \tag{5.6-16}$$

Example 5.20 *Steam flowing at a rate of* 8 kg/ s *is at* 7 MPa *and* 600 °C. *It is sent to a two-stage turbine. Part of the steam leaving the first stage at* 200 kPa *and* 450 °C *is sent to a heat exchanger to heat* 12 kg/ s *of air from* 20 °C *to* 135 °C. *The remaining part is expanded in the second stage to* 50 kPa. *What is the maximum possible power output of this turbine?*

Solution

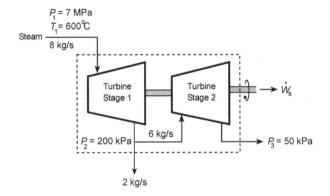

Power output will reach its maximum value if the turbine operates reversibly and adiabatically. It is expressed as

$$\dot{W}_s = (8)(\widehat{H}_{2,rev} - \widehat{H}_1) + (6)(\widehat{H}_{3,rev} - \widehat{H}_{2,rev})$$

From Table A.3 in Appendix A

$$\left. \begin{array}{l} P_1 = 7\,\text{MPa} \\ T_1 = 600\,°\text{C} \end{array} \right\} \begin{array}{l} \widehat{H}_1 = 3650.3\,\text{kJ/kg} \\ \widehat{S}_1 = 7.0894\,\text{kJ/kg.\,K} \end{array}$$

For a reversible and adiabatic expansion in the first stage of the turbine, $\widehat{S}_2 = \widehat{S}_1 = 7.0894\,\text{kJ/kg.\,K}$. *From Table A.2 in Appendix A*

$$P_2 = 200\,\text{kPa} \left\{ \begin{array}{ll} \widehat{H}^L = 504.7\,\text{kJ/kg} & \widehat{S}^L = 1.5301\,\text{kJ/kg.\,K} \\ \Delta\widehat{H} = 2201.9\,\text{kJ/kg} & \Delta\widehat{S} = 5.5970\,\text{kJ/kg.\,K} \\ & \widehat{S}^V = 7.1271\,\text{kJ/kg.\,K} \end{array} \right.$$

Since $1.5301 < \widehat{S}_2 < 7.1271$, *the steam at state 2 is a wet steam with a quality*

$$x = \frac{\widehat{S}_2 - \widehat{S}^L}{\Delta\widehat{S}} = \frac{7.0894 - 1.5301}{5.5970} = 0.993$$

The enthalpy at state 2 is

$$\widehat{H}_{2,rev} = 504.7 + (0.993)(2201.9) = 2691.2\,\text{kJ/kg}$$

Entropy remains the same during the expansion in the second stage, i.e.,
$\widehat{S}_3 = \widehat{S}_2 = \widehat{S}_1 = 7.0894\,\text{kJ/kg.K}$. *From Table A.2 in Appendix A*

$$P_3 = 50\,\text{kPa} \left.\begin{array}{l} \\ \\ \\ \end{array}\right\} \begin{array}{ll} \widehat{H}^L = 340.49\,\text{kJ/kg} & \widehat{S}^L = 1.0910\,\text{kJ/kg.K} \\ \Delta\widehat{H} = 2305.4\,\text{kJ/kg} & \Delta\widehat{S} = 6.5029\,\text{kJ/kg.K} \\ & \widehat{S}^V = 7.5939\,\text{kJ/kg.K} \end{array}$$

Since $1.0910 < \widehat{S}_3 < 7.5939$, *the steam at state 3 is a wet steam with a quality*

$$x = \frac{\widehat{S}_3 - \widehat{S}^L}{\Delta\widehat{S}} = \frac{7.0894 - 1.0910}{6.5029} = 0.922$$

The enthalpy at state 3 is

$$\widehat{H}_{3,rev} = 340.49 + (0.922)(2305.4) = 2466.1\,\text{kJ/kg}$$

Substitution of the numerical values into Eq. (1) gives the maximum power output as

$$\dot{W}_s = (8)(2691.2 - 3650.3) + (6)(2466.1 - 2691.2) = -9023.4\,\text{kW}$$

In actual applications, however, part of the work will be lost as a result of irreversibilities associated with the expansion process. As a result, as shown in Fig. 5.19-b, entropy will increase and the actual work is given by

$$-\widehat{W}_s = \widehat{H}_1 - \widehat{H}_2 \tag{5.6-17}$$

The *isentropic efficiency*, η, is defined by

$$\boxed{\eta = \frac{-\widehat{W}_s}{-\widehat{W}_{s,rev}} = \frac{\widehat{H}_1 - \widehat{H}_2}{\widehat{H}_1 - \widehat{H}_{2,rev}}} \tag{5.6-18}$$

From Eq. (5.6-18)

$$\boxed{\widehat{H}_2 = \widehat{H}_1 - \eta\left(\widehat{H}_1 - \widehat{H}_{2,rev}\right)} \tag{5.6-19}$$

Isentropic efficiencies of turbines range from 0.7 to 0.9. For example, a turbine with an efficiency of 0.85 will deliver 85% of the power of an ideal (isentropic) turbine operating under the same inlet conditions and exit pressure.

If the fluid expanding in the turbine is an ideal gas, then Eq. (5.6-18) takes the form

$$\eta = \frac{T_1 - T_2}{T_1 - T_{2,rev}} = \frac{1 - (T_2/T_1)}{1 - (T_{2,rev}/T_1)} = \frac{1 - (T_2/T_1)}{1 - (P_2/P_1)^{(\gamma-1)/\gamma}} \qquad (5.6\text{-}20)$$

Solving for T_2 yields

$$T_2 = T_1 \left\{ 1 + \eta \left[\left(\frac{P_2}{P_1} \right)^{(\gamma-1)/\gamma} - 1 \right] \right\} \qquad (5.6\text{-}21)$$

The power output of a turbine is

$$-\dot{W}_s = \dot{m} \left(\widehat{H}_1 - \widehat{H}_2 \right) = \dot{m}\, \widehat{C}_P^*(T_1 - T_2) \qquad (5.6\text{-}22)$$

The use of Eq. (5.6-21) in Eq. (5.6-22) leads to

$$\boxed{-\dot{W}_s = \dot{m}\, \widehat{C}_P^* T_1 \eta \left[1 - \left(\frac{P_2}{P_1} \right)^{(\gamma-1)/\gamma} \right]} \qquad \text{Ideal gas} \qquad (5.6\text{-}23)$$

Example 5.21 *Steam enters an adiabatic turbine at 9 MPa and 800 °C and leaves at 100 kPa and 150 °C. The power output of the turbine is 7 MW.*

a) *Calculate the mass flow rate of steam.*
b) *Calculate the isentropic efficiency of the turbine.*

Solution

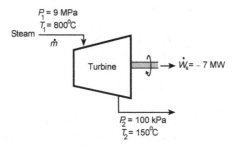

a) *From the first law*

$$\dot{m}(\widehat{H}_2 - \widehat{H}_1) = \dot{W}_s \qquad (1)$$

From Table A.3 in Appendix A

$$P_1 = 9\,\text{MPa} \atop T_1 = 800\,°\text{C}\Bigg\} \begin{array}{l} \widehat{H}_1 = 4119.3\,\text{kJ/kg} \\ \widehat{S}_1 = 7.4596\,\text{kJ/kg.K} \end{array}$$

$$P_2 = 100\,\text{kPa} \atop T_2 = 150\,°\text{C}\Bigg\} \widehat{H}_2 = 2776.4\,\text{kJ/kg}$$

Substitution of the numerical values into Eq. (1) gives

$$\dot{m} = \frac{-7000}{(2776.4 - 4119.3)} = 5.21\,\text{kg/s}$$

b) *If expansion takes place isentropically, $\widehat{S}_{2,rev} = \widehat{S}_1 = 7.4596\,\text{kJ/kg.K}$. From Table A.3 in Appendix A, at 100 kPa*

T (°C)	$\widehat{H}$ (kJ/kg)	$\widehat{S}$ (kJ/kg.K)
100	2676.2	7.3614
150	2776.4	7.6134

By interpolation

$$\widehat{H}_{2,rev} = 2676.2 + \left(\frac{2776.4 - 2676.2}{7.6134 - 7.3614}\right)(7.4596 - 7.3614) = 2715.2\,\text{kJ/kg}$$

Isentropic efficiency is calculated from Eq. (5.6-18) as

$$\eta = \frac{4119.3 - 2776.4}{4119.3 - 2715.2} = 0.956$$

Example 5.22 *The exit streams from two steam turbines operating in parallel, as shown in the figure below, are mixed adiabatically to produce saturated vapor at 1 MPa. Both turbines have an isentropic efficiency of 0.8. If the combined power output of the turbines is 15 MW, calculate the mass flow rate of the steam passing through each turbine.*

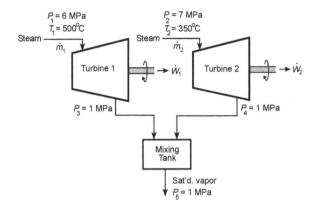

Solution

The power outputs of the turbines are expressed as

$$\dot{m}_1 \widehat{W}_1 + \dot{m}_2 \widehat{W}_2 = 15,000 \tag{1}$$

The energy balance around the mixing tank is

$$(\dot{m}_1 + \dot{m}_2)\widehat{H}_5 = \dot{m}_1 \widehat{H}_3 + \dot{m}_2 \widehat{H}_4 \tag{2}$$

Simultaneous solution of Eqs. (1) and (2) gives the flow rates $\dot{m}_1$ and $\dot{m}_2$.

- **Turbine 1:** *From Table A.3 in Appendix A*

$$\begin{array}{l} P_1 = 6\,\text{MPa} \\ T_1 = 500\,^\circ\text{C} \end{array} \left\} \begin{array}{l} \widehat{H}_1 = 3422.2\,\text{kJ/kg} \\ \widehat{S}_1 = 6.8803\,\text{kJ/kg.\,K} \end{array} \right.$$

If expansion takes place isentropically, $\widehat{S}_{3,rev} = \widehat{S}_1 = 6.8803\,\text{kJ/kg.\,K}$. From Table A.3 in Appendix A, at 1 MPa

T (°C)	$\widehat{H}$ (kJ/kg)	$\widehat{S}$ (kJ/kg.K)
200	2827.9	6.6940
250	2942.6	6.9247

By interpolation

$$\widehat{H}_{3,rev} = 2827.9 + \left(\frac{2942.6 - 2827.9}{6.9247 - 6.6940}\right)(6.8803 - 6.6940) = 2920.5\,\text{kJ/kg}$$

The isentropic work is

$$-\widehat{W}_{1,rev} = \widehat{H}_1 - \widehat{H}_{3,rev} = 3422.2 - 2920.5 = 501.7\,\text{kJ/kg}$$

The actual work is

$$-\widehat{W}_1 = \eta\left(-\widehat{W}_{1,rev}\right) = (0.8)(501.7) = 401.4\,\text{kJ}/\text{kg}$$

The enthalpy at the exit of turbine 1 is

$$\widehat{H}_3 = \widehat{H}_1 + \widehat{W}_1 = 3422.2 - 401.4 = 3020.8\,\text{kJ}/\text{kg}$$

• **Turbine 2:** *From Table A.3 in Appendix A*

$$\left.\begin{array}{l} P_2 = 7\,\text{MPa} \\ T_2 = 350\,^\circ\text{C} \end{array}\right\} \begin{array}{l} \widehat{H}_2 = 3016.0\,\text{kJ}/\text{kg} \\ \widehat{S}_2 = 6.2283\,\text{kJ}/\text{kg}\,.\,\text{K} \end{array}$$

If expansion takes place isentropically, $\widehat{S}_{4,rev} = \widehat{S}_2 = 6.2283\,\text{kJ}/\text{kg}\,.\,\text{K}$. *From Table A.3 in Appendix A, at 1 MPa*

$$\widehat{S}^L = 2.1387\,\text{kJ}/\text{kg}\,.\,\text{K} \quad \widehat{H}^L = 762.81\,\text{kJ}/\text{kg}$$
$$\Delta\widehat{S} = 4.4478\,\text{kJ}/\text{kg}\,.\,\text{K} \quad \Delta\widehat{H} = 2015.3\,\text{kJ}/\text{kg}$$
$$\widehat{S}^V = 6.5865\,\text{kJ}/\text{kg}\,.\,\text{K} \quad \widehat{H}^V = 2778.1\,\text{kJ}/\text{kg}$$

Since $2.1387 < \widehat{S}_{4,rev} < 6.5865$, *the steam at state 4 is a wet steam with a quality*

$$x = \frac{\widehat{S}_{4,rev} - \widehat{S}^L}{\Delta\widehat{S}} = \frac{6.2283 - 2.1387}{4.4478} = 0.919$$

Hence

$$\widehat{H}_{4,rev} = 762.81 + (0.919)(2015.3) = 2614.9\,\text{kJ}/\text{kg}$$

The isentropic work is

$$-\widehat{W}_{2,rev} = \widehat{H}_2 - \widehat{H}_{4,rev} = 3016.0 - 2614.9 = 401.1\,\text{kJ}/\text{kg}$$

The actual work is

$$-\widehat{W}_2 = \eta\left(-\widehat{W}_{2,rev}\right) = (0.8)(401.1) = 320.9\,\text{kJ}/\text{kg}$$

The enthalpy at the exit of turbine 2 is

$$\widehat{H}_4 = \widehat{H}_2 + \widehat{W}_2 = 3016.0 - 320.9 = 2695.1\,\text{kJ}/\text{kg}$$

Since saturated steam at 1 MPa leaves the mixing tank, $\widehat{H}_5 = 2778.1\,\text{kJ}/\text{kg}$. *Substitution of the numerical values into Eqs. (1) and (2) yields*

$$401.4\,\dot{m}_1 + 320.9\,\dot{m}_2 = 15,000 \tag{3}$$

$$2778.1(\dot{m}_1 + \dot{m}_2) = 3020.8\,\dot{m}_1 + 2695.1\,\dot{m}_2 \tag{4}$$

Simultaneous solution of Eqs. (3) and (4) gives

$$\dot{m}_1 = 11.2\,\text{kg/s} \qquad and \qquad \dot{m}_2 = 32.7\,\text{kg/s}$$

5.6.3   *Isentropic efficiency of a compressor*

Consider the adiabatic compressor shown in Fig. 5.20-a in which gas is compressed from P_1 to P_2. In compressors, as stated in Section 4.3.3, the changes in kinetic and potential energy terms are usually quite small in comparison to the change in enthalpy. Thus, the first law reduces to

$$\widehat{W}_s = \widehat{H}_2 - \widehat{H}_1 \tag{5.6-24}$$

Since the value of $\widehat{H}_1$ is fixed by the inlet conditions, $\widehat{H}_2$ should be kept as small as possible to minimize the work input. Referring to Fig. 5.20-b, the smallest value of $\widehat{H}_2$ corresponds to a reversible process in which $\widehat{S}_2 = \widehat{S}_1$. Thus,

$$\widehat{W}_{s,rev} = \widehat{H}_{2,rev} - \widehat{H}_1 \tag{5.6-25}$$

In actual applications, however, more work is required to overcome irreversibilities associated with the compression process. As a result, as shown in Fig. 5.20-b, entropy will increase and the actual work is given by

$$\widehat{W}_s = \widehat{H}_2 - \widehat{H}_1 \tag{5.6-26}$$

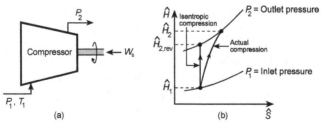

Fig. 5.20   Compression of a gas in an adiabatic compressor.

The *isentropic efficiency*, η, is defined by

$$\boxed{\eta = \frac{\widehat{W}_{s,rev}}{\widehat{W}_s} = \frac{\widehat{H}_{2,rev} - \widehat{H}_1}{\widehat{H}_2 - \widehat{H}_1}} \tag{5.6-27}$$

For an ideal gas, Eq. (5.6-27) takes the form

$$\eta = \frac{T_{2,rev} - T_1}{T_2 - T_1} = \frac{(T_{2,rev}/T_1) - 1}{(T_2/T_1) - 1} = \frac{(P_2/P_1)^{(\gamma-1)/\gamma} - 1}{(T_2/T_1) - 1} \tag{5.6-28}$$

Solving for T_2 yields

$$T_2 = T_1 \left\{ 1 + \frac{1}{\eta} \left[\left(\frac{P_2}{P_1} \right)^{(\gamma-1)/\gamma} - 1 \right] \right\} \tag{5.6-29}$$

Isentropic efficiencies of compressors range from 0.75 to 0.85. The power required to operate a compressor is

$$\dot{W}_s = \dot{m} \left(\widehat{H}_2 - \widehat{H}_1 \right) = \dot{m} \, \widehat{C}_P^*(T_2 - T_1) \tag{5.6-30}$$

The use of Eq. (5.6-29) in Eq. (5.6-30) leads to

$$\boxed{\dot{W}_s = \frac{\dot{m} \, \widehat{C}_P^* T_1}{\eta} \left[\left(\frac{P_2}{P_1} \right)^{(\gamma-1)/\gamma} - 1 \right]} \qquad \text{Ideal gas} \tag{5.6-31}$$

Compressors are one of the most energy consuming pieces of equipment in the chemical and process industries. Examination of Eq. (5.6-31) indicates that power consumption can be reduced by (*i*) increasing compressor efficiency (η), (*ii*) reducing inlet temperature (T_1), (*iii*) reducing pressure ratio (P_2/P_1), and (*iv*) reducing mass flow rate ($\dot{m}$) through the compressor.

Example 5.23 *Air flowing at a rate of* 1.2 kg/s *is at* 100 kPa *and* 20 °C. *It is compressed to* 700 kPa *in an adiabatic compressor. The isentropic efficiency of the compressor is* 0.85. *The exit stream from the compressor is heated at constant pressure in a heat exchanger with the addition of* 150 kW *of heat. Calculate the exit temperature of air from the heat exchanger.*

Solution

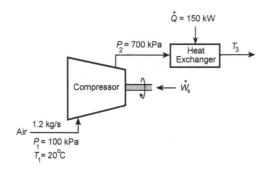

The exit temperature from the compressor can be calculated from Eq. (5.6-29) as

$$T_2 = (293)\left\{1 + \frac{1}{0.85}\left[\left(\frac{700}{100}\right)^{2/7} - 1\right]\right\} = 549.3\,\text{K}$$

Considering air as the system, the energy balance around the heat exchanger gives

$$\Delta\dot{H} = \dot{Q} \qquad \Rightarrow \qquad \dot{m}\,\widehat{C}_P^*(T_3 - T_2) = \dot{Q} \tag{1}$$

Substitution of the numerical values into Eq. (1) leads to

$$(1.2)\left(\frac{3.5 \times 8.314}{29}\right)(T_3 - 549.3) = 150 \qquad \Rightarrow \qquad T_3 = 673.9\,\text{K}$$

5.7 Second Law Analysis of Steady-State Flow Processes

The second law analysis presented in this section is limited to steady-state flow processes since the chemical and energy industries deal predominantly with them.

5.7.1 *Lost work*

The work associated with a completely reversible[10] process is called an *ideal work*, W_{ideal}. When the work is done by the system, the actual work done by the system, W_s, and W_{ideal} are both negative and

$$|W_{ideal}| > |W_s| \qquad \text{Work-producing process} \qquad (5.7\text{-}1)$$

When the work is done on the system, the actual work done on the system, W_s, and W_{ideal} are both positive and

$$W_s > W_{ideal} \qquad \text{Work-requiring process} \qquad (5.7\text{-}2)$$

The lost work, W_{lost}, is defined as

$$W_{lost} = \begin{cases} |W_{ideal}| - |W_s| & \text{Work-producing process} \\ \\ W_s - W_{ideal} & \text{Work-requiring process} \end{cases} \qquad (5.7\text{-}3)$$

Therefore, the lost work indicates the amount of work wasted as a result of irreversibilities in a process and is always positive. Also note that while the actual work and the lost work are path functions, the ideal work is a state function.

For a steady-state flow system, the first law of thermodynamics, Eq. (4.3-5), is expressed as

$$\Delta \dot{H} = \dot{Q} + \dot{W}_s \qquad (5.7\text{-}4)$$

in which kinetic and potential energy changes are considered negligible. The entropy balance, Eq. (5.4-46), on the other hand, is given by

$$\dot{S}_{gen} = \Delta \dot{S} - \frac{\dot{Q}}{T_{surr}} \qquad (5.7\text{-}5)$$

Elimination of $\dot{Q}$ between Eqs. (5.7-4) and (5.7-5) leads to

$$\dot{W}_s = \Delta(\dot{H} - T_{surr}\,\dot{S}) + T_{surr}\,\dot{S}_{gen} \qquad (5.7\text{-}6)$$

---

[10]Not only must all changes within the system be reversible, but also the heat transfer between the system and its surroundings must be reversible as well. Reversible heat transfer between the system and its surroundings implies either of the following:

• The part of the system involved in heat transfer must be at the temperature of the surroundings.

• The heat transfer must be accomplished by operating a reversible heat engine between the system and its surroundings.

In the literature, an *availability function, B,* is defined as

$$\boxed{B = H - T_{surr}\, S} \tag{5.7-7}$$

so that Eq. (5.7-6) is expressed as

$$\dot{W}_s = \dot{m}\, \Delta \widehat{B} + T_{surr}\, \dot{S}_{gen} \tag{5.7-8}$$

The ideal work can be obtained from Eq. (5.7-8) by simply letting $\dot{S}_{gen} = 0$, i.e.,

$$\dot{W}_{ideal} = \dot{m}\, \Delta \widehat{B} \tag{5.7-9}$$

The term *availability, $\mathcal{A}$,* is defined by

$$\boxed{\mathcal{A} = B - B_{surr}} \tag{5.7-10}$$

Since $\Delta B = \Delta \mathcal{A}$, Eq. (5.7-9) indicates that the change in availability, also known as the *exergy*[11], is the ideal work obtained when the system is brought to a state of thermodynamic equilibrium with the surroundings.

▶ **Work-producing processes**

When work is done by the system, Eqs. (5.7-8) and (5.7-9) become

$$\boxed{\begin{aligned} \left| \dot{W}_s \right| &= -\dot{W}_s = -\dot{m}\, \Delta(\widehat{B}) - T_{surr}\, \dot{S}_{gen} \\ &= \dot{m}\left[(\widehat{H}_{in} - \widehat{H}_{out}) - T_{surr}(\widehat{S}_{in} - \widehat{S}_{out}) \right] - T_{surr}\, \dot{S}_{gen} \end{aligned}} \tag{5.7-11}$$

and

$$\boxed{\left| \dot{W}_{ideal} \right| = -\dot{m}\, \Delta(\widehat{B}) = \dot{m}\left[(\widehat{H}_{in} - \widehat{H}_{out}) - T_{surr}(\widehat{S}_{in} - \widehat{S}_{out}) \right]} \tag{5.7-12}$$

▶ **Work-requiring processes**

When work is done on the system, Eqs. (5.7-8) and (5.7-9) become

$$\boxed{\begin{aligned} \dot{W}_s &= \dot{m}\, \Delta(\widehat{B}) + T_{surr}\, \dot{S}_{gen} \\ &= \dot{m}\left[(\widehat{H}_{out} - \widehat{H}_{in}) - T_{surr}(\widehat{S}_{out} - \widehat{S}_{in}) \right] + T_{surr}\, \dot{S}_{gen} \end{aligned}} \tag{5.7-13}$$

and

$$\boxed{\dot{W}_{ideal} = \dot{m}\, \Delta \widehat{B} = \dot{m}\left[(\widehat{H}_{out} - \widehat{H}_{in}) - T_{surr}(\widehat{S}_{out} - \widehat{S}_{in}) \right]} \tag{5.7-14}$$

---

[11]The word *exergy* comes from the combination of the Greek words *ex* = out and *erg* = work.

Substitution of Eqs. (5.7-11)-(5.7-14) into Eq. (5.7-3) gives

$$\boxed{\dot{W}_{lost} = T_{surr}\,\dot{S}_{gen}} \tag{5.7-15}$$

indicating that the generation of entropy corresponds to the decrease in the ability of energy to do work.

Example 5.24 *Consider steam at 5 MPa and 600 °C flowing at a mass flow rate of 1 kg/s. This stream is brought to equilibrium with the surroundings at 100 kPa and 25 °C by first expanding in a throttling valve to 100 kPa and then decreasing the temperature to 25 °C in a heat exchanger as shown in the figure below.*

The enthalpy and entropy values for the initial and final states are given by

$$\left. \begin{array}{l} 5\,\text{MPa} \\ 600\,°\text{C} \end{array} \right\} \begin{array}{l} \widehat{H}_1 = 3666.5\,\text{kJ/kg} \\ \widehat{S}_1 = 7.2589\,\text{kJ/kg.K} \end{array} \qquad \left. \begin{array}{l} 100\,\text{kPa} \\ 25\,°\text{C} \end{array} \right\} \begin{array}{l} \widehat{H}_3 = 104.89\,\text{kJ/kg} \\ \widehat{S}_3 = 0.3674\,\text{kJ/kg.K} \end{array}$$

Since $\Delta\widehat{H} = 0$ during the throttling process, $\widehat{H}_2 = \widehat{H}_1 = 3666.5\,\text{kJ/kg}$. In the heat exchanger, the rate of heat loss from steam to the surroundings is

$$\dot{Q} = \dot{m}\,(\widehat{H}_3 - \widehat{H}_2) = (1)(104.89 - 3666.5) = -3561.61\,\text{kW}$$

In other words, 3561.61 kW of energy is discarded to the atmosphere with no work output. In actual practice, part of this energy could have been converted into work. To calculate this amount, first it is necessary to determine the rate of entropy generation as a result of the overall process. From Eq. (5.7-5)

$$\dot{S}_{gen} = \dot{m}\,(\widehat{S}_3 - \widehat{S}_1) - \frac{\dot{Q}}{T_{surr}}$$

$$= (1)(0.3674 - 7.2589) + \frac{3561.61}{25 + 273} = 5.0602\,\text{kW/K}$$

Using Eq. (5.7-15), the rate of lost work is given by

$$\dot{W}_{lost} = T_{surr}\,\dot{S}_{gen} = (25 + 273)(5.0602) = 1507.94\,\text{kW}$$

Since the actual work is zero, from Eq. (5.7-3) $\left|\dot{W}_{ideal}\right| = 1507.94\,\text{kW}$. *In summary, by using* $3561.61\,\text{kW}$ *of energy, the maximum rate of work that could be produced is* $1507.94\,\text{kW}$. *This value can also be calculated from Eq. (5.7-12) as*

$$\left|\dot{W}_{ideal}\right| = \dot{m}\left[(\widehat{H}_1 - \widehat{H}_3) - T_{surr}(\widehat{S}_1 - \widehat{S}_3)\right]$$
$$= (1)\left[(3666.5 - 104.89) - (25 + 273)(7.2589 - 0.3674)\right]$$
$$= 1507.94\,\text{kW}$$

How to produce $1507.94\,\text{kW}$ *of work is the next question that needs to be answered. For this purpose, consider the arrangement shown in the figure below in which steam is first expanded in an isentropic turbine to* $100\,\text{kPa}$ *and then its temperature is decreased to* $25\,°\text{C}$ *by operating a reversible heat engine.*

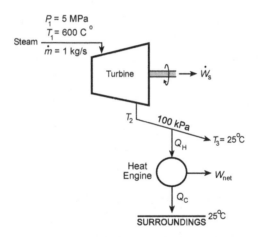

For an isentropic expansion $\widehat{S}_2 = \widehat{S}_1 = 7.2589\,\text{kJ}/\text{kg}.\,\text{K}$. *At* $100\,\text{kPa}$

$$100\,\text{kPa} \left.\begin{array}{l} \widehat{H}^L = 417.46\,\text{kJ}/\text{kg} \quad \widehat{S}^L = 1.3026\,\text{kJ}/\text{kg}.\,\text{K} \\ \Delta\widehat{H} = 2258.0\,\text{kJ}/\text{kg} \quad \Delta\widehat{S} = 6.0568\,\text{kJ}/\text{kg}.\,\text{K} \\ \widehat{H}^V = 2675.5\,\text{kJ}/\text{kg} \quad \widehat{S}^V = 7.3594\,\text{kJ}/\text{kg}.\,\text{K} \end{array}\right\}$$

Since $1.3026 < \widehat{S}_2 < 7.3594$, *the steam at state 2 is a wet steam with a quality*

$$7.2589 = 1.3026 + x(6.0568) \qquad \Rightarrow \qquad x = 0.9834$$

The enthalpy at state 2 is

$$\widehat{H}_2 = 417.46 + (0.9834)(2258.0) = 2638.0 \, \text{kJ/kg}$$

The power output of the turbine is

$$\left| \dot{W}_s \right| = -\dot{W}_s = \dot{m}\left(\widehat{H}_1 - \widehat{H}_2\right) = (1)(3666.5 - 2638.0) = 1028.5 \, \text{kW}$$

The rate of heat supplied to the heat engine is

$$\dot{Q}_H = \dot{m}\left(\widehat{H}_2 - \widehat{H}_3\right) = (1)(2638.0 - 104.89) = 2533.11 \, \text{kW}$$

For a reversible heat engine, the rate of entropy generation is zero

$$\dot{S}_{gen} = \Delta\dot{S}_{steam} + \Delta\dot{S}_{surroundings} = 0$$

The entropy changes of the steam and the surroundings are expressed as

$$\dot{m}(\widehat{S}_3 - \widehat{S}_2) + \frac{\dot{Q}_C}{T_{surr}} = 0$$

Substitution of the numerical values gives

$$(1)(0.3674 - 7.2589) + \frac{\dot{Q}_C}{25 + 273} = 0 \quad \Rightarrow \quad \dot{Q}_C = 2053.67 \, \text{kW}$$

Thus, the power output of the heat engine is

$$\dot{W}_{net} = 2533.11 - 2053.67 = 479.44 \, \text{kW}$$

The summation of the power outputs of the turbine and the heat engine is

$$1028.5 + 479.44 = 1507.94 \, \text{kW}$$

which is identical with the previously calculated rate of lost work. The following table shows the various quantities involved in this example:

Process	$\dot{Q}_{waste}$ (kW)	$\vert \dot{W}_s \vert$ (kW)	$\dot{S}_{gen}$ (kW/K)	$T_{surr}\dot{S}_{gen}$ (kW)
Throttling valve + Heat exchanger	3561.61	0	5.0602	1507.94
Isentropic turbine + Reversible heat engine	2053.67	1507.94	0	0

5.7.2 Second-law efficiency of separation processes

In a typical chemical plant, 40-80% of the total investment is for separation-related equipment. Therefore, thermodynamic analysis of separation processes is extremely important to minimize lost work. Consider a separation process with multiple inlet and outlet streams as shown in Fig. 5.21. Let $\dot{Q}_o$ be the rate of heat input to the system from the surroundings at T_o, and $\dot{Q}_i$ be the rate of heat input to the system from a heat reservoir at T_i.

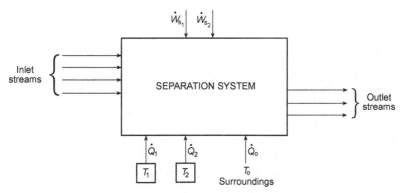

Fig. 5.21 Steady flow separation system exchanging heat and work with its surroundings.

For a separation system, the first and second laws of thermodynamics are expressed as

$$\sum_{out}(\dot{m}\,\widehat{H}) - \sum_{in}(\dot{m}\,\widehat{H}) = \dot{Q}_o + \sum_i \dot{Q}_i + \sum_j \dot{W}_{s_j} \tag{5.7-16}$$

$$\sum_{out}(\dot{m}\,\widehat{S}) - \sum_{in}(\dot{m}\,\widehat{S}) - \frac{\dot{Q}_o}{T_o} - \sum_i \frac{\dot{Q}_i}{T_i} = \dot{S}_{gen} \tag{5.7-17}$$

Elimination of $\dot{Q}_o$ between Eqs. (5.7-16) and (5.7-17) and the use of Eqs. (5.7-7) and (5.7-15) lead to

$$\boxed{\dot{W}_{lost} = \sum_{in}(\dot{m}\,\widehat{B}) - \sum_{out}(\dot{m}\,\widehat{B}) + \sum_i \left[\dot{Q}_i\left(1 - \frac{T_o}{T_i}\right)\right] + \sum_j \dot{W}_{s_j}} \tag{5.7-18}$$

Note that $\dot{Q}_i$ is considered positive when heat is added to the system and $\dot{W}_{s_j}$ is taken as positive when work is done on the system. Note that $\dot{W}_{lost}$ can also be calculated from Eq. (5.7-15).

In terms of molar quantities, Eq. (5.7-18) is expressed as

$$\boxed{\dot{W}_{lost} = \sum_{in}(\dot{n}\,\widetilde{B}) - \sum_{out}(\dot{n}\,\widetilde{B}) + \sum_i \left[\dot{Q}_i\left(1 - \frac{T_o}{T_i}\right)\right] + \sum_j \dot{W}_{s_j}} \tag{5.7-19}$$

The minimum work of separation is calculated from Eq. (5.7-14) as

$$\boxed{\dot{W}_{ideal} = \sum_{out}(\dot{m}\,\widehat{B}) - \sum_{in}(\dot{m}\,\widehat{B})} \text{ or } \boxed{\dot{W}_{ideal} = \sum_{out}(\dot{n}\,\widetilde{B}) - \sum_{in}(\dot{n}\,\widetilde{B})}$$
$$\tag{5.7-20}$$

In other words, the difference between the availability of products and feed streams gives the minimum work for a reversible separation.

The *second-law efficiency*, η^*, is defined by

$$\boxed{\eta^* = \frac{\dot{W}_{ideal}}{\dot{W}_{ideal} + \dot{W}_{lost}}} \tag{5.7-21}$$

Example 5.25 *Consider distillation of a mixture of ethane, propane, n-butane, n-pentane, and n-hexane as shown in the figure below.*

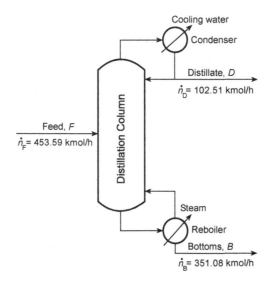

Enthalpy and entropy values are given as (Demirel, 2004)

	Feed	Distillate	Bottoms
$\widetilde{H}$ (kJ/ kmol)	134, 570	102, 840	143, 460
$\widetilde{S}$ (kJ/ kmol. K)	− 403.93	− 277.61	− 443.76

The column has a condenser duty of 3431.88 kW and a reboiler duty of 3395.34 kW. The condenser and the reboiler temperatures are 319.3 K and 400.2 K, respectively. Assuming $T_o = 300$ K, calculate:

a) *The lost work,*
b) *The minimum work of separation,*
c) *The thermodynamic efficiency.*

Solution

a) *The lost work can be calculated by two different methods:*

▶ **Method 1**

The use of Eq. (5.7-19) gives

$$\dot{W}_{lost} = \left(\frac{453.59}{3600}\right)\left[134,570 - (300)(-403.93)\right]$$

$$- \left(\frac{102.51}{3600}\right)\left[102,840 - (300)(-277.61)\right]$$

$$- \left(\frac{351.08}{3600}\right)\left[143,460 - (300)(-443.76)\right]$$

$$+ 3395.34\left(1 - \frac{300}{400.2}\right) - 3431.88\left(1 - \frac{300}{319.3}\right) = 593\,\text{kW}$$

▶ **Method 2**

An entropy balance around the entire distillation column gives

$$\dot{S}_{gen} = \dot{n}_D \tilde{S}_D + \dot{n}_B \tilde{S}_B - \dot{n}_F \tilde{S}_F + \frac{\dot{Q}_C}{T_C} - \frac{\dot{Q}_R}{T_R}$$

where $\dot{Q}_C$ is the rate of heat transfer to cooling water in the condenser and $\dot{Q}_R$ is the rate of heat transfer from the condensing steam in the reboiler. Substitution of the numerical values yields

$$\dot{S}_{gen} = \left(\frac{102.51}{3600}\right)(-277.61) + \left(\frac{351.08}{3600}\right)(-443.76)$$

$$- \left(\frac{453.59}{3600}\right)(-403.93) + \frac{3431.88}{319.3} - \frac{3395.34}{400.2} = 1.977\,\text{kW/K}$$

The lost work is calculated from Eq. (5.7-15) as

$$\dot{W}_{lost} = T_o \dot{S}_{gen} = (300)(1.977) = 593.1\,\text{kW}$$

b) *The minimum work of separation is calculated from Eq. (5.7-20) as*

$$\dot{W}_{ideal} = \left(\frac{102.51}{3600}\right)\left[102,840 - (300)(-277.61)\right]$$

$$+ \left(\frac{351.08}{3600}\right)\left[143,460 - (300)(-443.76)\right]$$

$$- \left(\frac{453.59}{3600}\right)\left[134,570 - (300)(-403.93)\right] = 49.7\,\text{kW}$$

c) *The use of Eq. (5.7-21) gives the thermodynamic efficiency as*

$$\eta^* = \frac{49.7}{49.7 + 593} = 0.077 \quad or \quad 7.7\%$$

Comment: *The second-law efficiency of distillation systems is generally low. Seider et al. (1999) discuss innovative ways of decreasing the lost work in separation systems.*

5.8 What is Entropy?

As defined originally by Clausius in 1856, entropy is the heat transferred in a reversible process divided by the absolute temperature, i.e., Eq. (5.3-10). In the example problems presented in Sections 5.3, 5.4, 5.5, and 5.6, the integrated form of this equation, Eq. (5.3-11), has been used to calculate entropy changes during various processes without giving a second thought to the meaning of entropy. Since Eq. (5.3-10) has no evident meaning, the focus in the solution of these problems was on the use of entropy change to calculate heat and work interactions in a given process.

What do we understand by the term "entropy?" What is the physical significance of this term? The interpretation of entropy has been the subject of hot debates over the years[12]. The jury is still out on the exact meaning of entropy. First of all, it should be kept in mind that entropy, like temperature and pressure, is a macroscopic variable and is defined only for systems at equilibrium.

5.8.1 *Is entropy a measure of disorder?*

Based on the conjecture that spontaneous processes move from "ordered" to "disordered" state, up until the beginning of the 21st century, entropy was considered a measure of disorder (or chaos). Two typical examples to go along with this interpretation of entropy are as follows:

• Shuffling a new deck by a dealer results in an increase in entropy of the cards,

• As the papers and books on a professor's desk become more disorganized as the semester progresses, the entropy of the papers and books increases.

As pointed out by Lambert (1999), the movement of cards, papers, and books in the above examples is not spontaneous, i.e., these processes never

---

[12]The interested reader may refer to the following websites:

• http://www.energyandentropy.com
• http://entropysite.oxy.edu
• http://www.ariehbennaim.com/index.html

take place by themselves. An external agent, dealer or professor, is responsible for the movement of these macro objects. As a result, the entropy change of cards, papers, and books is zero! However, the entropy of the agent increases in the process. For example, entropy increases in the dealer's muscles.

Moreover, like "beauty", "disorder" is in the eye of the beholder and the terms "order" and "disorder", as stated by Ben-Naim (2012), are very subjective, sometimes ambiguous, and at times totally misleading. For a pure substance one can easily claim $S_{gas} > S_{liquid}$, on the basis that the molecules in the gas phase are more disordered than the molecules in the liquid phase. Similarly, the molecules in the liquid phase are more disordered than the molecules in the solid phase. Hence, $S_{liquid} > S_{solid}$. In these examples there is no ambiguity in defining "disordered" and "ordered" states. Now, consider the following example given by Lambert (2002). Let us vigorously shake olive oil and water to form an emulsion. When this "disorderly" emulsion is left alone, it separates into two "orderly" layers. Is the direction of this spontaneous process from "disordered" to "ordered" state? Lambert (2002) and Ben-Naim (2011) give various examples in which "ordered" and "disordered" states cannot be identified.

In conclusion, interpretation of entropy as a measure of disorder should be abandoned. Entropy is neither disorder nor a driving force.

5.8.2 Molecular interpretation of entropy

In 1877, Boltzmann[13] defined entropy as

$$\boxed{S = k \ln W} \tag{5.8-1}$$

where k is the Boltzmann's constant and W stands for "Wahrscheinlichkeit", meaning *probability* in German. The term W represents the number of accessible microstates (or configurations) of a thermodynamic system (or macrostate). The entropy change is given by

$$\Delta S = k \ln \left(\frac{W_{final}}{W_{initial}} \right) \tag{5.8-2}$$

Boltzmann's definition of entropy has paved the way to molecular interpretation of entropy.

---

[13]Ludwig Eduard Boltzmann (1844-1906), Austrian physicist and philosopher. Equation (5.8-1) is engraved on his tombstone in the Zentralfriedhof (German for "Central Cemetery"), Vienna.

To understand what is meant by accessible "microstates of a macro state", let us consider 4 molecules of an ideal gas in a rigid tank. These molecules move around randomly in a tank, colliding with each other and bouncing off the walls. Suppose that the tank is divided into two equal chambers, right and left, by an imaginary partition. How many different equally likely possibilities are there for these 4 molecules to be distributed between the right and left chambers? For each molecule, the possibility of being in either the right or left chamber is 2. Since there are 4 molecules, then the total number of possibilities for the arrangement of these molecules is

$$2 \times 2 \times 2 \times 2 = 2^4 = 16$$

Among these 16 possibilities, the possibilities of having 3 molecules in the right chamber and 1 molecule in the left chamber are

$$\frac{4!}{3!\,1!} = 4$$

Thus, the probability of having 3 molecules in the right chamber and 1 molecule in the left is $4/16 = 0.25$. The molecule in the left chamber can be any one of the 4 molecules as shown in Fig. 5.22. In other words, the **macrostate** "with 3 molecules in the right chamber and 1 molecule in the left" has 4 **microstates**.

Now let us consider the possibilities of having 2 molecules in each chamber. The number of such possibilities is

$$\frac{4!}{2!\,2!} = 6$$

The probability of having 2 molecules in each chamber is $6/16 = 0.375$. The possibilities (or microstates) for having 2 molecules in each chamber are shown in Fig. 5.23.

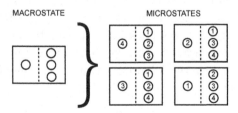

Fig. 5.22   The macrostate with 3 molecules in the right chamber and 1 molecule in the left chamber has 4 microstates.

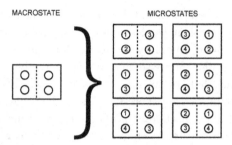

Fig. 5.23   The macrostate with 2 molecules in each chamber has 6 microstates.

According to Eq. (5.8-1), entropy is a monotonic function of W, number of accessible microstates. Therefore, a macrostate with higher number of microstates (or entropy) is more likely (or probable) than one with smaller number of microstates (or entropy). Since entropy of the universe increases continuously, it is possible to conclude that nature seeks to maximize the number of accessible microstates. In other words, the change in the number of accessible microstates is solely responsible for entropy change. Entropy change should not and cannot be explained with the change in the pattern of a group of macro objects.

Entropy is a measure of energy dispersal

The leading advocate of this approach, Frank L. Lambert[14], states that energy spontaneously changes from being localized (or concentrated) to becoming more dispersed (or spread out) if it is not hindered from doing so. Entropy is a measure of this tendency and it quantifies how much energy is spread out among molecules in microstates in a particular process, or how widely spread out it becomes (at a specific temperature). In other words, there are more ways for energy to be spread out than for it to be concentrated. Consider the following examples presented by Lambert to clarify this concept.

---

[14]http://entropysite.oxy.edu

▶ **Spontaneous cooling off a hot copper block on the laboratory counter:** Since the atoms in the hot copper block are rapidly vibrating, initially the block has localized (or concentrated) energy among all of its atoms. When the less rapidly moving molecules in the cooler air of the laboratory hit the copper block, copper atoms transfer some of their energy to air molecules. Thus, the block's overall localized energy is dispersed (or spread out) to molecules in the laboratory air. Eventually, the copper block and laboratory air reach equilibrium at a state of maximum entropy, in which energy is uniformly distributed.

▶ **Spontaneous rusting of an iron nail:** Iron atoms plus oxygen molecules of the air have more energy localized within their bonds than the product, iron oxide. Thus, whenever iron is exposed to air, it spontaneously reacts with oxygen and each spreads out some of its bond energy to the surroundings, i.e., an exothermic reaction.

▶ **Spontaneous mixing of liquids under isothermal condition:** Liquid water and ethanol spontaneously mix with a resulting increase in volume. In the larger volume, the original motional energy of each liquid is spread out more widely.

Entropy is a measure of missing information

While searching for the means to transmit information in the most efficient way, Shannon published a paper entitled "The Mathematical Theory of Communication" in 1948, which is considered the Magna Carta of communication. Shannon's equation states that the *amount of missing information* (or *uncertainty*), H, in a given experiment with N possible outcomes is given by

$$H = -\sum_{i=1}^{N} p_i \log_2 p_i \qquad (5.8\text{-}3)$$

where p_i is the probability of the ith outcome. The lower the H value, the more information is known about the outcomes of an experiment.

According to Eq. (5.8-3), any nonuniformity in the distribution leads to an increase in the information on the experiment. For example, consider a coin tossing experiment. The number of possible outcomes is 2, i.e., either heads or tails. Let the probabilities associated with heads and tails be p and $(1-p)$, respectively, and consider the following three cases:

- **Case 1:** If the coin is fair, the probability distribution is uniform, i.e., $p = 1/2$. Application of Eq. (5.8-3) gives

$$H = -\left(\frac{1}{2}\log_2\frac{1}{2} + \frac{1}{2}\log_2\frac{1}{2}\right) = 1$$

and the chance of guessing the outcome is 50%.

- **Case 2:** Consider an "unfair" coin that is heavier on the heads side so that $p = 9/10$. From Eq. (5.8-3)

$$H = -\left(\frac{9}{10}\log_2\frac{9}{10} + \frac{1}{10}\log_2\frac{1}{10}\right) = 0.469$$

As a result of the nonuniformity in probability distribution, the amount of information about the outcome is increased with a concomitant decrease in H. In this case the outcome will be heads most of the time.

- **Case 3:** Consider a "fake" coin with two heads so that $p = 1$ and $H = 0$. In this case, the maximum information is provided and one can predict the outcome at all times.

When Shannon asked von Neumann[15] what he should call the function H, von Neumann replied[16] as follows:

> "You should call it entropy, for two reasons. In the first place your uncertainty function has been used in statistical mechanics under that name, so it already has a name. In the second place, and more important, no one knows what entropy really is, so in a debate you will always have the advantage!"

Unfortunately, the interchangeable use of "information entropy" and "thermodynamic entropy" has been causing a great deal of confusion in the literature.

Ben-Naim[17] called the term H "Shannon's measure of information" (SMI). He also pointed out the fact that the SMI of tossing a coin has nothing to do with thermodynamic entropy. However, Ben-Naim (2012) was able to derive thermodynamic entropy from the SMI for an ideal gas.

---

[15] John von Neumann (1903-1957), Hungarian-born American mathematician.
[16] This quote is of von Neumann to Shannon [*Scientific Amarican*, **225** (3), 180 (1971)].
[17] http://www.ariehbennaim.com/index.html

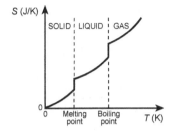

Fig. 5.24 Variation in absolute entropy as a function of temperature.

5.8.3 The third law of thermodynamics

Absolute zero (or zero degrees Kelvin) is the lowest limit on temperature. The third law of thermodynamics states that the entropy of a pure crystal is zero at $0\,K$. In other words, the movements of all molecules cease at $0\,K$ and $W = 1$. As a result, Eq. (5.8-1) gives $S = 0\,J/K$. In this way, the third law of thermodynamics provides the reference state for calculating absolute entropies. Fig. 5.24 shows qualitative variation in absolute entropy as a function of temperature for a pure substance.

5.9 Interpretation of Adages, Songs, and Poems in Terms of the Second Law of Thermodynamics

The following adages (Smith, 1975) express the concepts of the unidirectionality of events, the trend toward increasing entropy, and eventual attainment of the equilibrium state:

- "It's no use crying over spilled milk", i.e., you shouldn't worry about something that has already happened and/or you cannot change.

- "Burning your bridges behind you", i.e., you are ruining any chance you have of going back to where you were.

- "Water doesn't run uphill."

- "Water seeks its own level."

- "What you get is less than what you expect."

The following songs express the ideas of the irreversibility of spontaneous processes:

- "All Mixed Up" by Pete Seeger
- "I'm Gonna Fade Away" by the Rolling Stones
- "The Times They Are a-Changin'" by Bob Dylan

In the "Hollow Men", T.S. Eliot expresses the idea of gradually increasing entropy and the eventual "heat death" of the universe as

> This is the way the world ends
> Not with a bang
> But a whimper.

In "Essay on Man" by Alexander Pope, one finds

> Man is governed by two forces, "self-love" and "reason."
> Self-love drives man, reason restraints him.

These two lines may be interpreted as representing the balance between entropy and enthalpy and the drive toward equilibrium.

Problems

Problems related to Section 5.1

5.1 It is proposed to heat a house using a heat pump. The heat transfer from the house is 15 kW. The house is to be maintained at 24 °C while the outside air is at a temperature of -7 °C. What is the minimum power required to drive the heat pump?

(**Answer:** 1.57 kW)

5.2 As a source of fresh water, Kabel (1979) considered moving icebergs, initially weighing 10^{11} kg, from the Antarctic to Saudi Arabia by towing with large tug boats. According to his calculations, the trip takes about 7 months and 51% of the iceberg melts as a result of heat transfer from the surrounding water at 15 °C. If a Carnot heat engine is operated using the surrounding water as a heat source and the iceberg as a cold sink, what is the work output that could be generated during this trip? Assume that the iceberg's temperature is 0 °C and the heat of fusion for ice is 334 kJ/kg.

(**Answer:** 9.36×10^{11} kJ)

5.3 A Carnot heat engine using air as working fluid receives 175 kJ of heat from a high temperature reservoir and rejects 70 kJ of heat to a low temperature reservoir. The pressure after isothermal expansion is 280 kPa. If the maximum molar volume in the cycle is $0.18\,\mathrm{m^3/mol}$, calculate the high and low reservoir temperatures.

(**Answer:** 613.3 K and 245.3 K)

Problems related to Section 5.2

5.4 Is it possible to convert work completely into heat? If yes, think of an example. If no, why?

5.5 A heat pump removes 30 kW of heat from a 250 K thermal reservoir and exhausts 50 kW of heat to a 350 K thermal reservoir.

a) Calculate the coefficient of performance of the heat pump,
b) Calculate the coefficient of performance of a Carnot heat pump operating between the same temperature limits.

(**Answer:** a) 1.5, b) 2.5)

5.6 The work output of a Carnot heat engine is used to drive a Carnot heat pump as shown in the figure below. If $Q_H = \overline{Q}_C$, calculate T.

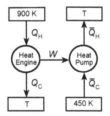

(**Answer:** 600 K)

5.7 Ocean thermal energy conversion (OTEC) generates electricity indirectly from solar energy by harnessing the temperature difference between the sun-warmed surface of tropical oceans and the colder deep waters (Masutani and Takahashi, 2001). Use the surface of the ocean as the heat source at 300 K and the bottom of the ocean as the heat sink at 275 K, and calculate the maximum efficiency of an OTEC heat engine.

(**Answer:** 8.3%)

5.8 Although ambient air temperature fluctuates with the changing seasons, underground temperatures do not change dramatically due to the insulating properties of the earth. The temperature of the soil remains fairly constant at a depth of greater than 9 m. Depending on the latitude, the temperature varies between 10 °C and 16 °C. Geothermal heat pumps use the constant temperature of the earth as the heat source (or sink) instead of the outside air. A non-freezing, non-corrosive fluid, like propylene glycol, is circulated through a coil of pipe buried in the ground to transfer thermal energy to and from the ground.

In the city where you live, people like to keep the interior temperature of their homes constant at 22 °C all year long. The ground temperature is 12 °C. As an expert, you are asked to compare the amount of heat transferred per unit of electrical work consumed for geothermal and conventional, i.e., air-to-air, heat pump systems operating under the following conditions:

• Summer day when the outside temperature is 30 °C,
• Winter day when the outside temperature is − 15 °C.

Which heat pump performs better based on your analysis?

5.9 Is it possible to cool your kitchen by leaving the refrigerator door open?

5.10 An inventor claims to have developed a heat pump with a coefficient of performance of 7 that operates between − 5 °C and 27 °C. Is the inventor serious?

Problems related to Section 5.3

5.11 The temperature of a system is increased reversibly from 25 °C to 75 °C. If the rate of heat transfer per unit temperature rise is constant and is equal to 1.2 kJ/ K, calculate the increase in entropy of the system.

(**Answer:** 0.186 kJ/ K)

5.12 A reversible heat engine operates between 800 °C and 150 °C. The work output of the engine is 85 kW. Calculate the entropy change of each reservoir after quarter of an hour.

(**Answer:** $\Delta S_H = -117.7$ kJ/ K, $\Delta S_C = 117.7$ kJ/ K)

5.13 A cylinder fitted with a frictionless piston contains 2.5 kg of steam at 0.2 MPa and 225 °C. If the steam is to be compressed reversibly and adiabatically to a final pressure of 1.2 MPa, calculate the work required to accomplish this compression.

(**Answer:** 979 kJ)

5.14 2 kg of steam at 0.2 MPa and 250 °C (state 1) undergoes the following reversible changes in a series of nonflow processes:

- Compressed adiabatically to 400 °C (state 2),
- Heated at constant pressure to 500 °C (state 3),
- Expanded isothermally until the specific volume reaches that of state 1 (state 4),
- Cooled at constant volume to state 1.

a) Show each process in a single *P-V* diagram,

b) Calculate Q, W, and ΔU for each process and for the entire cycle.

(**Answer:** For the entire cycle $\Delta U = 0$, $Q = -W = 132.59\,\text{kJ}$)

5.15 A rigid tank of $1\,\text{m}^3$ volume containing steam at $25\,\text{MPa}$ and $1100\,°\text{C}$ explodes. Estimate the maximum work associated with the explosion for damage assessment purposes.

(**Answer:** $57.3\,\text{MJ}$)

5.16 Steam flowing at a mass flow rate of $0.02\,\text{kg/s}$ enters an adiabatic and reversible turbine at $3\,\text{MPa}$ and $400\,°\text{C}$. The exit stream from the turbine is at $500\,\text{kPa}$ and it is sent to a heat exchanger, where it is condensed at constant pressure to some temperature. The cooled water is then throttled to $100\,\text{kPa}$, where 15% of the stream is now vapor. In the heat exchanger, cooling is accomplished by a reversible heat pump. Calculate the power requirement of the heat pump if its coefficient of performance (COP) is 1.5.

(**Answer:** $27.2\,\text{kW}$)

Problems related to Section 5.4

5.17 For the Carnot heat engine shown in the figure below, calculate the net work output.

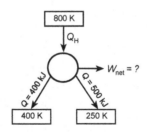

(**Answer:** $1500\,\text{kJ}$)

5.18 For the Carnot heat pump shown in the figure below, calculate the temperature of the hot thermal reservoir.

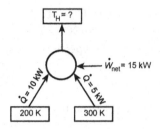

(**Answer:** 450 K)

5.19 Consider the following heat engines operating between 1200 K and 300 K. For each case determine whether the heat engine operates reversibly, operates irreversibly, or is impossible.

a) $Q_H = 850$ kJ, $Q_C = 300$ kJ, $W_{net} = 550$ kJ,
b) $Q_H = 900$ kJ, $Q_C = 400$ kJ, $W_{net} = 550$ kJ,
c) $Q_H = 700$ kJ, $Q_C = 140$ kJ, $W_{net} = 560$ kJ,
d) $Q_H = 800$ kJ, $Q_C = 200$ kJ, $W_{net} = 600$ kJ.

Problems related to Section 5.5

5.20 Consider an ideal gas undergoing a reversible and adiabatic process.

a) Under these circumstances show that Eq. (5.5-3) reduces to

$$0 = \frac{\widetilde{C}_V^*}{T} dT + \frac{R}{\widetilde{V}} d\widetilde{V} \tag{1}$$

b) Integrate Eq. (1) from state 1 to state 2 to obtain

$$\frac{T_2}{T_1} = \left(\frac{\widetilde{V}_1}{\widetilde{V}_2}\right)^{\gamma-1} \tag{2}$$

which is equivalent to Eq. (4.1-75).

5.21 A 3 kg aluminum block is taken out of a furnace at 175 °C and quenched in a 15 L oil bath ($\rho = 850$ kg/m^3, $\widehat{C}_P = 1.67$ kJ/kg. K) at an initial temperature of 25 °C. The heat capacity of aluminum is given as

$$\widehat{C}_P = 0.537 + 1.916 \times 10^{-3}T - 2.938 \times 10^{-6}T^2 + 1.838 \times 10^{-9}T^3$$

where $\widehat{C}_P$ is in kJ/kg. K and T is in K. Calculate the entropy change of the universe once the thermal equilibrium is achieved.

(**Answer:** 0.235 kJ/K)

5.22 A rigid tank of $2\,m^3$ volume contains air at $20\,°C$ and $200\,kPa$. Calculate the entropy change of air for the following processes:

a) $100\,kJ$ of heat is transferred to the air,
b) The tank is fitted with a stirrer that is rotated by a shaft and $720\,kJ$ of work is done on the air during an adiabatic process.

(**Answer:** a) $0.325\,kJ/K$, b) $1.851\,kJ/K$)

5.23 A cylinder fitted with a piston initially contains 2 moles of air at $1\,bar$ and $10\,°C$. Air is compressed adiabatically but irreversibly until the pressure reaches $7\,bar$. The work required is estimated to be 20% more than that of an isentropic compression from the same initial state to the same final pressure. Calculate the entropy change of air.

(**Answer:** $4.76\,J/K$)

5.24 A cylinder fitted with a piston initially contains 3 moles of air at $2\,bar$ and $380\,K$. The air undergoes a series of reversible processes as shown in the figure below. Calculate the entropy change for each process.

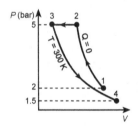

5.25 An insulated cylinder fitted with a frictionless piston is divided into two chambers, A and B, by a metal partition as shown in the figure below. The piston does not conduct heat. Initially the metal partition is covered with an adiabatic film. While chamber A contains 3 moles of helium at $100\,kPa$ and $1000\,K$, chamber B contains 2 moles of helium at $100\,kPa$ and $400\,K$. The adiabatic film is then removed and the helium in chamber A is compressed reversibly and isothermally until the temperature of the helium in chamber B reaches $1000\,K$.

a) Calculate the final pressures in chambers A and B.
b) Calculate the entropy change of the universe associated with this process.
c) Although the compression takes place reversibly, $\Delta S_{universe} \neq 0$. Why?

Metal partition (Initially covered
with an adiabatic film)

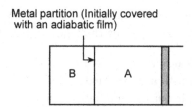

(**Answer:** a) $P_A = 182.2\,\text{kPa}$, $P_B = 250\,\text{kPa}$, b) $7.89\,\text{J/K}$)

5.26 A cylinder fitted with a frictionless piston contains 2 kg of steam at 0.1 MPa and 100 °C. It is first compressed isentropically to 1 MPa and then heated reversibly at constant pressure until the temperature reaches 400 °C. Calculate the heat transfer and work for the overall process.

(**Answer:** $Q = 134.28\,\text{kJ}$, $W = 766.92\,\text{kJ}$)

5.27 A 40 m³ rigid tank initially contains steam at 2 MPa and 400 °C. A valve at the top of the tank is opened, allowing steam to escape. Assume that at any instant the steam that remains in the tank has undergone a reversible adiabatic process. Determine the amount of steam that has escaped when the steam remaining in the tank is saturated vapor.

(**Answer:** 219.4 kg)

5.28 Two well-insulated rigid tanks, each having a volume of 0.1 m³, are connected by an adiabatic turbine. Initially, one of the tanks contains air at 10 bar and 300 K while the other is evacuated. Calculate the maximum work that can be obtained from this system if air is allowed to flow between the two tanks until equilibrium is reached.

(**Answer:** 60, 511 J)

5.29 An insulated rigid tank A of 0.5 m³ volume is connected to another rigid tank B of 0.25 m³ volume. Initially, while tank A contains steam at 1.8 MPa and 650 °C, tank B contains steam at 100 kPa and 200 °C. The valve between the tanks is opened and steam flows from A to B until the temperature in tank A reaches 400 °C, at which point the valve is closed. The cooling coils placed in tank B keep the temperature of tank B's contents constant at 200 °C throughout the process.

a) Calculate the final pressures of tanks A and B. (Hint: Assume that the steam remaining in tank A undergoes an adiabatic and reversible expansion.)
b) Calculate the amount of heat transferred from tank B.

(**Answer:** a) 0.422 MPa, 1.273 MPa, b) − 1408.8 kJ)

5.30 An insulated cylinder fitted with a frictionless piston is divided into two chambers by a metal partition as shown in the figure below. Initially chamber A contains 20 kg of wet steam at 200 °C and 10% quality, and chamber B contains 2 mol of air at 1.2 MPa and 200 °C. The piston moves slowly to the left and compresses the air until the pressure reaches 2 MPa. Calculate the final temperature. State your assumptions.

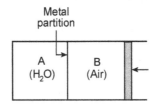

(**Answer:** 218.7 °C)

5.31 125 mol/s of an air stream at a pressure of 200 kPa is to be heated from 15 °C to 85 °C, the heat being supplied by condensing saturated steam at 550 kPa. Steam leaves the exchanger as saturated liquid.

a) How much steam would be required if a conventional heat exchanger were used?
b) How much steam would be required if heating of the air were carried out by using the heat transferred from the steam to drive a reversible heat engine and by using the air stream as the heat sink? What would be the power output of the engine?

(**Answer:** a) 0.121 kg/s, b) 0.162 kg/s, 85.1 kW)

5.32 Estimate the minimum amount of work required to make 10 kg of ice cubes from water initially at 5 °C. The heat of fusion of water at 0 °C is 334 kJ/kg. Assume that the ambient air temperature is 25 °C.

(**Answer:** 323 kJ)

5.33 Steam flowing at a mass flow rate of 3 kg/s enters pipe A at 200 kPa and 700 °C. Water flowing at a mass flow rate of 4 kg/s enters pipe B at 200 kPa and 25 °C, and exits from the pipe as a wet steam with a quality of 70%. Assume negligible pressure drop in both pipes.

It is proposed to operate a reversible Carnot engine by using the streams flowing in pipes A and B as a heat source and a heat sink, respectively.

a) Calculate the condition of the steam exiting pipe A,

b) Calculate the power output of the engine.

(**Answer:** a) Wet steam with $x = 0.128$, b) $1658.8\,\text{kW}$)

5.34   You are given a thermal reservoir at $300\,^{\circ}\text{C}$, a reversible heat engine, and a $1\,\text{m}^3$ rigid tank of air at $100\,\text{kPa}$ and $20\,^{\circ}\text{C}$. Arrange these elements to produce the maximum amount of work and calculate the amount.

(**Answer:** $89\,\text{kJ}$)

5.35   A reversible heat pump transfers heat from the water contained in a cylinder A fitted with a frictionless piston and rejects heat to the water contained in a rigid tank B of $0.8\,\text{m}^3$ volume. The schematic of the process is shown in the figure below.

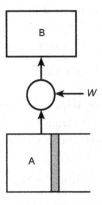

The ambient pressure is $100\,\text{kPa}$. Cylinder A initially contains $1\,\text{kg}$ of wet steam with a quality of 50% while tank B contains wet steam at $200\,\text{kPa}$ and 60% quality. If the heat pump operates until cylinder A contains saturated liquid, determine the final temperature in tank B.

(**Answer:** $138.1\,^{\circ}\text{C}$)

5.36   Levenspiel (1993) posed the following problem and asked readers to respond. Consider a column of isothermal ideal gas, such as air, at equilibrium as shown in the figure below. What happens to the pressure if a chunk of this gas is raised reversibly and isothermally from z_1 to z_2?

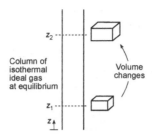

To analyze the problem, Levenspiel started with the first law of thermodynamics, i.e.,

$$\Delta \widehat{U} + \Delta \widehat{E}_K + \Delta \widehat{E}_P = \widehat{Q} + \widehat{W} \tag{1}$$

and simplified Eq. (1) to

$$g(z_2 - z_1) = -\int_{\widehat{V}_1}^{\widehat{V}_2} P \, d\widehat{V} \tag{2}$$

The use of the identities

$$P = \frac{MRT}{\widehat{V}} \quad \text{and} \quad \frac{\widehat{V}_2}{\widehat{V}_1} = \frac{P_1}{P_2} \tag{3}$$

in Eq. (2) leads to

$$g(z_2 - z_1) = MRT \ln\left(\frac{P_2}{P_1}\right) \tag{4}$$

Since the left-hand side of Eq. (4) is positive,

$$P_2 > P_1 \tag{5}$$

In other words, as you climb the mountain, the air pressure increases, contrary to experience! Where is the error in the development of Eq. (4)? The reader may refer to Levenspiel (1994) for different opinions on this simple problem.

5.37 A cylinder fitted with a piston initially contains $0.1 \, \text{m}^3$ of air at $100 \, \text{kPa}$ and $35 \, ^\circ\text{C}$. An inventor claims to have designed a device that compresses air adiabatically to $550 \, \text{kPa}$ and $180 \, ^\circ\text{C}$. Is this process feasible?

5.38 Steam at $3.5 \, \text{MPa}$ and $700 \, ^\circ\text{C}$ is flowing at a mass flow rate of $450 \, \text{kg/h}$. It is expanded in a turbine to produce work. Two exit streams are removed from the turbine. One of the streams is at $1.4 \, \text{MPa}$ and $200 \, ^\circ\text{C}$ and has a mass flow rate of $150 \, \text{kg/h}$. The other stream is at $0.1 \, \text{MPa}$ and $150 \, ^\circ\text{C}$. The rate of heat loss from the turbine is estimated to be $19.43 \, \text{kW}$.

The temperature of the surroundings is $20\,°C$. The power output of the turbine is expected to be $121\,kW$. Determine if this process is possible.

5.39 Steam is first cooled by mixing with liquid water and then expanded in an isentropic turbine as shown in the figure below. Calculate the power output of the turbine.

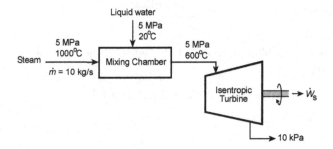

(**Answer:** $17{,}328\,kW$)

5.40 Consider a rigid, well-insulated tank divided into two chambers, A and B, by a partition. Initially, chamber A contains n_A moles of ideal gas A and chamber B contains n_B moles of ideal gas B. Temperature and pressure in each chamber are the same. If the partition is removed, mixing occurs and gas molecules of A and B fill the entire volume.

a) Use Eq. (5.5-19) and show that

$$\Delta S_A = n_A R \ln\left(\frac{V_A + V_B}{V_A}\right) \quad \text{and} \quad \Delta S_B = n_B R \ln\left(\frac{V_A + V_B}{V_B}\right)$$

b) Conclude that the entropy change of the overall system is given by

$$\Delta S = \Delta S_A + \Delta S_B = - n_T R(x_A \ln x_A + x_B \ln x_B) \tag{1}$$

where the total number of moles, n_T, and mole fraction of species i, x_i, are defined as

$$n_T = n_A + n_B \quad \text{and} \quad x_i = \frac{n_i}{n_T}$$

c) In the literature, the term ΔS in Eq. (1) is called the *entropy change on mixing* (or *entropy of mixing*). Ben-Naim (2012), however, states that the entropy change given by Eq. (1) is due to expansion of the two gases (not the mixing) and ΔS in this process should be referred to as the *entropy of expansion* and not as the *entropy of mixing*. Discuss.

d) Calculate the entropy change associated with mixing 2 moles of argon and 1 mole of neon at 1 bar and 25 °C.

e) Calculate ΔS if the two chambers originally contain the same ideal gas.

(**Answer:** d) 15.876 J/ K, e) 0 J/ K)

Problems related to Section 5.6

5.41 Steam enters an adiabatic turbine at $10\,MPa$ and 700 °C and leaves at 100 kPa. If the isentropic efficiency of the turbine is 0.8, calculate the work output.

(**Answer:** -1013 kJ)

5.42 Air flowing at a mass flow rate of 8 kg/ s expands in an adiabatic turbine from 1.4 MPa and 500 °C to 100 kPa. If the isentropic efficiency of the turbine is 0.9, calculate the power output of the turbine.

(**Answer:** -2957.3 kW)

5.43 Steam enters an adiabatic turbine at 550 °C and exits as a saturated vapor at 75 kPa. If the isentropic efficiency of the turbine is 0.85, calculate the inlet pressure of steam.

(**Answer:** 6 MPa)

5.44 Steam enters an adiabatic turbine at 12.5 MPa and 700 °C and leaves as a saturated vapor. If the isentropic efficiency of the turbine is 0.93, what is the minimum allowable exit pressure?

(**Answer:** 125 kPa)

5.45 Steam enters an adiabatic turbine at 600 kPa and 400 °C and leaves as a saturated vapor at 25 kPa. Calculate the isentropic efficiency of the turbine.

(**Answer:** 0.94)

5.46 The exit streams from the two adiabatic turbines operating in parallel are mixed and sent to a third adiabatic turbine as shown in the figure below. All turbines have an isentropic efficiency of 0.85. Calculate the total power output of the three turbines.

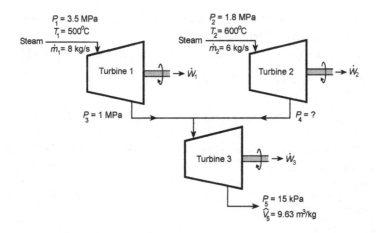

(**Answer:** $14,666\,\text{kW}$)

5.47   An adiabatic compressor receives steam from a supply line at $800\,\text{kPa}$ and $300\,^\circ\text{C}$. The exit stream at $3.5\,\text{MPa}$ is sent to an insulated cylindrical tank fitted with a frictionless piston. The mass of the piston is such that it maintains a constant pressure of $3.5\,\text{MPa}$ inside the tank. The tank is initially empty and the isentropic efficiency of the compressor is 0.70. Estimate the work supplied to the compressor when the tank volume reaches $0.5\,\text{m}^3$.

(**Answer:** $2835.3\,\text{kJ}$)

5.48   The work output of a Carnot heat engine is used to drive an adiabatic air compressor as shown in the figure below. The exit stream of the compressor is used as the heat sink for the heat engine. If the isentropic efficiency of the compressor is 0.75, calculate the reservoir temperature T_H.

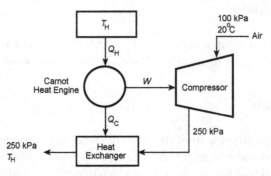

(**Answer:** $756.4\,\text{K}$)

5.49 Consider a two-stage turbine with heat addition between the stages as shown in the figure below. Steam at 8 MPa and 500 °C enters the first stage and exits at 0.8 MPa. Before entering the second stage, steam passes through a heat exchanger and its temperature is increased to 400 °C. The pressure of steam at the exit of the second stage is 10 kPa. Both turbines are adiabatic and have isentropic efficiencies of 0.9. The power output of the second stage turbine is 5 MW. Calculate the power output of the first stage turbine.

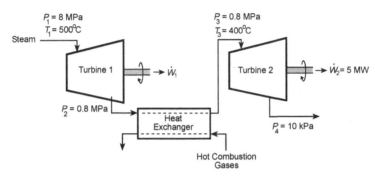

(**Answer:** 3.47 MW)

References

Battino, R., L.E. Strong and S.E. Wood, 1997, *J. Chem. Ed.*, **74** (3), 304-305.

Ben-Naim, A., 2011, *J. Chem. Ed.*, **88**, 594-596.

Ben-Naim, A., 2012, *Entropy and the Second Law: Interpretation and Misss-Interpretationsss*, World Scientific, Singapore.

Demirel, Y., 2004, *Sep. Sci. Tech.*, **39** (16), 3897-3942.

De Nevers, N. and J.D. Seader, 1984, *Chem. Eng. Ed.*, **18** (3), 128-131; 146-148.

Howard, K.I., 2001, *J. Chem. Ed.*, **78** (4), 505-508.

Kabel, R.L., 1979, *Chem. Eng. Ed.*, **13** (2), 70-72.

Lambert, F.L., 1999, *J. Chem. Ed.*, **76** (10), 1385-1387.

Lambert, F.L., 2002, *J. Chem. Ed.*, **79** (2), 187-192.

Levenspiel, O., 1993, *Chem. Eng. Ed.*, **27** (4), 206-207.

Levenspiel, O., 1994, *Chem. Eng. Ed.*, **28** (3), 183.

Masutani, S. M. and P.K.Takahashi, 2001, *Encyclopedia of Ocean Sciences*, 2nd Ed., Vol. 4, pp. 1993-1999, Elsevier.

Seider, W.D., J.D. Seader, D.R. Lewin and S. Widagdo, 2009, *Product and Process Design Principles: Synthesis, Analysis and Design*, 3rd Ed., Wiley, New York, N.Y.

Smith, W.L., 1975, *J. Chem. Ed.*, **52** (2), 97-98.

Tester, J.W. and M. Modell, 1997, *Thermodynamics and Its Applications*, 3rd Ed., Prentice Hill, Upper Saddle River, N.J.

Van Ness, H.C., 1983, *Understanding Thermodynamics*, Dover Pub., New York, N.Y.

Chapter 6

Power and Refrigeration Cycles

Thermodynamic cycles are broadly classified as power and refrigeration cycles. A power cycle (or heat engine) generates work using an energy source at a high temperature while a refrigeration cycle (or heat pump) provides cooling from a work input.

Power cycles are divided into two categories: *vapor power cycles* in which the working fluid is alternately vaporized and condensed, and *gas power cycles* in which the working fluid remains in the gas phase throughout the entire cycle.

Like power cycles, refrigeration cycles can be divided into two categories depending on the type of working fluid, known as the *refrigerant*: *vapor-compression refrigeration cycles* in which the refrigerant is alternately vaporized and condensed, and *gas compression refrigeration cycles* in which the refrigerant remains in the gas phase throughout the entire cycle.

The purpose of this chapter is to provide the basic principles for the analysis of power and refrigeration cycles.

6.1 Vapor Power Cycles

In vapor power cycles, the selection of a working fluid is dependent on various factors, such as physical properties, availability and cost, stability, safety and compatibility, and effect on the environment. Steam, by far, is the most commonly used working fluid. For low temperature heat sources, such as geothermal and solar, low boiling organic fluids are preferred[1].

A typical vapor power cycle is the steam power plant shown in Fig. 6.1. Today, much of the electricity used in the world is produced in steam power

---

[1]For a comprehensive review on the various working fluids used in vapor power cycles, see Chen *et al.* (2010).

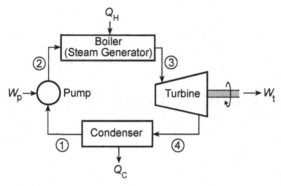

Fig. 6.1   A typical steam power plant.

plants consisting of four major components: boiler (or steam generator), turbine, condenser, and pump.

In the boiler, different energy resources, such as fossil fuels (coal[2], oil, natural gas), nuclear fuels, and geothermal steam, are used to convert liquid water into high-pressure and high-temperature steam. The steam is then fed to a turbine where part of its energy is converted to mechanical energy, which is transmitted by a rotating shaft to drive an electrical generator. In the turbine, expansion of steam occurs adiabatically and is as nearly reversible as possible. Low pressure steam flowing out of the turbine condenses to liquid water in the condenser. For this purpose, a separate cooling water loop is used to transfer the heat removed in the condenser to a nearby lake or river. In some power plants, cooling towers are used for cooling the water circulated through the condenser. The pressure of the liquid condensate is increased by a pump and is fed to the boiler to complete the cycle. A small fraction of the work obtained from the turbine is used to operate the pump.

Consider a typical 500 MW coal power plant[3] to have some idea about the quantities involved in such a plant. It produces 500 MW h of energy per hour. If the plant has a capacity factor of 90%, i.e., operates 90% of

---

[2]Coal is the most abundant fossil energy resource throughout the world. Due to a series of continuing advances in coal mining and coal utilization technology, it is expected to remain a major fuel of choice for electric power generation in many developed and developing countries in the next 3-4 decades (Lior, 2007).

[3]As a rule of thumb, keep in mind that a power plant with a capacity of 1000 MW produces enough energy for 1 million homes.

the time over a year, kW h produced per year is

$$\left(500\frac{\text{MW h}}{\text{h}}\right)\left(1000\frac{\text{kW}}{\text{MW}}\right)\left(24\frac{\text{h}}{\text{day}}\right)\left(0.9\times365\frac{\text{day}}{\text{year}}\right) = 3.94\times10^9\frac{\text{kW h}}{\text{year}}$$

Suppose that the steam generated in the boiler is at 15 MPa and 550 °C, and the condenser operates at 7.5 kPa ($T^{sat} \simeq 40$ °C). The most efficient cycle that can operate between 550 °C and 40 °C is the Carnot heat engine and its thermal efficiency is given by

$$\eta = \frac{\dot{W}_{net}}{\dot{Q}_H} = 1 - \frac{T_C}{T_H} = 1 - \frac{40+273}{550+273} = 0.62$$

Thus, at best the power plant converts 62% of the heat received by the boiler into work, and discards 38% to the surroundings as waste. In practice, the efficiency of an actual power plant will be much less than that of a Carnot heat engine, varying between 32% and 42%. If the thermal efficiency is taken as 35%, then

$$\dot{Q}_H = \frac{\dot{W}_{net}}{\eta} = \frac{5\times10^5}{0.35} = 1.43\times10^6\,\text{kW}$$

The latent heat of vaporization of water is approximately 2260 kJ/kg. Therefore, the steam circulation rate is

$$\dot{m}_{steam} = \frac{1.43\times10^6}{2260} \simeq 633\,\text{kg/s} = 2280\,\text{ton/h}$$

The specific volume of steam at 15 MPa and 550 °C is 0.02293 m³/kg and a reasonable velocity for high-pressure steam in a pipe is 15 m/s. Hence, the diameter of the pipe carrying steam to the turbine is

$$D = \sqrt{\frac{4\dot{m}_{steam}\widehat{V}}{\pi v}} = \sqrt{\frac{(4)(633)(0.02293)}{\pi(15)}} = 1.11\,\text{m}$$

If the efficiency of the boiler and combustion is 88% and the higher heating value of coal is taken as 20,000 kJ/kg, then the amount of coal required is

$$\frac{(1.43\times10^6)(3600)}{(0.88)(20,000)} = 2.925\times10^5\,\text{kg/h} = 292.5\,\text{ton/h}$$

The rate of heat removal in the condenser is

$$\dot{Q}_C = \dot{Q}_H - \dot{W}_{net} = 1.43\times10^6 - 5\times10^5 = 9.3\times10^5\,\text{kW}$$

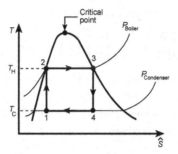

Fig. 6.2   The Carnot power cycle.

Within the condenser, the increase in the cooling water temperature is limited to 3-4 °C to prevent any detrimental effect on the environment[4], i.e., fish, algal blooms, etc.   Taking $\Delta T = 4$ °C, the mass flow rate of cooling water is

$$\dot{m}_{H_2O} = \frac{(9.3 \times 10^5)(3600)}{(4.2)(4)} = 2 \times 10^8 \text{ kg/h} = 200{,}000 \text{ ton/h}$$

In other words, $200{,}000 \text{ m}^3$/h of cooling water must be circulated from lakes, rivers, or the sea.

6.1.1   *Carnot cycle*

The most efficient cycle for the steam power plant shown in Fig. 6.1 is the Carnot cycle. The Carnot cycle, represented in a $T\text{-}\widehat{S}$ diagram in Fig. 6.2, is composed of the following four processes:

- **Process 1-2:** Adiabatic and reversible compression in a pump,
- **Process 2-3:** Isothermal and reversible heat addition in a boiler,
- **Process 3-4:** Adiabatic and reversible expansion in a turbine,
- **Process 4-1:** Isothermal and reversible heat rejection in a condenser.

The Carnot cycle, however, has the following two major drawbacks:

- Heat addition (process 2-3) should take place in the two-phase region to keep the temperature constant. The critical point is the upper limit of the two-phase region. Since the critical temperature of water is 374 °C, the

---

[4]Keep in mind that cold water dissolves more oxygen than warm water does.

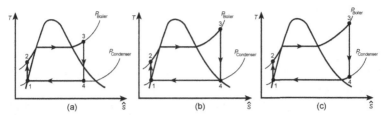

Fig. 6.3 Representation of the ideal Rankine cycle in a T-$\widehat{S}$ diagram.

temperature of the saturated steam leaving the boiler cannot be greater than 374 °C.

• Steam exiting the turbine (state 4) is partially condensed in the condenser and condensation must be stopped at state 1, which is a mixture of vapor and liquid. Pumping a two-phase mixture causes operational difficulties.

6.1.2 *Rankine cycle*

Impracticalities encountered in the Carnot cycle can be eliminated by superheating the steam in the boiler and condensing it completely in the condenser. The resulting cycle is called the *Rankine*[5] *cycle*. Today, steam power plants all over the world operate on different variants of the Rankine cycle.

The T-$\widehat{S}$ diagram of an ideal[6] Rankine cycle in its simplest form is shown in Fig. 6.3. It is composed of the following four processes:

• **Process 1-2:** Pressure of the saturated liquid at state 1 is increased reversibly and adiabatically to boiler pressure in a pump. For an adiabatic pump, the energy balance becomes

$$\widehat{W}_{p,rev} = \widehat{H}_2 - \widehat{H}_1 \tag{6.1-1}$$

The pump work is calculated from Eq. (5.6-9) as

$$\widehat{W}_{p,rev} = \int_{P_1}^{P_2} \widehat{V} \, dP \tag{6.1-2}$$

Liquid water is incompressible, i.e., density $(1/\widehat{V})$ is independent of pressure. Taking $\widehat{V} = \widehat{V}_1 = \text{constant}$, Eq. (6.1-2) becomes

$$\widehat{W}_{p,rev} = \widehat{V}_1(P_2 - P_1) \tag{6.1-3}$$

---

[5]William John Macquorn Rankine (1820-1872), Scottish physicist and engineer.
[6]A cycle is *ideal* if all processes taking place within the cycle are reversible.

Substitution of Eq. (6.1-3) into Eq. (6.1-1) gives

$$\widehat{H}_2 = \widehat{H}_1 + \widehat{V}_1(P_2 - P_1) \qquad (6.1\text{-}4)$$

• **Process 2-3:** A boiler (or steam generator) is a heat exchanger in which hot combustion gases flow outside of the water-filled tubes. Water enters the boiler as a compressed liquid at state 2 and leaves as a superheated vapor at state 3. In a well-designed heat exchanger, pressure drop may be considered negligible. The energy balance around the boiler gives

$$\widehat{Q}_H = \widehat{H}_3 - \widehat{H}_2 \qquad (6.1\text{-}5)$$

• **Process 3-4:** Superheated steam enters the turbine and part of its thermal energy is converted to work as a result of an isentropic expansion. Depending on the degrees of superheat of the steam entering the turbine and condenser pressure, the steam leaves the turbine as a wet steam, saturated steam, or superheated steam as shown in Figures 6.3-a, 6.3-b, and 6.3-c, respectively. For an adiabatic turbine, the energy balance simplifies to

$$-\widehat{W}_{t,rev} = \widehat{H}_3 - \widehat{H}_4 \qquad (6.1\text{-}6)$$

• **Process 4-1:** A condenser is another heat exchanger in which condensation of low-pressure steam leaving the turbine is achieved on the external surface of the tubes by passing external cooling water through them. Again, pressure drop may be neglected. The energy balance around the condenser is given by

$$\widehat{Q}_C = \widehat{H}_1 - \widehat{H}_4 \qquad (6.1\text{-}7)$$

The thermal efficiency of the Rankine cycle is defined by

$$\eta = \frac{\widehat{W}_{net}}{\widehat{Q}_H} = \frac{(-\widehat{W}_{t,rev}) - \widehat{W}_{p,rev}}{\widehat{Q}_H} \qquad (6.1\text{-}8)$$

Substitution of Eqs. (6.1-1), (6.1-5), and (6.1-6) into Eq. (6.1-8) and rearrangement result in

$$\boxed{\eta = 1 - \frac{\widehat{H}_4 - \widehat{H}_1}{\widehat{H}_3 - \widehat{H}_2}} \qquad (6.1\text{-}9)$$

Example 6.1 *In an ideal Rankine cycle steam leaves the boiler and enters the turbine at* 4 MPa *and* 425 °C. *The condenser pressure is* 10 kPa. *Determine the thermal efficiency of the cycle.*

Solution

• **State: 1**

From Table A.2 in Appendix A

$$P_1 = 10 \, \text{kPa} \left.\right\} \begin{array}{l} \widehat{V}_1 = 0.00101 \, \text{m}^3/\text{kg} \\ \widehat{H}_1 = 191.83 \, \text{kJ}/\text{kg} \end{array}$$
$$\text{Sat. Liquid}$$

• **State: 2**

The pressure is 4 MPa (4000 kPa). *From Eq. (6.1-4)*

$$\widehat{H}_2 = 191.83 + (0.00101)(4000 - 10) = 195.86 \, \text{kJ}/\text{kg}$$

• **State: 3**

From Table A.3 in Appendix A

$$P_3 = 4 \, \text{MPa} \left.\right\} \begin{array}{l} \widehat{H}_3 = 3271.95 \, \text{kJ}/\text{kg} \\ \widehat{S}_3 = 6.85265 \, \text{kJ}/\text{kg.K} \end{array}$$
$$T_3 = 425 \, °\text{C}$$

• **State: 4**

Since the turbine operates isentropically, $\widehat{S}_4 = \widehat{S}_3 = 6.85265 \, \text{kJ}/\text{kg.K}$. *From Table A.2 in Appendix A*

$$P_4 = 10 \, \text{kPa} \left\{\begin{array}{ll} \widehat{H}^L = 191.83 \, \text{kJ}/\text{kg} & \widehat{S}^L = 0.6493 \, \text{kJ}/\text{kg.K} \\ \Delta\widehat{H} = 2392.8 \, \text{kJ}/\text{kg} & \Delta\widehat{S} = 7.5009 \, \text{kJ}/\text{kg.K} \\ \widehat{H}^V = 2584.7 \, \text{kJ}/\text{kg} & \widehat{S}^V = 8.1502 \, \text{kJ}/\text{kg.K} \end{array}\right.$$

Since $0.6493 < \widehat{S}_4 < 8.1502$, *the steam at state 4 is a wet steam with a quality*

$$x = \frac{\widehat{S}_4 - \widehat{S}^L}{\Delta\widehat{S}} = \frac{6.85265 - 0.6493}{7.5009} = 0.827$$

The enthalpy at state 4 is

$$\widehat{H}_4 = 191.83 + (0.827)(2392.8) = 2170.7 \, \text{kJ}/\text{kg}$$

Thus, the thermal efficiency is calculated from Eq. (6.1-9) as

$$\eta = 1 - \frac{2170.7 - 191.83}{3271.95 - 195.86} = 0.357$$

Comment: *The T-$\widehat{S}$ diagram of the cycle is similar to the one shown in Figure 6.3-a.*

In analyzing the Rankine cycle, it is helpful to think of the thermal efficiency depending on the average temperature during heat addition in the boiler and the condenser temperature. Since heat addition in the boiler (process 2-3) is reversible,

$$\widehat{Q}_H = \int_{\widehat{S}_2}^{\widehat{S}_3} T \, d\widehat{S} = \langle T \rangle_H \left(\widehat{S}_3 - \widehat{S}_2 \right) \tag{6.1-10}$$

where $\langle T \rangle_H$ represents the average temperature during heat addition. On a T-$\widehat{S}$ diagram, the shaded area under the process path during heat addition gives $\widehat{Q}_H$ as shown in Fig. 6.4-a.

The temperature remains constant at T_C during heat rejection in the condenser (process 4-1). Therefore,

$$\left| \widehat{Q}_C \right| = \int_{\widehat{S}_1}^{\widehat{S}_4} T \, d\widehat{S} = T_C \left(\widehat{S}_4 - \widehat{S}_1 \right) = T_C \left(\widehat{S}_3 - \widehat{S}_2 \right) \tag{6.1-11}$$

On a T-$\widehat{S}$ diagram, the shaded area under the process path during heat rejection gives $\widehat{Q}_C$ as shown in Fig. 6.4-b.

The energy balance for the overall cycle is expressed as

$$\underbrace{\Delta \widehat{U}}_{0} = \widehat{Q} + \widehat{W} \qquad \Rightarrow \qquad \widehat{W}_{net} = \widehat{Q}_H - \left| \widehat{Q}_C \right| \tag{6.1-12}$$

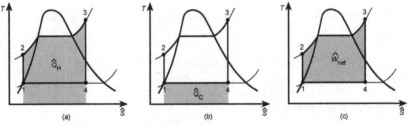

Fig. 6.4 T-$\widehat{S}$ diagram of the ideal Rankine cycle.

Thus, the shaded area in Figure 6.4-c shows the net work output of the Rankine cycle. Substitution of Eqs. (6.1-10)-(6.1-12) into Eq. (6.1-8) gives the thermal efficiency of the cycle as

$$\eta = 1 - \frac{T_C}{\langle T \rangle_H} \qquad (6.1\text{-}13)$$

which is similar to the thermal efficiency of a Carnot heat engine.

Example 6.2 *Use Eq. (6.1-13) and recalculate the thermal efficiency of the Rankine cycle given in Example 6.1.*

Solution

The average temperature during heat addition is given by Eq. (6.1-10) as

$$\langle T \rangle_H = \frac{\widehat{Q}_H}{\widehat{S}_3 - \widehat{S}_2} = \frac{\widehat{Q}_H}{\widehat{S}_3 - \widehat{S}_1} \qquad (1)$$

The heat added to the boiler is

$$\widehat{Q}_H = \widehat{H}_3 - \widehat{H}_2 = 3271.95 - 195.86 = 3076.1 \, \text{kJ} / \text{kg}$$

For a saturated liquid at $10\,\text{kPa}$, $T_C = 45.81\,°\text{C}$ and $\widehat{S}_1 = 0.6493\,\text{kJ}/\text{kg.K}$. Substitution of the values into Eq. (1) gives

$$\langle T \rangle_H = \frac{3076.1}{6.85265 - 0.6493} = 495.9 \, \text{K}$$

Hence, the thermal efficiency is

$$\eta = 1 - \frac{45.81 + 273}{495.9} = 0.357$$

How to increase the efficiency of a Rankine cycle?

To achieve maximum efficiency, examination of Eq. (6.1-13) indicates that the average fluid temperature should be as high as possible during heat addition and as low as possible during heat rejection. The average temperature during heat addition can be increased either by superheating the steam to high temperatures or by raising the boiler pressure. Condenser temperature, on the other hand, can be decreased by lowering condenser pressure.

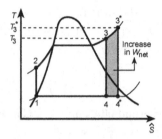

Fig. 6.5   The effect of superheating the steam to a higher temperature on the ideal Rankine cycle.

▶ Effect of increasing steam temperature

As shown in Fig. 6.5, superheating the steam to higher temperatures leads to higher efficiencies as a result of an increase in the net work output. However, the maximum steam temperature exiting the boiler is limited by metallurgical considerations of turbine blades since they cannot sustain temperatures above 650-700 °C. The use of ceramic turbine blades increases the upper limit of temperature to 1200 °C.

▶ Effect of increasing boiler pressure

The effect of increasing boiler pressure on the ideal Rankine cycle is shown in Fig. 6.6. Although the maximum temperature is kept constant at T_3, the average temperature at which heat is added increases with an increase in boiler pressure, leading to a higher thermal efficiency. However, the increase in boiler pressure causes a decrease in the steam quality of the vapor leaving the turbine. Since the presence of moisture leads to turbine blade erosion, steam quality at the turbine exit should be higher than 90%.

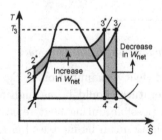

Fig. 6.6   The effect of increasing boiler pressure on the ideal Rankine cycle.

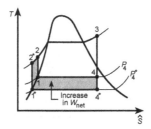

Fig. 6.7 The effect of lowering condenser pressure on the ideal Rankine cycle.

▶ **Effect of lowering condenser pressure**

By lowering condenser pressure, less energy is lost to the surroundings with a concomitant increase in the net work output of the cycle as shown in Fig. 6.7. However, the moisture content in the steam leaving the turbine also increases ($x_{4*} > x_4$). Therefore, once the condenser pressure is lowered, one should check the quality of steam at the exit of the turbine. Moreover, there is a limit to which the condenser pressure can be decreased. For effective heat transfer between the condensing steam and the cooling water, the temperature difference must be at least $10\,°C$. Thus, the condenser pressure should be adjusted so that the temperature of the condensing steam is at least $10\,°C$ greater than the available cooling water temperature. If the condenser operates below atmospheric pressure, one should also be careful about air leakage to the system.

Deviations from the ideal Rankine cycle

In the analysis of an ideal power cycle, pressure drops are not taken into consideration. In reality, pressure losses do occur between the cycle elements as a result of fluid friction. Moreover, the compression in a pump and expansion in a turbine take place irreversibly. As a result, a pump requires more work input, and a turbine produces a smaller work output compared to reversible operation. These irreversibilities lead to an increase in entropy during compression and expansion stages as shown in Fig. 6.8.

The isentropic efficiency of a pump, η_p, is defined by Eq. (5.6-27) as

$$\eta_p = \frac{\widehat{W}_{p,rev}}{\widehat{W}_p} = \frac{\widehat{V}_1(P_2 - P_1)}{\widehat{H}_2 - \widehat{H}_1} \tag{6.1-14}$$

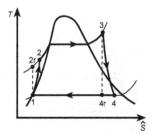

Fig. 6.8 T-$\widehat{S}$ diagram showing the effects of pump and turbine
irreversibilities.

Rearrangement of Eq. (6.1-14) gives

$$\widehat{H}_2 = \widehat{H}_1 + \frac{\widehat{V}_1(P_2 - P_1)}{\eta_p} \qquad (6.1\text{-}15)$$

The isentropic efficiency of a turbine, η_t, is defined by Eq. (5.6-18) as

$$\eta_t = \frac{-\widehat{W}_t}{-\widehat{W}_{t,rev}} = \frac{\widehat{H}_3 - \widehat{H}_4}{\widehat{H}_3 - \widehat{H}_{4r}} \qquad (6.1\text{-}16)$$

Rearrangement of Eq. (6.1-16) gives

$$\widehat{H}_4 = \widehat{H}_3 - \eta_t\left(\widehat{H}_3 - \widehat{H}_{4r}\right) \qquad (6.1\text{-}17)$$

The thermal efficiency of the cycle is again calculated from Eq. (6.1-9).

Example 6.3 *Consider a power plant operating on a Rankine cycle using steam as the working fluid. The boiler pressure is 2.5 MPa and the steam leaving the boiler is superheated to a temperature 126 °C above its saturation temperature. The condenser temperature is 50 °C and it discharges saturated liquid. The turbine and pump have isentropic efficiencies of 0.90 and 0.80, respectively. Calculate the thermal efficiency of the cycle.*

Solution

• **State: 1**

From Table A.1 in Appendix A

$$\left.\begin{array}{l} T_1 = 50\,^\circ\text{C} \\ \text{Sat. Liquid} \end{array}\right\} \quad \begin{array}{l} P_1 = 12.349\,\text{kPa} \\ \widehat{V}_1 = 0.001012\,\text{m}^3/\,\text{kg} \\ \widehat{H}_1 = 209.33\,\text{kJ}/\,\text{kg} \end{array}$$

• **State: 2**

The pressure is 2.5 MPa (2500 kPa). *From Eq. (6.1-15)*

$$\widehat{H}_2 = 209.33 + \frac{(0.001012)(2500 - 12.349)}{0.8} = 212.48 \, \text{kJ/kg}$$

• **State: 3**

The saturation temperature of steam at 2.5 MPa *is* ∼ 224 °C. *Therefore, the temperature of superheated steam at the exit of the boiler is* 224 + 126 = 350 °C. *From Table A.3 in Appendix A*

$$\left. \begin{array}{l} P_3 = 2.5 \, \text{MPa} \\ T_3 = 350 \, °\text{C} \end{array} \right\} \begin{array}{l} \widehat{H}_3 = 3126.3 \, \text{kJ/kg} \\ \widehat{S}_3 = 6.8403 \, \text{kJ/kg.K} \end{array}$$

• **State: 4**

If the turbine operates isentropically, $\widehat{S}_{4r} = \widehat{S}_3 = 6.8403 \, \text{kJ/kg.K}$. *From Table A.1 in Appendix A*

$$T = 50 \, °\text{C} \left\{ \begin{array}{ll} \widehat{H}^L = 209.33 \, \text{kJ/kg} & \widehat{S}^L = 0.7038 \, \text{kJ/kg.K} \\ \Delta\widehat{H} = 2382.7 \, \text{kJ/kg} & \Delta\widehat{S} = 7.3725 \, \text{kJ/kg.K} \\ \widehat{H}^V = 2592.1 \, \text{kJ/kg} & \widehat{S}^V = 8.0763 \, \text{kJ/kg.K} \end{array} \right.$$

Since $0.7038 < \widehat{S}_{4r} < 8.0763$, *the steam at state 4r is a wet steam with a quality*

$$x = \frac{\widehat{S}_{4r} - \widehat{S}^L}{\Delta\widehat{S}} = \frac{6.8403 - 0.7038}{7.3725} = 0.832$$

The enthalpy at state 4r is

$$\widehat{H}_{4r} = 209.33 + (0.832)(2382.7) = 2191.7 \, \text{kJ/kg}$$

The enthalpy at state 4 is calculated from Eq. (6.1-17) as

$$\widehat{H}_4 = 3126.3 - (0.9)(3126.3 - 2191.7) = 2285.2 \, \text{kJ/kg}$$

Thus, the thermal efficiency is calculated from Eq. (6.1-9) as

$$\eta = 1 - \frac{2285.2 - 209.33}{3126.3 - 212.48} = 0.288$$

Comment: *The T-$\widehat{S}$ diagram of the cycle is similar to the one shown in Fig. 6.8.*

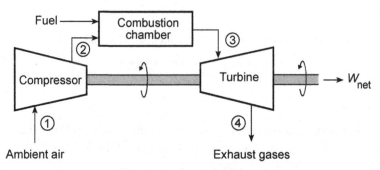

Fig. 6.9   A gas turbine.

6.2   Gas Power Cycles

In gas power cycles, heat is supplied to the system by burning fuel with air. As a result, gas power cycles are different from vapor cycles in two aspects: (i) gas composition does not remain constant, i.e., while the gas consists of a mixture of air and fuel at the beginning of the cycle, a mixture of air and combustion products (exhaust gases) is present at the end of the cycle; (ii) exhaust gases are expelled from the system and the working fluid is renewed at the end of each cycle, i.e., the working fluid is not recirculated. Gas turbine power plants and internal combustion engines are typical examples of gas power cycles. Since the composition of the working fluid does not remain constant, an accurate analysis of gas power cycles is rather complicated.

6.2.1   *Gas turbine power plant*

Gas turbines comprise a compressor, a combustion chamber, and a turbine as shown in Fig. 6.9. Ambient air is compressed and fed to a combustion chamber, where the fuel is burned at constant pressure. Once the combustion products are expanded in the turbine, they are discharged to the atmosphere. Gas turbine power plants are mainly used as backup for steam and hydroelectric power plants. They are also used to power aircraft and ships.

The complications in the analysis of the gas turbine given in Fig. 6.9 can be avoided by considering the so-called *ideal air-standard cycle* with the following assumptions:

- The working fluid is air,
- Air behaves as an ideal gas with constant heat capacities,
- The processes are reversible,
- The combustion process is modeled as a heat addition process from an external heat source, and the exhaust process is replaced by a heat rejection process so as to restore the working fluid to its initial state.

Idealized air-standard Brayton[7] cycle

An ideal air-standard Brayton cycle, shown in Fig. 6.10, consists of the following processes:

- **Process 1-2:** Adiabatic and reversible compression,
- **Process 2-3:** Isobaric and reversible heat addition,
- **Process 3-4:** Adiabatic and reversible expansion,
- **Process 4-1:** Isobaric and reversible heat rejection.

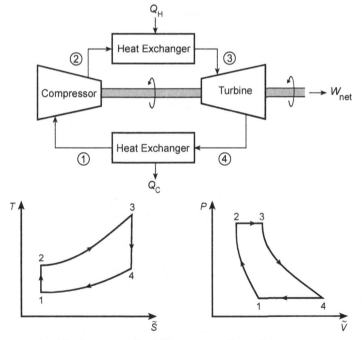

Fig. 6.10 An ideal air-standard Brayton cycle and its representations on T-$\widetilde{S}$ and P-$\widetilde{V}$ diagrams.

---

[7]George Brayton (1830-1892), American mechanical engineer.

The thermal efficiency of the cycle is defined by

$$\eta = \frac{\widetilde{W}_{net}}{\widetilde{Q}_H} = \frac{\widetilde{Q}_H - \left|\widetilde{Q}_C\right|}{\widetilde{Q}_H} = 1 - \frac{\left|\widetilde{Q}_C\right|}{\widetilde{Q}_H} \tag{6.2-1}$$

The energy balances around the heat exchangers give

$$\widetilde{Q}_H = \widetilde{H}_3 - \widetilde{H}_2 = \widetilde{C}_P^* \left(T_3 - T_2\right) \tag{6.2-2}$$

$$\left|\widetilde{Q}_C\right| = \widetilde{H}_4 - \widetilde{H}_1 = \widetilde{C}_P^* \left(T_4 - T_1\right) \tag{6.2-3}$$

Substitution of Eqs. (6.2-2) and (6.2-3) into Eq. (6.2-1) and rearrangement give

$$\eta = 1 - \frac{T_1 \left[(T_4/T_1) - 1\right]}{T_2 \left[(T_3/T_2) - 1\right]} \tag{6.2-4}$$

Processes 1-2 and 3-4 are both isentropic. The use of Eq. (4.1-77) leads to

$$\frac{T_2}{T_1} = \left(\frac{P_2}{P_1}\right)^{(\gamma-1)/\gamma} \tag{6.2-5}$$

and

$$\frac{T_4}{T_3} = \left(\frac{P_4}{P_3}\right)^{(\gamma-1)/\gamma} \tag{6.2-6}$$

Since $P_2 = P_3$ and $P_4 = P_1$, combination of Eqs. (6.2-5) and (6.2-6) results in

$$\left(\frac{T_2}{T_1}\right)\left(\frac{T_4}{T_3}\right) = 1 \quad \Rightarrow \quad \frac{T_4}{T_1} = \frac{T_3}{T_2} \tag{6.2-7}$$

Hence, Eq. (6.2-4) simplifies to

$$\boxed{\eta = 1 - \left(\frac{1}{r_p}\right)^{(\gamma-1)/\gamma}} \tag{6.2-8}$$

where r_p is the *pressure ratio* defined by

$$r_p = \frac{P_2}{P_1} = \frac{P_3}{P_4} \tag{6.2-9}$$

A plot of the thermal efficiency versus the pressure ratio for $\gamma = 1.4$ is shown in Fig. 6.11. Note that the thermal efficiency of the cycle is 0.62 for

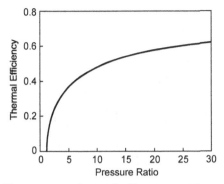

Fig. 6.11 Variation in thermal efficiency with pressure ratio.

the pressure ratio of 30. For higher efficiencies, the pressure ratio must be extremely high.

In an air-standard Brayton cycle, T_3 is the maximum turbine inlet temperature (design constraint), T_1 is the ambient air temperature, and T_4 is the compressor exit temperature (design variable). The practical question to ask at this stage is "what should the value of T_4 be to maximize the work output?"

With the help of Eqs. (6.2-2) and (6.2-3), the net work output is expressed as

$$\widetilde{W}_{net} = \widetilde{Q}_H - \left|\widetilde{Q}_C\right| = \widetilde{C}_P^* \left[(T_3 - T_2) - (T_4 - T_1)\right] \tag{6.2-10}$$

The value of T_4 to maximize work can be found from the following expression:

$$\frac{d\widetilde{W}_{net}}{dT_4} = \widetilde{C}_P^* \left(-\frac{dT_2}{dT_4} - 1\right) = 0 \tag{6.2-11}$$

From Eq. (6.2-7)

$$T_2 = \frac{T_1 T_3}{T_4} \qquad \Rightarrow \qquad \frac{dT_2}{dT_4} = -\frac{T_1 T_3}{T_4^2} \tag{6.2-12}$$

Substitution of Eq. (6.2-12) into Eq. (6.2-11) yields

$$T_4 = \sqrt{T_1 T_3} \tag{6.2-13}$$

Using Eqs. (6.2-6) and (6.2-9), the pressure ratio is expressed in terms of temperature as

$$r_p = \frac{P_2}{P_1} = \frac{P_3}{P_4} = \left(\frac{T_3}{T_4}\right)^{\gamma/(\gamma-1)} \tag{6.2-14}$$

The use of Eq. (6.2-13) in Eq. (6.2-14) results in

$$r_p = \left(\frac{T_3}{T_1}\right)^{\gamma/[2(\gamma-1)]} \tag{6.2-15}$$

Finally, substitution of Eqs. (6.2-12) and (6.2-13) into Eq. (6.2-10) and rearrangement lead to

$$(\widetilde{W}_{net})_{max} = \widetilde{C}_P^* T_1 \left(\frac{T_3}{T_1} - 2\sqrt{\frac{T_3}{T_1}} + 1\right) \tag{6.2-16}$$

6.2.2 *Internal combustion engines*

Internal combustion engines are either reciprocating or rotary. Among the reciprocating internal combustion engines the Otto[8] cycle and the Diesel[9] cycle are the two principal types. As shown in Figure 6.12, these cycles consist of four basic steps:

• **Intake stroke:** Downward movement of the piston creates a vacuum. The intake valve opens and a mixture of fuel and air fills the cylinder. The intake valve closes at the end of this stroke.
• **Compression stroke:** With all the valves closed, the piston moves up and compresses the fuel-air mixture.
• **Power stroke:** In the Otto cycle, the spark plug located on the cylinder head initiates the combustion of the fuel-air mixture. The resulting expansion of the combustion gases forces the piston down. In the Diesel cycle, on the other hand, the fuel-air mixture is compressed to high pressure so that the temperature at the end of the compression stroke is sufficient for self-ignition.

---

[8]Nikolaus August Otto (1832-1891), German developer of the internal combustion engine.
[9]Rudolf Christian Karl Diesel (1858-1913), German inventor of the diesel engine.

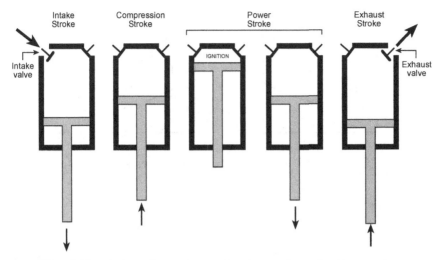

Fig. 6.12   A 4-stroke reciprocating internal combustion engine.

• **Exhaust stroke:** The exhaust valve opens at the end of the power stroke and the upward movement of the piston pushes the combustion gases out of the cylinder. At the end of this stroke, the exhaust valve closes, the intake valve opens, and the sequence repeats in the next cycle.

Since the variation in gas composition complicates the analysis, the air-standard Otto cycle and air-standard Diesel cycle will be considered in the thermodynamic analysis of these internal combustion engines.

Air-standard Otto cycle

The air-standard Otto cycle, shown in Fig. 6.13, consists of the following processes[10]:

• **Process 1-2:** Adiabatic and reversible compression (compression stroke),
• **Process 2-3:** Constant volume heat transfer from an external source (this process corresponds to the ignition of the fuel-air mixture by the spark plug and subsequent rapid increase in temperature and pressure),
• **Process 3-4:** Adiabatic and reversible expansion (power stroke),
• **Process 4-1:** Constant volume heat rejection (this process corresponds to the opening of the exhaust valve as the piston reaches the bottom of its travel).

---

[10]Note that intake and exhaust strokes are not considered in the analysis.

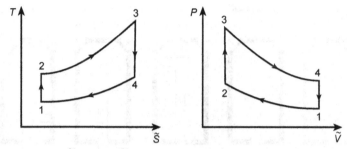

Fig. 6.13 T-$\tilde{S}$ and P-$\tilde{V}$ diagrams of the air-standard Otto cycle.

The thermal efficiency of the cycle is defined by

$$\eta = \frac{\widetilde{W}_{net}}{\widetilde{Q}_H} = \frac{\widetilde{Q}_H - \left|\widetilde{Q}_C\right|}{\widetilde{Q}_H} = 1 - \frac{\left|\widetilde{Q}_C\right|}{\widetilde{Q}_H} \qquad (6.2\text{-}17)$$

Process 2-3 represents constant volume heat addition:

$$\widetilde{Q}_H = \widetilde{H}_3 - \widetilde{H}_2 = \widetilde{C}_V^* \left(T_3 - T_2\right) \qquad (6.2\text{-}18)$$

Process 4-1 represents constant volume heat rejection:

$$\left|\widetilde{Q}_C\right| = \widetilde{H}_4 - \widetilde{H}_1 = \widetilde{C}_V^* \left(T_4 - T_1\right) \qquad (6.2\text{-}19)$$

Substitution of Eqs. (6.2-18) and (6.2-19) into Eq. (6.2-17) and rearrangement give

$$\eta = 1 - \frac{T_1 \left[(T_4/T_1) - 1\right]}{T_2 \left[(T_3/T_2) - 1\right]} \qquad (6.2\text{-}20)$$

Processes 1-2 and 3-4 are both isentropic. The use of Eq. (4.1-75) leads to

$$\frac{T_2}{T_1} = \left(\frac{\widetilde{V}_1}{\widetilde{V}_2}\right)^{\gamma-1} \qquad (6.2\text{-}21)$$

and

$$\frac{T_4}{T_3} = \left(\frac{\widetilde{V}_3}{\widetilde{V}_4}\right)^{\gamma-1} \qquad (6.2\text{-}22)$$

Since $\widetilde{V}_2 = \widetilde{V}_3$ and $\widetilde{V}_4 = \widetilde{V}_1$, combination of Eqs. (6.2-21) and (6.2-22) results in

$$\left(\frac{T_2}{T_1}\right)\left(\frac{T_4}{T_3}\right) = 1 \quad \Rightarrow \quad \frac{T_4}{T_1} = \frac{T_3}{T_2} \tag{6.2-23}$$

Hence, Eq. (6.2-20) simplifies to

$$\boxed{\eta = 1 - \frac{1}{r_c^{\gamma-1}}} \tag{6.2-24}$$

where r_c is the *compression ratio* defined by

$$r_c = \frac{\widetilde{V}_1}{\widetilde{V}_2} = \frac{\widetilde{V}_4}{\widetilde{V}_3} \tag{6.2-25}$$

Example 6.4 *The compression ratio in an air-standard Otto cycle is 8. The pressure and temperature at the beginning of the compression stroke are 1 bar and 290 K, respectively. The maximum temperature during the cycle is 2300 K.*

a) *Calculate the heat added to the cycle,*
b) *Calculate the thermal efficiency of the cycle,*
c) *Calculate the net work of the cycle.*

Solution

a) *From Eqs. (6.2-21) and (6.2-25)*

$$T_2 = T_1\, r_c^{\gamma-1} = (290)(8)^{0.4} = 666.2\,\text{K}$$

The amount of heat added is calculated from Eq. (6.2-18) as

$$\widetilde{Q}_H = \widetilde{C}_V^* (T_3 - T_2) = (2.5 \times 8.314)(2300 - 666.2) = 33,959\,\text{kJ}/\,\text{kmol}$$

b) *From Eq. (6.2-24)*

$$\eta = 1 - \frac{1}{(8)^{0.4}} = 0.5647$$

c) *Note that*

$$\eta = \frac{\widetilde{W}_{net}}{\widetilde{Q}_H} \quad \Rightarrow \quad \widetilde{W}_{net} = (0.5647)(33,959) = 19,177\,\text{kJ}/\,\text{kmol}$$

Alternative solution: *From Eqs. (6.2-22) and (6.2-25)*

$$T_4 = \frac{T_3}{r_c^{\gamma-1}} = \frac{2300}{(8)^{0.4}} = 1001.1\,\text{K}$$

From Eq. (6.2-19)

$$\left|\widetilde{Q}_C\right| = \widetilde{C}_V^* \left(T_4 - T_1\right) = (2.5 \times 8.314)(1001.1 - 290) = 14,780\,\text{kJ/kmol}$$

Therefore

$$\widetilde{W}_{net} = \widetilde{Q}_H - \left|\widetilde{Q}_C\right| = 33,959 - 14,780 = 19,179\,\text{kJ/kmol}$$

The difference comes from the round-off errors.

Air-standard Diesel cycle

An air-standard Diesel cycle, shown in Fig. 6.14, consists of the following processes[11]:

- **Process 1-2:** Adiabatic and reversible compression (compression stroke),
- **Process 2-3:** Constant pressure heat transfer from an external source (this process corresponds to the spontaneous ignition of the fuel-air mixture and part of the power stroke),
- **Process 3-4:** Adiabatic and reversible expansion (remainder of the power stroke),
- **Process 4-1:** Constant volume heat rejection (this process corresponds to the opening of the exhaust valve as the piston reaches the bottom of its travel).

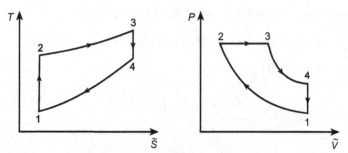

Fig. 6.14 T-$\widetilde{S}$ and P-$\widetilde{V}$ diagrams of the air-standard Diesel cycle.

---

[11]Note that intake and exhaust strokes are not considered in the analysis.

The thermal efficiency of the cycle is defined by

$$\eta = \frac{\widetilde{W}_{net}}{\widetilde{Q}_H} = \frac{\widetilde{Q}_H - \left|\widetilde{Q}_C\right|}{\widetilde{Q}_H} = 1 - \frac{\left|\widetilde{Q}_C\right|}{\widetilde{Q}_H} \tag{6.2-26}$$

Process 2-3 represents constant pressure heat addition:

$$\widetilde{Q}_H = \widetilde{H}_3 - \widetilde{H}_2 = \widetilde{C}_P^* (T_3 - T_2) \tag{6.2-27}$$

Process 4-1 represents constant volume heat rejection:

$$\left|\widetilde{Q}_C\right| = \widetilde{H}_4 - \widetilde{H}_1 = \widetilde{C}_V^* (T_4 - T_1) \tag{6.2-28}$$

Substitution of Eqs. (6.2-27) and (6.2-28) into Eq. (6.2-26) and rearrangement give

$$\eta = 1 - \frac{1}{\gamma} \frac{T_1}{T_2} \frac{\left[(T_4/T_1) - 1\right]}{\left[(T_3/T_2) - 1\right]} \tag{6.2-29}$$

Processes 1-2 and 3-4 are both isentropic. The use of Eqs. (4.1-75) and (4.1-77) leads to

$$\frac{T_2}{T_1} = \left(\frac{\widetilde{V}_1}{\widetilde{V}_2}\right)^{\gamma-1} = \left(\frac{P_2}{P_1}\right)^{(\gamma-1)/\gamma} \tag{6.2-30}$$

and

$$\frac{T_4}{T_3} = \left(\frac{\widetilde{V}_3}{\widetilde{V}_4}\right)^{\gamma-1} = \left(\frac{P_4}{P_3}\right)^{(\gamma-1)/\gamma} \tag{6.2-31}$$

Since $\widetilde{V}_4 = \widetilde{V}_1$, combination of Eqs. (6.2-30) and (6.2-31) results in

$$\left(\frac{T_2}{T_1}\right)\left(\frac{T_4}{T_3}\right) = \left(\frac{\widetilde{V}_3}{\widetilde{V}_2}\right)^{\gamma-1} \tag{6.2-32}$$

The entropy changes for processes 2-3 and 4-1 are given by

$$\widetilde{S}_3 - \widetilde{S}_2 = \widetilde{C}_P^* \ln\left(\frac{T_3}{T_2}\right) \quad \text{and} \quad \widetilde{S}_1 - \widetilde{S}_4 = \widetilde{C}_V^* \ln\left(\frac{T_1}{T_4}\right) \tag{6.2-33}$$

Since $\widetilde{S}_1 = \widetilde{S}_2$ and $\widetilde{S}_3 = \widetilde{S}_4$, it follows that

$$\frac{T_4}{T_1} = \left(\frac{T_3}{T_2}\right)^{\gamma} \tag{6.2-34}$$

The use of Eq. (6.2-34) in Eq. (6.2-32) results in

$$\frac{T_3}{T_2} = \frac{\widetilde{V}_3}{\widetilde{V}_2} = r_e \qquad (6.2\text{-}35)$$

where the *cutoff ratio*, r_e, is the ratio of the cylinder volumes after and before the combustion process. If the *compression ratio*, r_c, is defined by

$$r_c = \frac{\widetilde{V}_1}{\widetilde{V}_2} \qquad (6.2\text{-}36)$$

Eq. (6.2-29) becomes

$$\boxed{\eta = 1 - \frac{1}{r_c^{\gamma-1}}\left[\frac{r_e^{\gamma} - 1}{\gamma\,(r_e - 1)}\right]} \qquad (6.2\text{-}37)$$

In Eq. (6.2-37), the term

$$\left[\frac{r_e^{\gamma} - 1}{\gamma\,(r_e - 1)}\right] > 1 \qquad (6.2\text{-}38)$$

is always greater than unity. Thus, comparison of Eqs. (6.2-24) and (6.2-37) indicates that when Otto and Diesel cycles operate at the same compression ratio, $\eta_{otto} > \eta_{diesel}$. However, the conclusions drawn from the comparison of these two cycles must always be related to the basis on which the comparison has been made. For example, Diesel cycles have the following advantages over Otto cycles:

• Diesel cycles operate at much higher compression ratios and thus are usually more efficient than Otto cycles.
• Diesel cycles burn the fuel more completely since they operate at lower revolutions per minute (rpm) than Otto cycles.

Example 6.5 *The compression ratio in an air-standard Diesel cycle is* 16. *The pressure and temperature at the beginning of the compression stroke are* 1 bar *and* 290 K, *respectively. Calculate the thermal efficiency of the cycle if the heat transfer to the air per cycle is* 50,000 kJ/kmol.

Solution

From Eqs. (6.2-30) and (6.2-36)

$$T_2 = T_1\,r_c^{\gamma-1} = (290)(16)^{0.4} = 879.1\,\text{K}$$

From Eq. (6.2-27)

$$50,000 = (3.5 \times 8.314)(T_3 - 879.1) \qquad \Rightarrow \qquad T_3 = 2597.4\,\text{K}$$

The cutoff ratio is calculated from Eq. (6.2-35) as

$$r_e = \frac{T_3}{T_2} = \frac{2597.4}{879.1} = 2.95$$

Hence, the thermal efficiency of the cycle is calculated from Eq. (6.2-37) as

$$\eta = 1 - \frac{1}{(16)^{0.4}} \left[\frac{2.95^{1.4} - 1}{(1.4)(2.95 - 1)} \right] = 0.57$$

6.3 Refrigeration Cycles

A refrigeration cycle is a reversed heat engine cycle. Heat is transferred from a low temperature level to a high temperature level and this, according to the second law of thermodynamics, cannot be accomplished without the use of external energy. Vapor-compression refrigeration cycles are widely used for refrigeration and air-conditioning purposes in domestic and industrial applications. Gas compression refrigeration cycles are less efficient but more lightweight than vapor-compression refrigeration cycles. They are mainly used in mobile refrigeration and air-conditioning applications, such as in aircraft.

6.3.1 Refrigerants

While circulating within the refrigeration system, refrigerants absorb heat from areas to be cooled and transfer it to where the heat can be dissipated. They must possess the following properties:

- Low boiling point and high latent heat of vaporization,
- Nontoxic, noncorrosive, nonflammable,
- Low miscibility with oil,
- Low cost,
- Chemically stable.

Since the chemical names of refrigerants are generally long and cumbersome, an R number is assigned to each refrigerant related to its chemical composition, and the system has been formalized as ASHRAE[12] Standard 34. The first digit on the right is the number of fluorine (F) atoms in the refrigerant. The second digit from the right is one more than the number of hydrogen (H) atoms present. The third digit from the right is one less than the number of carbon (C) atoms, but when this digit is zero it is omitted. The fourth digit from the right is equal to the number of unsaturated carbon-carbon bonds in the compound, but when this digit is zero it is omitted. For example, dichlorodifluoromethane (CCl_2F_2) is known as R-12:

$$
\begin{array}{l}
\text{R-12} \quad CCl_2F_2 \\
\quad\;\; \llcorner L\; \text{Number of fluorine atoms} \quad = 2 \\
\quad\;\; \llcorner\!\!\text{Number of hydrogen atoms} + 1 = 1
\end{array}
$$

Ammonia and sulfur dioxide were the early refrigerants for commercial use as a result of their high latent heats of vaporization. However, they have the obvious drawbacks of being highly toxic and corrosive, and with NH_3, flammable. Research efforts in the 1920's supported by Frigidaire, the company owned by General Motors (GM) from 1919 to 1979, to identify safe, effective refrigerants led to the conclusion that small molecules having C-F bonds (fluorocarbons) were suitable choices, with R-12 having the best properties. Dupont, by virtue of its close ties with GM, was the first producer of fluorocarbons, beginning in 1931. Chlorofluorocarbon (CFC) and hydro-chlorofluorocarbon (HCFC) refrigerants produced by Dupont were sold under the trade name *Freon* and R-12 quickly became the refrigerant of choice for most applications.

With the expanding markets since their 1931 introduction, the world-wide production of all CFCs and HCFCs grew to a maximum of 1.25 million metric tons/year by 1988. As a result, besides Dupont, Allied Signal, Pennsalt, Union Carbide, ICI, Hoechst, Monticatini, and Daikin started producing similar refrigerants.

CFCs were widely used as refrigerants until the 1980's, when it was confirmed that they were the main source of depletion of the earth's protective ozone layer[13]. HCFCs were developed as "transitional" refrigerants, and

---

[12] American Society of Heating, Refrigeration, and Air-conditioning Engineers (www.ashrae.org)

[13] CFCs accumulate in the earth's atmosphere and dissociate with the help of the sun's radiation to produce the highly reactive chlorine radical (Cl). The Cl radical so generated in the stratosphere, on the other hand, destroys O_3 in a chain reaction.

chlorodifluoromethane (R-22) started to replace R-12 in various applications. HCFCs are less damaging to the ozone layer but are highly potent greenhouse gases leading to global warming.

To protect the stratospheric ozone layer, the Montreal Protocol, an international environmental agreement, was originally signed in 1987 and substantially amended in 1990 and 1992. In accordance with this treaty, CFCs, halons, carbon tetrachloride, and methyl chloroform were phased out by 2000 (2005 for methyl chloroform). The regulations allow the use of HCFCs in new equipment until the year 2015 and in existing equipment until 2020.

Researchers interested in comparing the impact of various fluorinated compounds on O_3 developed a ranking scale known as the *ozone depletion potential* (ODP). The ODP is the ratio of the impact on ozone of a chemical compared to the impact of a similar mass of R-11 ($CFCl_3$). Thus, the ODP of R-11 is defined to be 1.0.

Hydrofluorocarbons (HFCs) were developed as alternatives to the ozone-depleting refrigerants and R-134a (1,1,1,2-tetrafluoroethane - CF_3CH_2F) has become the refrigerant of choice to replace R-12 in most refrigeration and auto air conditioning systems. The properties of R-134a are given in Appendix B[14].

6.3.2 *Carnot vapor-compression refrigeration cycle*

A typical Carnot vapor-compression refrigeration cycle is shown in Fig. 6.15. In the evaporator, the refrigerant evaporates at constant temperature and pressure as heat is absorbed from the low temperature region (process 4-1). The vapor leaving the evaporator is then compressed to a higher pressure (process 1-2). An external work is required for this purpose. In the condenser, the refrigerant condenses at constant temperature and pressure as heat is rejected to the high temperature region (process 2-3). To complete the cycle, the liquid from the condenser is returned to its original state by an expansion process (process 3-4). The overall process is also represented in a $T\text{-}\widehat{S}$ diagram in Fig. 6.15.

As stated in Section 5.2, the performance of a refrigeration cycle (or heat pump) is measured in terms of the *coefficient of performance*, COP. It is the ratio of the heat absorbed from the low temperature region to the net work supplied to the cycle, i.e.,

---

[14]Property tables for R-134a can also be found at:
www.peacesoftware.de/einigewerte/r134a_e.html.

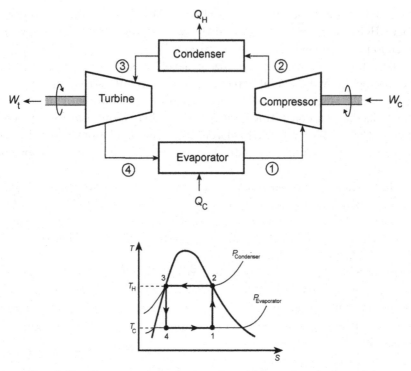

Fig. 6.15   Carnot vapor-compression refrigeration cycle and its
representation in a T-$\tilde{S}$ diagram.

$$\text{COP} = \frac{\text{Cooling effect}}{\text{Net power input}} = \frac{\dot{Q}_C}{\dot{W}_{net}} \qquad (6.3\text{-}1)$$

The COP of a Carnot refrigeration cycle is given by

$$\text{COP}_{carnot} = \frac{\dot{Q}_C}{\dot{W}_{net}} = \frac{T_C}{T_H - T_C} \qquad (6.3\text{-}2)$$

Thus, higher COP can be achieved with higher evaporator temperature and lower condenser temperature. The following rules of thumb should be taken into consideration in the design of refrigerating systems[15]:

---

[15] *Detailed Assessment: Energy Efficiency*, Chapter 5, United Nations Environment Programme Publication (www.unep.org).

• Refrigeration capacity decreases by 6% for every 3.5 °C increase in condensing temperature.

• Reducing condensing temperature by 5.5 °C results in a 20-25% decrease in compressor power consumption.

• A reduction of 0.55 °C in cooling water temperature at the condenser inlet reduces compressor power consumption by 3%.

• A 5.5 °C increase in evaporator temperature reduces compressor power consumption by 20-25%.

A common unit used in practice to describe the refrigeration effect is the *ton*. *One ton of refrigeration* is the term used to refer to $12,000 \, \text{Btu/h}$[16]. Thus, a refrigeration unit with a cooling capacity of $72,000 \, \text{Btu/h}$ is said to have a capacity of 6 tons. Note that

$$1 \text{ ton of refrigeration } = 12,000 \, \text{Btu/h} = 12,661 \, \text{kJ/h} = 3.517 \, \text{kW}$$

Both the compression and expansion steps in the Carnot refrigeration cycle occur within the two-phase region. The compression of a two-phase mixture, however, is very difficult in practice and is generally avoided. As a result, although the Carnot refrigeration cycle requires the least amount of energy, it is not used in industrial applications.

Example 6.6 *A Carnot refrigeration cycle is used to maintain a room at 23 °C by removing heat from groundwater at 15 °C. Refrigerant R-134a enters the condenser as saturated vapor at 40 °C and leaves as saturated liquid at the same temperature. The evaporator pressure is 351 kPa.*

a) *If the room is to receive 2 kW, calculate the power input to the compressor.*

b) *Calculate the net power input to the cycle.*

Solution

From Table B.1 in Appendix B, the saturation temperature corresponding to 351 kPa is 5 °C. The T-$\widehat{S}$ diagram of the cycle is shown in the figure below:

---

[16] Historically, this value was developed by the ice manufacturers in the US, who wanted an easy way of understanding the size of a refrigeration system in terms of the production of ice. To make one ton ($2000 \, \text{lb}_m$) of ice, $288,000 \, \text{Btu}$ is required. Divide this value by 24 hours to obtain $12,000 \, \text{Btu/h}$ required to make one ton of ice in one day.

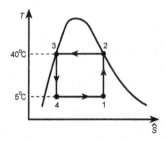

- **State: 2**

From Table B.1 in Appendix B

$$\left.\begin{array}{l} T_2 = 40\,^\circ\text{C} \\ \text{Sat. Vapor} \end{array}\right\} \begin{array}{l} \widehat{H}_2 = 419.821\,\text{kJ/kg} \\ \widehat{S}_2 = 1.7123\,\text{kJ/kg. K} \end{array}$$

- **State: 3**

From Table B.1 in Appendix B

$$\left.\begin{array}{l} T_3 = 40\,^\circ\text{C} \\ \text{Sat. Liquid} \end{array}\right\} \begin{array}{l} \widehat{H}_3 = 256.539\,\text{kJ/kg} \\ \widehat{S}_3 = 1.1909\,\text{kJ/kg. K} \end{array}$$

From Table B.1 in Appendix B, the enthalpy and entropy values at the evaporator pressure of 351 kPa are given as

$$P = 351\,\text{kPa} \left\{\begin{array}{ll} \widehat{H}^L = 206.751\,\text{kJ/kg} & \widehat{S}^L = 1.0243\,\text{kJ/kg. K} \\ \Delta\widehat{H} = 194.572\,\text{kJ/kg} & \Delta\widehat{S} = 0.6995\,\text{kJ/kg. K} \\ \widehat{H}^V = 401.323\,\text{kJ/kg} & \widehat{S}^V = 1.7239\,\text{kJ/kg. K} \end{array}\right.$$

- **State: 1**

For an isentropic compression in the compressor $\widehat{S}_1 = \widehat{S}_2 = 1.7123\,\text{kJ/kg. K}$. *The quality at state 1 is*

$$1.7123 = 1.0243 + x(0.6995) \qquad \Rightarrow \qquad x = 0.984$$

Therefore, the enthalpy is

$$\widehat{H}_1 = 206.751 + (0.984)(194.572) = 398.21\,\text{kJ/kg}$$

• **State: 4**

For an isentropic expansion in the turbine $\widehat{S}_4 = \widehat{S}_3 = 1.1909\,\text{kJ}/\text{kg. K.}$ *The quality at state 4 is*

$$1.1909 = 1.0243 + x(0.6995) \qquad \Rightarrow \qquad x = 0.238$$

Therefore, the enthalpy is

$$\widehat{H}_4 = 206.751 + (0.238)(194.572) = 253.06\,\text{kJ}/\text{kg}$$

a) *The energy balance around the condenser gives the circulation rate of the refrigerant as*

$$\left| \dot{Q}_H \right| = \dot{m}\,(\widehat{H}_2 - \widehat{H}_3) \qquad \Longrightarrow \qquad \dot{m} = \frac{2}{419.821 - 256.539} = 0.0122\,\text{kg}/\text{s}$$

The energy balance around the compressor gives

$$\dot{W}_c = \dot{m}\,\widehat{W}_c = \dot{m}\,(\widehat{H}_2 - \widehat{H}_1) = (0.0122)(419.821 - 398.21) = 0.264\,\text{kW}$$

b) *The energy balance around the turbine gives*

$$-\dot{W}_t = \dot{m}\,(-\widehat{W}_t) = \dot{m}\,(\widehat{H}_3 - \widehat{H}_4) = (0.0122)(256.539 - 253.06) = 0.042\,\text{kW}$$

Thus, the net power input to the cycle is

$$\dot{W}_{net} = 0.264 - 0.042 = 0.222\,\text{kW}$$

Alternative solution: *The coefficient of performance of a Carnot refrigeration cycle is*

$$\text{COP}_{carnot} = \frac{\dot{Q}_C}{\dot{W}_{net}} = \frac{T_C}{T_H - T_C} \tag{1}$$

The energy balance for the overall cycle gives

$$\dot{Q}_C = 2 - \dot{W}_{net} \tag{2}$$

Combination of Eqs. (1) and (2) leads to

$$\dot{W}_{net} = 2\left[1 - \left(\frac{T_C}{T_H}\right)\right] = 2\left[1 - \left(\frac{5 + 273}{40 + 273}\right)\right] = 0.224\,\text{kW}$$

The difference comes from the round-off errors.

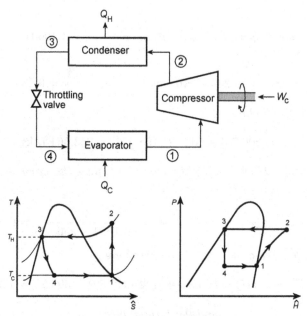

Fig. 6.16 The ideal vapor-compression refrigeration cycle and its representations on T-$\widehat{S}$ and P-$\widehat{H}$ diagrams.

6.3.3 *The ideal vapor-compression refrigeration cycle*

Impracticalities encountered in the Carnot refrigeration cycle can be eliminated by simply allowing the refrigerant to evaporate completely in the evaporator, producing a saturated vapor. On the other hand, the fluid passing through the turbine is mostly liquid and its specific volume is relatively low. Consequently, the work output of the turbine is not appreciable. For this reason, much less expensive and almost maintenance-free throttling valves are preferred over turbines in practice. The modified cycle and its representation on T-$\widehat{S}$ and P-$\widehat{H}$ diagrams are shown in Fig. 6.16. It is composed of the following four processes:

• **Process 1-2:** Saturated vapor at state 1 is compressed reversibly and adiabatically to condenser pressure. The energy balance around the compressor gives

$$\widehat{W}_{c,rev} = \widehat{H}_2 - \widehat{H}_1 \tag{6.3-3}$$

• **Process 2-3:** Refrigerant enters the condenser as a superheated vapor and leaves as a saturated liquid while pressure remains constant during condensation. The energy balance around the condenser becomes

$$\widehat{Q}_H = \widehat{H}_3 - \widehat{H}_2 \qquad (6.3\text{-}4)$$

• **Process 3-4:** Pressure of the saturated liquid coming out of the condenser is lowered to the evaporator pressure as it passes through the throttling valve. Since the expansion process is isenthalpic

$$\widehat{H}_4 = \widehat{H}_3 \qquad (6.3\text{-}5)$$

• **Process 4-1:** Refrigerant evaporates in the evaporator by absorbing heat from the low temperature region. The energy balance around the evaporator gives

$$\widehat{Q}_C = \widehat{H}_1 - \widehat{H}_4 \qquad (6.3\text{-}6)$$

The COP of the refrigeration cycle, Eq. (6.3-1), then becomes

$$\boxed{\text{COP} = \frac{\text{Cooling effect}}{\text{Power input to compressor}} = \frac{\widehat{Q}_C}{\widehat{W}_{c,rev}} = \frac{\widehat{H}_1 - \widehat{H}_4}{\widehat{H}_2 - \widehat{H}_1}} \qquad (6.3\text{-}7)$$

Example 6.7 *An ideal vapor-compression refrigeration cycle operates with R-134a as the working fluid. The evaporator temperature is 15 °C and the condenser pressure is 1.6 MPa. Saturated vapor enters the compressor and saturated liquid exits the condenser. The mass flow rate of the refrigerant is 0.1 kg/ s.*

a) *Calculate the coefficient of performance of the cycle,*
b) *Determine the power input to the compressor,*
c) *Calculate the refrigeration capacity in tons.*

Solution

The T-$\widehat{S}$ diagram of the cycle is shown below:

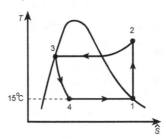

•**State: 1**

From Table B.1 in Appendix B

$$\left.\begin{array}{l} T_1 = 15\,^\circ\text{C} \\ \text{Sat. Vapor} \end{array}\right\} \begin{array}{l} \widehat{H}_1 = 407.075\,\text{kJ/kg} \\ \widehat{S}_1 = 1.7200\,\text{kJ/kg.K} \end{array}$$

• **State: 2**

For an isentropic compression, $\widehat{S}_2 = \widehat{S}_1 = 1.7200\,\text{kJ/kg.K}$. *From Table B.2 in Appendix B at 1.6 MPa*

$T\,(^\circ\text{C})$	$\widehat{H}\,(\text{kJ/kg})$	$\widehat{S}\,(\text{kJ/kg.K})$
60	429.322	1.71349
70	441.888	1.75066

By interpolation

$$\widehat{H}_2 = 429.322 + \left(\frac{441.888 - 429.322}{1.75066 - 1.71349}\right)(1.7200 - 1.71349) = 431.5\,\text{kJ/kg}$$

• **State: 3**

From Table B.1 in Appendix B, by interpolation

$$\left.\begin{array}{l} P_3 = 1.6\,\text{MPa} \\ \text{Sat. Liquid} \end{array}\right\} \widehat{H}_3 = 284.323\,\text{kJ/kg}$$

• **State: 4**

Since the expansion through the valve is isenthalpic

$$\widehat{H}_4 = \widehat{H}_3 = 284.323\,\text{kJ/kg}$$

a) *The COP is calculated from Eq. (6.3-7) as*

$$\text{COP} = \frac{407.075 - 284.323}{431.5 - 407.075} = 5.03$$

b) *The power of the compressor is calculated from Eq. (6.3-3) as*

$$\dot{W}_{c,rev} = \dot{m}\,\widehat{W}_{c,rev} = \dot{m}\,(\widehat{H}_2 - \widehat{H}_1) = (0.1)(431.5 - 407.075) = 2.44\,\text{kW}$$

c) *The refrigeration capacity is determined from Eq. (6.3-6) as*

$$\dot{Q}_C = \dot{m}\,(\widehat{H}_1 - \widehat{H}_4) = (0.1)(407.075 - 284.323)\left(\frac{1\,\text{ton}}{3.517\,\text{kW}}\right) = 3.49\,\text{ton}$$

6.3.4 Deviations from idealities

The isentropic efficiency of a compressor, η_c, is defined by Eq. (5.6-27) as

$$\eta_c = \frac{\widehat{W}_{c,rev}}{\widehat{W}_c} = \frac{\widehat{H}_{2r} - \widehat{H}_1}{\widehat{H}_2 - \widehat{H}_1} \tag{6.3-8}$$

Rearrangement of Eq. (6.3-8) gives

$$\widehat{H}_2 = \widehat{H}_1 + \frac{\widehat{H}_{2r} - \widehat{H}_1}{\eta_c} \tag{6.3-9}$$

Example 6.8 *A vapor-compression refrigeration cycle with R-134a as the working fluid has an evaporator temperature of* $-20\,^\circ$C *and a condenser pressure of* 1.2 MPa. *Saturated vapor enters the compressor and saturated liquid exits the condenser. The isentropic efficiency of the compressor is* 0.85.

a) *Calculate the coefficient of performance of the cycle,*
b) *Determine the power input to the compressor if* $\dot{Q}_C = 2\,$kW,
c) *Calculate the circulation rate of the refrigerant.*

Solution

The P-$\widehat{H}$ diagram of the cycle is shown in the figure below:

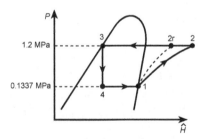

• **State: 1**

From Table B.1 in Appendix B

$$\left.\begin{array}{l} T_1 = -20\,^\circ\text{C} \\ \text{Sat. Vapor} \end{array}\right\} \begin{array}{l} P_1 = 0.1337\,\text{MPa} \\ \widehat{H}_1 = 386.083\,\text{kJ}/\,\text{kg} \\ \widehat{S}_1 = 1.7395\,\text{kJ}/\,\text{kg. K} \end{array}$$

• **State: 2**

If the compressor operates isentropically, $\widehat{S}_{2r} = \widehat{S}_1 = 1.7395\,\text{kJ/kg. K.}$
From Table B.2 in Appendix B at 1.2 MPa

$T\,(°C)$	$\widehat{H}\,(\text{kJ/kg})$	$\widehat{S}\,(\text{kJ/kg. K})$
50	426.845	1.72373
60	438.210	1.75837

By interpolation

$$\widehat{H}_{2r} = 426.845 + \left(\frac{438.210 - 426.845}{1.75837 - 1.72373}\right)(1.7395 - 1.72373)$$

$$= 432.02\,\text{kJ/kg}$$

The use of Eq. (6.3-9) gives

$$\widehat{H}_2 = 386.083 + \frac{432.02 - 386.083}{0.85} = 440.127\,\text{kJ/kg}$$

• **State: 3**

From Table B.1 in Appendix B

$$\left.\begin{array}{l} P_3 = 1.2\,\text{MPa} \\ \text{Sat. Liquid} \end{array}\right\} \widehat{H}_3 = 266.06\,\text{kJ/kg}$$

• **State: 4**

Since the expansion through the valve is isenthalpic

$$\widehat{H}_4 = \widehat{H}_3 = 266.06\,\text{kJ/kg}$$

a) *The COP is calculated from Eq. (6.3-7) as*

$$\text{COP} = \frac{386.083 - 266.06}{440.127 - 386.083} = 2.221$$

b) *The power input to the compressor is*

$$\text{COP} = \frac{\dot{Q}_C}{\dot{W}} = 2.221 = \frac{2}{\dot{W}} \quad \Longrightarrow \quad \dot{W} = 0.9\,\text{kW}$$

c) The circulation rate of the refrigerant is

$$\dot{Q}_C = \dot{m}\,(\widehat{H}_1 - \widehat{H}_4) \quad \Longrightarrow \quad \dot{m} = \frac{(2)(60)}{386.083 - 266.06} = 1\,\text{kg/min}$$

Example 6.9 *The capacity of a R-134a vapor-compression refrigeration system is 35 kW. The cycle is slightly modified by installing a countercurrent heat exchanger to subcool the liquid from the condenser with a vapor stream exiting the evaporator as shown in the figure below. Saturated vapor exits the evaporator at − 10 °C and the condenser pressure is 1.2 MPa. The isentropic efficiency of the compressor is 0.80. The minimum temperature difference that can be attained in the heat exchanger is 6 °C.*

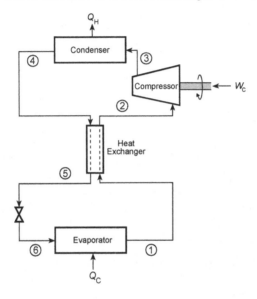

a) *Calculate the circulation rate of the refrigerant,*

b) *Determine the power input to the compressor.*

Solution

The P-$\widehat{H}$ and T-$\widehat{S}$ diagrams of the cycle are shown in the figure below:

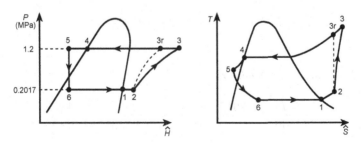

• **State: 1**

From Table B.1 in Appendix B

$$T_1 = -10\,°C \\ \text{Sat. Vapor} \left.\right\} \begin{array}{l} P_1 = 0.2017\,\text{MPa} \\ \widehat{H}_1 = 392.285\,\text{kJ/kg} \\ \widehat{S}_1 = 1.7319\,\text{kJ/kg.K} \end{array}$$

• **State: 4**

Saturated liquid is at 1.2 MPa. From Table B.1 in Appendix B

$T\,(°C)$	$P\,(\text{MPa})$	$\widehat{H}^L\,(\text{kJ/kg})$
45	1.1602	264.110
50	1.3180	271.830

By interpolation

$$\widehat{H}_4 = 264.110 + \left(\frac{271.830 - 264.110}{1.3180 - 1.1602}\right)(1.2 - 1.1602) = 266.057\,\text{kJ/kg}$$

$$T_4 = 45 + \left(\frac{50 - 45}{1.3180 - 1.1602}\right)(1.2 - 1.1602) = 46.3\,°C$$

The energy balance around the heat exchanger gives

$$\dot{m}(\widehat{H}_2 - \widehat{H}_1) + \dot{m}(\widehat{H}_5 - \widehat{H}_4) = 0 \;\Rightarrow\; \dot{m}\,\widehat{C}_P^V(T_2 - T_1) + \dot{m}\,\widehat{C}_P^L(T_5 - T_4) = 0 \quad (1)$$

Simplification of Eq. (1) leads to

$$\widehat{C}_P^V(T_2 - T_1) = \widehat{C}_P^L(T_4 - T_5) \qquad (2)$$

Since $\widehat{C}_P^L > \widehat{C}_P^V$, then one can conclude from Eq. (2) that

$$T_2 - T_1 > T_4 - T_5 \quad\Rightarrow\quad T_5 - T_1 > T_4 - T_2 \qquad (3)$$

Since the minimum temperature difference is 6 °C

$$T_4 - T_2 = 6 \quad\Rightarrow\quad T_2 = 46.3 - 6 \simeq 40\,°C$$

• **State: 2**

From Table B.2 in Appendix B

$$P_2 = P_1 = 0.2017\,\text{MPa} \\ T_2 = 40\,°C \left.\right\} \begin{array}{l} \widehat{H}_2 = 435.708\,\text{kJ/kg} \\ \widehat{S}_2 = 1.88357\,\text{kJ/kg.K} \end{array}$$

● **State: 3**

If the compressor operates isentropically, $\widehat{S}_{3r} = \widehat{S}_2 = 1.88357\,\text{kJ/kg.K}$. *From Table B.2 in Appendix B at* $1.2\,\text{MPa}$

$T\,(^\circ\text{C})$	$\widehat{H}\,(\text{kJ/kg})$	$\widehat{S}\,(\text{kJ/kg.K})$
100	481.128	1.88009
110	491.702	1.90805

By interpolation

$$\widehat{H}_{3r} = 481.128 + \left(\frac{491.702 - 481.128}{1.90805 - 1.88009}\right)(1.88357 - 1.88009) = 482.44\,\text{kJ/kg}$$

The use of Eq. (6.3-9) gives

$$\widehat{H}_3 = 435.708 + \frac{482.44 - 435.708}{0.80} = 494.12\,\text{kJ/kg}$$

● **State: 5**

From Eq. (1)

$$\widehat{H}_5 = \widehat{H}_1 + \widehat{H}_4 - \widehat{H}_2 = 392.285 + 266.057 - 435.708 = 222.634\,\text{kJ/kg}$$

● **State: 6**

Since the expansion through the valve is isenthalpic

$$\widehat{H}_6 = \widehat{H}_5 = 222.634\,\text{kJ/kg}$$

a) *The energy balance around the evaporator gives*

$$\dot{Q}_C = \dot{m}\,(\widehat{H}_1 - \widehat{H}_6) \quad \Rightarrow \quad \dot{m} = \frac{35}{392.285 - 222.634} = 0.206\,\text{kg/s}$$

b) *The energy balance around the compressor gives*

$$\dot{W}_c = \dot{m}\,(\widehat{H}_3 - \widehat{H}_2) = (0.206)(494.12 - 435.708) = 12.03\,\text{kW}$$

6.3.5 *Cascade refrigeration cycles*

In some applications, the temperature difference between the heat source and the sink is extremely high. For example, in order to liquefy air under atmospheric pressure, it must be cooled to $78\,\text{K}$ ($-195\,^\circ\text{C}$). In such cases, it is impossible to achieve extremely low temperatures in a single vapor-

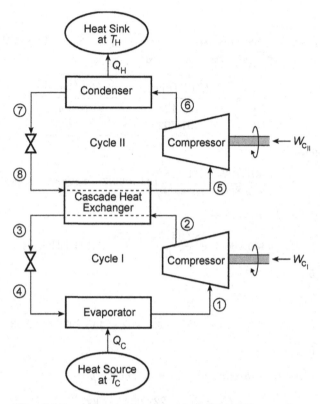

Fig. 6.17 A two-stage cascade refrigeration cycle.

compression cycle. This problem can be avoided by means of a cascade cycle as shown in Fig. 6.17. The so-called "cascade heat exchanger" acts as the condenser of cycle I and the evaporator of cycle II. Although there is no upper limit for the number of cycles to be used, three or four appears to be the maximum practical number.

In the design of cascade cycles, the refrigerants in each cycle should be determined based on the evaporator and condenser temperatures. For the heat transfer to take place in the evaporator, the boiling point of a refrigerant must be less than the temperature of the heat source, T_C. On the other hand, the critical temperature of a refrigerant must be less than the temperature of the heat sink, T_H. Otherwise, condensation cannot be achieved in the condenser.

The coefficient of performance of the two-stage cascade refrigeration cycle shown in Fig. 6.17 is given by

$$\text{COP} = \frac{\dot{Q}_C}{\dot{W}_{c_I} + \dot{W}_{c_{II}}}$$

$$= \frac{(\widehat{H}_1 - \widehat{H}_4)(\widehat{H}_5 - \widehat{H}_8)}{(\widehat{H}_2 - \widehat{H}_3)(\widehat{H}_6 - \widehat{H}_5) + (\widehat{H}_2 - \widehat{H}_1)(\widehat{H}_5 - \widehat{H}_8)} \qquad (6.3\text{-}10)$$

Example 6.10 *Devise a two-stage cascade refrigeration system to transfer heat from $-60\,°C$ to $30\,°C$ using the following refrigerants:*

Refrigerant	NBP[17]($°C$)	T_c ($°C$)	P_c (MPa)	ODP
Ethane (R-170)	-88.80	32.20	4.89	0
Propane (R-290)	-42.09	96.70	4.25	0

Solution

Consider the two-stage cascade refrigeration system shown in Figure 6.17. Since heat must be removed from a low temperature heat source at $-60\,°C$, the boiling point of refrigerant for Cycle I should be less than $-60\,°C$. Therefore, R-170 is an appropriate refrigerant for Cycle I. Since $T_c = 32.2\,°C$ for R-170, condensation temperature of R-170 in the cascade heat exchanger must be less than $32.2\,°C$. Let us adjust the pressure so that the condensation of R-170 takes place at $-20\,°C$. If the temperature difference between the flowing streams in the cascade heat exchanger is taken as $10\,°C$, then the boiling point of refrigerant for Cycle II should be less than $-30\,°C$. Since heat must be rejected to the region at $30\,°C$, the temperature in the condenser of Cycle II should be greater than $30\,°C$. Therefore, the refrigerant for Cycle II must have a critical temperature greater than $30\,°C$ and a boiling point temperature less than $-30\,°C$. Hence, R-290 is a suitable refrigerant for Cycle II.

---

[17]Normal **B**oiling **P**oint, i.e., boiling point temperature under atmospheric pressure.

Problems

Problems related to Section 6.1

6.1   In 2012, the United States generated about 4054 billion kW h of electricity. About 68% of the electricity generated was from fossil fuel (coal, natural gas, and petroleum), with 37% from coal[18]. The amount of coal consumption was estimated as 742 billion kg. Assuming that each kg of coal produces 22,500 kJ of heat at a temperature of 1600 K and the power plant rejects heat to the atmosphere at 20 °C, calculate the maximum amount of electricity that could have been generated from the coal.

(**Answer:** 3789 billion kW h)

6.2   An ideal Rankine cycle operates between the pressure limits of 15 MPa in the boiler and 50 kPa in the condenser. The boiler generates steam at a mass flow rate of 50 kg/s and it enters the turbine at a temperature of 1100 °C.

a) Calculate the net power output of the cycle,
b) Calculate the thermal efficiency of the cycle.

(**Answer:** a) 105 MW, b) 0.47)

6.3   Steam enters the turbine of an ideal Rankine cycle at 8 MPa and 550 °C, and expands adiabatically to 10 kPa. The net power output of the cycle is 18 MW.

a) Calculate the thermal efficiency of the cycle,
b) Determine the mass flow rate of steam.

(**Answer:** a) 0.402, b) 13.5 kg/s)

6.4   Steam enters the turbine of an ideal Rankine cycle at 8 MPa and expands adiabatically to 10 kPa. If the moisture content of the steam at the turbine exit is not to exceed 0.1, calculate the thermal efficiency of the cycle.

(**Answer:** 0.41)

6.5   Superheated steam at 5 MPa and 700 °C leaves the boiler of a Rankine cycle. The pressure at the exit of the turbine is 100 kPa. The mass flow rate of steam is 80 kg/s and the power output of the turbine is 90 MW. Calculate the isentropic efficiency of the turbine.

(**Answer:** 0.97)

---

[18]U.S. Energy Information Administration (www.eia.gov).

6.6 The tubes of a steam generator in an ideal Rankine cycle withstand a maximum pressure of 12.5 MPa. The condenser is designed for a 10 °C temperature difference and cooling water is available at 35 °C. If the steam exiting the turbine is saturated vapor, calculate the thermal efficiency of the cycle.

(**Answer:** 0.52)

6.7 In a steam power plant, 65 kg/s of liquid water at 6 MPa is heated to 500 °C in a boiler with a heat input rate of 200 MW. The pressure at the exit of the turbine is 100 kPa. The isentropic efficiency of the turbine is 0.8.

a) Calculate the power output of the turbine,
b) If the work produced by the pump is 9 kJ/kg, calculate the thermal efficiency of the power plant.

(**Answer:** a) 48.1 MW, b) 0.24)

6.8 Consider the steam power plant shown in the figure below.

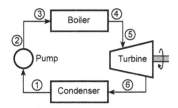

The following operating data are recorded in the plant:

State	T (°C)	P (kPa)
1	35	7.5
2		5500
3	35	5200
4	425	4250
5	400	4000
6		7.5

The turbine and pump have isentropic efficiencies of 0.9 and 0.8, respectively. They both operate adiabatically. Calculate the thermal efficiency of the cycle.

(**Answer:** 0.32)

6.9   One of the most common modifications of a Rankine cycle is the *reheat cycle* as shown in the figure below. In this cycle, steam coming out of a boiler is first expanded in the high-pressure (HP) turbine. Steam exiting the HP turbine is reheated in the boiler before passing through the low-pressure (LP) turbine. In this way, not only is the moisture content of the steam leaving the turbine reduced but also the cycle efficiency is increased by raising the average temperature during heat addition.

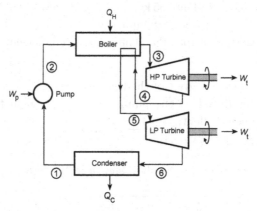

In an ideal Rankine cycle with reheat, steam leaves the boiler and enters the turbine at 4 MPa and 450 °C. After expansion in the high-pressure turbine to 400 kPa, the steam is reheated to 450 °C and then expanded in the low-pressure turbine to 7.5 kPa. Calculate the thermal efficiency of the cycle.

(**Answer:** 0.38)

Problems related to Section 6.2

6.10   The maximum and minimum temperatures in an ideal air-standard Brayton cycle are 310 K and 900 K, respectively. The net power output of the cycle is 14 MW and the compression ratio is 8.

a) Calculate the mass flow rate of air through the cycle,
b) Calculate the thermal efficiency of the cycle.

(**Answer:** a) 92 kg/s, b) 0.45)

6.11   An ideal air-standard Brayton cycle produces 1000 kW of power by compressing air at 100 kPa and 20 °C to 900 kPa. If the maximum temperature is 840 °C, calculate the molar flow rate of air.

(**Answer:** 0.114 kmol/s)

6.12 The compression ratio in an air-standard Otto cycle is 9. The pressure and temperature at the beginning of the compression stroke are 1 bar and 300 K, respectively. The heat supplied during the constant-volume process is 24 MJ/ kmol.

a) Calculate the net work output of the cycle,
b) Calculate the thermal efficiency of the cycle.

(**Answer:** a) 14,034 kJ/ kmol, b) 0.58)

6.13 The conditions at the beginning of the compression stroke in an air-standard Diesel cycle are 1 bar and 310 K. The compression ratio is 16 and the cutoff ratio is 2.

a) Calculate the maximum temperature in the cycle,
b) Calculate the pressure after the isentropic expansion,
c) Calculate the net work per cycle and thermal efficiency.

(**Answer:** a) 1880 K, b) 2.635 bar, c) 16,794 J/ mol, 0.61)

6.14 The Ericsson cycle employs air as a working fluid and consists of two isothermal and two isobaric processes as shown in the figure below. Show that the thermal efficiency of this cycle is given by

$$\eta = 1 - \frac{T_C}{T_H}$$

where T_C and T_H represent compressor and turbine temperatures, respectively.

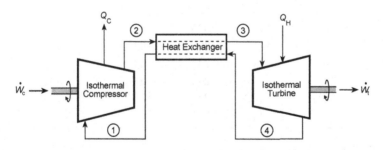

Problems related to Section 6.3

6.15 An ideal vapor-compression refrigeration cycle with a capacity of 3 tons has R-134a entering the compressor as saturated vapor at $-20\,°C$ and exiting at 1 MPa.

a) Determine the coefficient of performance of the cycle,

b) Determine the power input to the compressor.

(**Answer:** a) 3.1, b) 3.4 kW)

6.16 A vapor-compression refrigeration cycle using R-134a has evaporator and condenser pressures of 201.7 kPa and 1 MPa, respectively. The isentropic efficiency of the compressor is 0.85. Calculate the coefficient of performance.

(**Answer:** 3.45)

6.17 Derive Eq. (6.3-10).

6.18 The Linde[19] process is used to liquefy gases. It is similar to a vapor-compression refrigeration cycle. As shown in the figure below, it includes a compressor, an after cooler, a heat exchanger, a throttling valve, and a flash drum (separator). Consider that the cycle is used to produce liquefied propane at 172 kPa and 244 K from a feed of propane gas at 172 kPa and 322 K. The isentropic efficiency of the compressor is 0.8. Cooling water is assumed to be available at 10.13 kPa and 305 K. The power input to the process is 16 kW. All heat exchangers are assumed to have a minimum ΔT of 5 K. Leipziger and Huang (1984) provided the enthalpy and entropy values of the streams as follows:

Stream #	P (kPa)	T (K)	$\widehat{H}$ (J/g)	$\widehat{S}$ (J/g.K)
1	172	322	-1478.7	6.1358
2	172	301	-1515.8	6.0153
3	1137	383	-1377.8	6.0902
4	1137	305	-1544.2	5.5974
5	1137	305	-1629.4	5.3185
6	172	244	-1629.4	5.5853
7	172	244	-2017.7	3.9938
8	172	244	-1608.5	5.6711
9	172	300	-1517.9	6.0086

a) Calculate the mass flow rate of propane passing through the compressor.

b) Calculate the rate of heat transfer to the cooling water in the after cooler.

c) Calculate the production rate of liquefied propane by carrying out an energy balance around the dotted system shown in the figure below.

[19] Carl Paul Gottfried von Linde (1842-1934), German engineer.

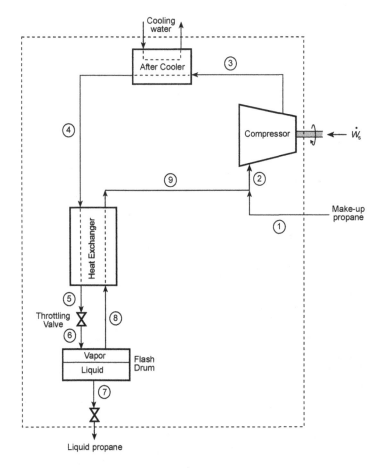

(**Answer:** a) 115.9 g/s, b) 19,286 W, c) 6.1 g/s)

References

Chen, H., D.Y. Goswami, and E.K. Stefanakos, 2010, *Renewable and Sustainable Energy Reviews*, **14**, 3059-3067.

Leipziger, S. and C.C. Huang, 1984, *Can. J. Chem. Eng.*, **62** (4), 278-283.

Lior, N., 2007, *Energy*, **32**, 254-260.

Appendix A

Steam Tables

Table A.1 - Saturated Steam: Temperature Table

Table A.2 - Saturated Steam: Pressure Table

Table A.3 - Superheated Vapor

Table A.4 - Compressed Liquid

Source: G.J. Van Wylen and R.E. Sonntag, *Fundamentals of Classical Thermodynamics, S.I. Version,* 2nd Ed., John Wiley & Sons, Inc., New York, 1978. Used with permission.

Table A.1 Saturated Steam: Temperature Table

T (°C)	P (kPa)	Specific Volume (m³/kg)		Internal Energy (kJ/kg)			Enthalpy (kJ/kg)			Entropy (kJ/kg·K)		
		$\hat{V}^L$	$\hat{V}^V$	$\hat{U}^L$	$\Delta\hat{U}$	$\hat{U}^V$	$\hat{H}^L$	$\Delta\hat{H}$	$\hat{H}^V$	$\hat{S}^L$	$\Delta\hat{S}$	$\hat{S}^V$
0.01	0.6113	0.001000	206.14	0.00	2375.3	2375.3	0.01	2501.3	2501.4	0.0000	9.1562	9.1562
5	0.8721	0.001000	147.12	20.97	2361.3	2382.3	20.98	2489.6	2510.6	0.0761	8.9496	9.0257
10	1.2276	0.001000	106.38	42.00	2347.2	2389.2	42.01	2477.7	2519.8	0.1510	8.7498	8.9008
15	1.7051	0.001001	77.93	62.99	2333.1	2396.1	62.99	2465.9	2528.9	0.2245	8.5569	8.7814
20	2.339	0.001002	57.79	83.95	2319.0	2402.9	83.96	2454.1	2538.1	0.2966	8.3706	8.6672
25	3.169	0.001003	43.36	104.88	2304.9	2409.8	104.89	2442.3	2547.2	0.3674	8.1905	8.5580
30	4.246	0.001004	32.89	125.78	2290.8	2416.6	125.79	2430.5	2556.3	0.4369	8.0164	8.4533
35	5.628	0.001006	25.22	146.67	2276.7	2423.4	146.68	2418.6	2565.3	0.5053	7.8478	8.3531
40	7.384	0.001008	19.52	167.56	2262.6	2430.1	167.57	2406.7	2574.3	0.5725	7.6845	8.2570
45	9.593	0.001010	15.26	188.44	2248.4	2436.8	188.45	2394.8	2583.2	0.6387	7.5261	8.1648
50	12.349	0.001012	12.03	209.32	2234.2	2443.5	209.33	2382.7	2592.1	0.7038	7.3725	8.0763
55	15.758	0.001015	9.568	230.21	2219.9	2450.1	230.23	2370.7	2600.9	0.7679	7.2234	7.9913
60	19.940	0.001017	7.671	251.11	2205.5	2456.6	251.13	2358.5	2609.6	0.8312	7.0784	7.9096
65	25.03	0.001020	6.197	272.02	2191.1	2463.1	272.06	2346.2	2618.3	0.8935	6.9375	7.8310
70	31.19	0.001023	5.042	292.95	2176.6	2469.6	292.98	2333.8	2626.8	0.9549	6.8004	7.7553
75	38.58	0.001026	4.131	313.90	2162.0	2475.9	313.93	2321.4	2635.3	1.0155	6.6669	7.6824
80	47.39	0.001029	3.407	334.86	2147.4	2482.2	334.91	2308.8	2643.7	1.0753	6.5369	7.6122
85	57.83	0.001033	2.828	355.84	2132.6	2488.4	355.90	2296.0	2651.9	1.1343	6.4102	7.5445
90	70.14	0.001036	2.361	376.85	2117.7	2494.5	376.92	2283.2	2660.1	1.1925	6.2866	7.4791
95	84.55	0.001040	1.982	397.88	2102.7	2500.6	397.96	2270.2	2668.1	1.2500	6.1659	7.4159

Table A.1 (continued)   Saturated Steam: Temperature Table

T (°C)	P (MPa)	Specific Volume (m³/kg)		Internal Energy (kJ/kg)			Enthalpy (kJ/kg)			Entropy (kJ/kg·K)		
		$\hat{V}^L$	$\hat{V}^V$	$\hat{U}^L$	$\Delta\hat{U}$	$\hat{U}^V$	$\hat{H}^L$	$\Delta\hat{H}$	$\hat{H}^V$	$\hat{S}^L$	$\Delta\hat{S}$	$\hat{S}^V$
100	0.10135	0.001044	1.6729	418.94	2087.6	2506.5	419.04	2257.0	2676.1	1.3069	6.0480	7.3549
105	0.12082	0.001048	1.4194	440.02	2072.3	2512.4	440.15	2243.7	2683.8	1.3630	5.9328	7.2958
110	0.14327	0.001052	1.2102	461.14	2057.0	2518.1	461.30	2230.2	2691.5	1.4185	5.8202	7.2387
115	0.16906	0.001056	1.0366	482.30	2041.4	2523.7	482.48	2216.5	2699.0	1.4734	5.7100	7.1833
120	0.19853	0.001060	0.8919	503.50	2025.8	2529.3	503.71	2202.6	2706.3	1.5276	5.6020	7.1296
125	0.2321	0.001065	0.7706	524.74	2009.9	2534.6	524.99	2188.5	2713.5	1.5813	5.4962	7.0775
130	0.2701	0.001070	0.6685	546.02	1993.9	2539.9	546.31	2174.2	2720.5	1.6344	5.3925	7.0269
135	0.3130	0.001075	0.5822	567.35	1977.7	2545.0	567.69	2159.6	2727.3	1.6870	5.2907	6.9777
140	0.3613	0.001080	0.5089	588.74	1961.3	2550.0	589.13	2144.7	2733.9	1.7391	5.1908	6.9299
145	0.4154	0.001085	0.4463	610.18	1944.7	2554.9	610.63	2129.6	2740.3	1.7907	5.0926	6.8833
150	0.4758	0.001091	0.3928	631.68	1927.9	2559.5	632.20	2114.3	2746.5	1.8418	4.9960	6.8379
155	0.5431	0.001096	0.3468	653.24	1910.8	2564.1	653.84	2098.6	2752.4	1.8925	4.9010	6.7935
160	0.6178	0.001102	0.3071	674.87	1893.5	2568.4	675.55	2082.6	2758.1	1.9427	4.8075	6.7502
165	0.7005	0.001108	0.2727	696.56	1876.0	2572.5	697.34	2066.2	2763.5	1.9925	4.7153	6.7078
170	0.7917	0.001114	0.2428	718.33	1858.1	2576.5	719.21	2049.5	2768.7	2.0419	4.6244	6.6663
175	0.8920	0.001121	0.2168	740.17	1840.0	2580.2	741.17	2032.4	2773.6	2.0909	4.5347	6.6256
180	1.0021	0.001127	0.19405	762.09	1821.6	2583.7	763.22	2015.0	2778.2	2.1396	4.4461	6.5857
185	1.1227	0.001134	0.17409	784.10	1802.9	2587.0	785.37	1997.1	2782.4	2.1879	4.3586	6.5465
190	1.2544	0.001141	0.15654	806.19	1783.8	2590.0	807.62	1978.8	2786.4	2.2359	4.2720	6.5079
195	1.3978	0.001149	0.14105	828.37	1764.4	2592.8	829.98	1960.0	2790.0	2.2835	4.1863	6.4698
200	1.5538	0.001157	0.12736	850.65	1744.7	2595.3	852.45	1940.7	2793.2	2.3309	4.1014	6.4323
205	1.7230	0.001164	0.11521	873.04	1724.5	2597.5	875.04	1921.0	2796.0	2.3780	4.0172	6.3952
210	1.9062	0.001173	0.10441	895.53	1703.9	2599.5	897.76	1900.7	2798.5	2.4248	3.9337	6.3585

Table A.1 (continued)   Saturated Steam: Temperature Table

T (°C)	P (MPa)	Specific Volume (m³/kg)		Internal Energy (kJ/kg)			Enthalpy (kJ/kg)			Entropy (kJ/kg·K)		
		$\hat{V}^L$	$\hat{V}^V$	$\hat{U}^L$	$\Delta\hat{U}$	$\hat{U}^V$	$\hat{H}^L$	$\Delta\hat{H}$	$\hat{H}^V$	$\hat{S}^L$	$\Delta\hat{S}$	$\hat{S}^V$
215	2.104	0.001181	0.09479	918.14	1682.9	2601.1	920.62	1879.9	2800.5	2.4714	3.8507	6.3221
220	2.318	0.001190	0.08619	940.87	1661.5	2602.4	943.62	1858.5	2802.1	2.5178	3.7683	6.2861
225	2.548	0.001199	0.07849	963.73	1639.6	2603.3	966.78	1836.5	2803.3	2.5639	3.6863	6.2503
230	2.795	0.001209	0.07158	986.74	1617.2	2603.9	990.12	1813.8	2804.0	2.6099	3.6047	6.2146
235	3.060	0.001219	0.06537	1009.89	1594.2	2604.1	1013.62	1790.5	2804.2	2.6558	3.5233	6.1791
240	3.344	0.001229	0.05976	1033.21	1570.8	2604.0	1037.32	1766.5	2803.8	2.7015	3.4422	6.1437
245	3.648	0.001240	0.05471	1056.71	1546.7	2603.4	1061.23	1741.7	2803.0	2.7472	3.3612	6.1083
250	3.973	0.001251	0.05013	1080.39	1522.0	2602.4	1085.36	1716.2	2801.5	2.7927	3.2802	6.0730
255	4.319	0.001263	0.04598	1104.28	1496.7	2600.9	1109.73	1689.8	2799.5	2.8383	3.1992	6.0375
260	4.688	0.001276	0.04221	1128.39	1470.6	2599.0	1134.37	1662.5	2796.9	2.8838	3.1181	6.0019
265	5.081	0.001289	0.03877	1152.74	1443.9	2596.6	1159.28	1634.4	2793.6	2.9294	3.0368	5.9662
270	5.499	0.001302	0.03564	1177.36	1416.3	2593.7	1184.51	1605.2	2789.7	2.9751	2.9551	5.9301
280	6.412	0.001332	0.03017	1227.46	1358.7	2586.1	1235.99	1543.6	2779.6	3.0668	2.7903	5.8571
290	7.436	0.001366	0.02557	1278.92	1297.1	2576.0	1289.07	1477.1	2766.2	3.1594	2.6227	5.7821
300	8.581	0.001404	0.02167	1332.0	1231.0	2563.0	1344.0	1404.9	2749.0	3.2534	2.4511	5.7045
310	9.856	0.001447	0.018350	1387.1	1159.4	2546.4	1401.3	1326.0	2727.3	3.3493	2.2737	5.6230
320	11.274	0.001499	0.015488	1444.6	1080.9	2525.5	1461.5	1238.6	2700.1	3.4480	2.0882	5.5362
330	12.845	0.001561	0.012996	1505.3	993.7	2498.9	1525.3	1140.6	2665.9	3.5507	1.8909	5.4417
340	14.586	0.001638	0.010797	1570.3	894.3	2464.6	1594.2	1027.9	2622.0	3.6594	1.6763	5.3357
350	16.513	0.001740	0.008813	1641.9	776.6	2418.4	1670.6	893.4	2563.9	3.7777	1.4335	5.2112
360	18.651	0.001893	0.006945	1725.2	626.3	2351.5	1760.5	720.5	2481.0	3.9147	1.1379	5.0526
370	21.03	0.002213	0.004925	1844.0	384.5	2228.5	1890.5	441.6	2332.1	4.1106	0.6865	4.7971
374.14	22.09	0.003155	0.003155	2029.6	0	2029.6	2099.3	0	2099.3	4.4298	0	4.4298

Table A.2 Saturated Steam: Pressure Table

P (kPa)	T (°C)	Specific Volume (m³/kg)		Internal Energy (kJ/kg)			Enthalpy (kJ/kg)			Entropy (kJ/kg·K)		
		$\hat{V}^L$	$\hat{V}^V$	$\hat{U}^L$	$\Delta\hat{U}$	$\hat{U}^V$	$\hat{H}^L$	$\Delta\hat{H}$	$\hat{H}^V$	$\hat{S}^L$	$\Delta\hat{S}$	$\hat{S}^V$
0.6113	0.01	0.001000	206.14	0.00	2375.3	2375.3	0.01	2501.3	2501.4	0.0000	9.1562	9.1562
1.0	6.98	0.001000	129.21	29.30	2355.7	2385.0	29.30	2484.9	2514.2	0.1059	8.8697	8.9756
1.5	13.03	0.001001	87.98	54.71	2338.6	2393.3	54.71	2470.6	2525.3	0.1957	8.6322	8.8279
2.0	17.50	0.001001	67.00	73.48	2326.0	2399.5	73.48	2460.0	2533.5	0.2607	8.4629	8.7237
2.5	21.08	0.001002	54.25	88.48	2315.9	2404.4	88.49	2451.6	2540.0	0.3120	8.3311	8.6432
3.0	24.08	0.001003	45.67	101.04	2307.5	2408.5	101.05	2444.5	2545.5	0.3545	8.2231	8.5776
4.0	28.96	0.001004	34.80	121.45	2293.7	2415.2	121.46	2432.9	2554.4	0.4226	8.0520	8.4746
5.0	32.88	0.001005	28.19	137.81	2282.7	2420.5	137.82	2423.7	2561.5	0.4764	7.9187	8.3951
7.5	40.29	0.001008	19.24	168.78	2261.7	2430.5	168.79	2406.0	2574.8	0.5764	7.6750	8.2515
10	45.81	0.001010	14.67	191.82	2246.1	2437.9	191.83	2392.8	2584.7	0.6493	7.5009	8.1502
15	53.97	0.001014	10.02	225.92	2222.8	2448.7	225.94	2373.1	2599.1	0.7549	7.2536	8.0085
20	60.06	0.001017	7.649	251.38	2205.4	2456.7	251.40	2358.3	2609.7	0.8320	7.0766	7.9085
25	64.97	0.001020	6.204	271.90	2191.2	2463.1	271.93	2346.3	2618.2	0.8931	6.9383	7.8314
30	69.10	0.001022	5.229	289.20	2179.2	2468.4	289.23	2336.1	2625.3	0.9439	6.8247	7.7686
40	75.87	0.001027	3.993	317.53	2159.5	2477.0	317.58	2319.2	2636.8	1.0259	6.6441	7.6700
50	81.33	0.001030	3.240	340.44	2143.4	2483.9	340.49	2305.4	2645.9	1.0910	6.5029	7.5939
75	91.78	0.001037	2.217	384.31	2112.4	2496.7	384.39	2278.6	2663.0	1.2130	6.2434	7.4564
(MPa)												
0.100	99.63	0.001043	1.6940	417.36	2088.7	2506.1	417.46	2258.0	2675.5	1.3026	6.0568	7.3594
0.125	105.99	0.001048	1.3749	444.19	2069.3	2513.5	444.32	2241.0	2685.4	1.3740	5.9104	7.2844
0.150	111.37	0.001053	1.1593	466.94	2052.7	2519.7	467.11	2226.5	2693.6	1.4336	5.7897	7.2233

Table A.2 (continued)   **Saturated Steam: Pressure Table**

P (MPa)	T (°C)	Specific Volume (m³/kg)		Internal Energy (kJ/kg)			Enthalpy (kJ/kg)			Entropy (kJ/kg·K)		
		$\hat{V}^L$	$\hat{V}^V$	$\hat{U}^L$	$\Delta\hat{U}$	$\hat{U}^V$	$\hat{H}^L$	$\Delta\hat{H}$	$\hat{H}^V$	$\hat{S}^L$	$\Delta\hat{S}$	$\hat{S}^V$
0.175	116.06	0.001057	1.0036	486.80	2038.1	2524.9	486.99	2213.6	2700.6	1.4849	5.6868	7.1717
0.200	120.23	0.001061	0.8857	504.49	2025.0	2529.5	504.70	2201.9	2706.7	1.5301	5.5970	7.1271
0.225	124.00	0.001064	0.7933	520.47	2013.1	2533.6	520.72	2191.3	2712.1	1.5706	5.5173	7.0878
0.250	127.44	0.001067	0.7187	535.10	2002.1	2537.2	535.37	2181.5	2716.9	1.6072	5.4455	7.0527
0.275	130.60	0.001070	0.6573	548.59	1991.9	2540.5	548.89	2172.4	2721.3	1.6408	5.3801	7.0209
0.300	133.55	0.001073	0.6058	561.15	1982.4	2543.6	561.47	2163.8	2725.3	1.6718	5.3201	6.9919
0.325	136.30	0.001076	0.5620	572.90	1973.5	2546.4	573.25	2155.8	2729.0	1.7006	5.2646	6.9652
0.350	138.88	0.001079	0.5243	583.95	1965.0	2548.9	584.33	2148.1	2732.4	1.7275	5.2130	6.9405
0.375	141.32	0.001081	0.4914	594.40	1956.9	2551.3	594.81	2140.8	2735.6	1.7528	5.1647	6.9175
0.40	143.63	0.001084	0.4625	604.31	1949.3	2553.6	604.74	2133.8	2738.6	1.7766	5.1193	6.8959
0.45	147.93	0.001088	0.4140	622.77	1934.9	2557.6	623.25	2120.7	2743.9	1.8207	5.0359	6.8565
0.50	151.86	0.001093	0.3749	639.68	1921.6	2561.2	640.23	2108.5	2748.7	1.8607	4.9606	6.8213
0.60	158.85	0.001101	0.3157	669.90	1897.5	2567.4	670.56	2086.3	2756.8	1.9312	4.8288	6.7600
0.7	164.97	0.001108	0.2729	696.44	1876.1	2572.5	697.22	2066.3	2763.5	1.9922	4.7158	6.7080
0.8	170.43	0.001115	0.2404	720.22	1856.6	2576.8	721.11	2048.0	2769.1	2.0462	4.6166	6.6628
0.9	175.38	0.001121	0.2150	741.83	1838.6	2580.5	742.83	2031.1	2773.9	2.0946	4.5280	6.6226
1.00	179.91	0.001127	0.19444	761.68	1822.0	2583.6	762.81	2015.3	2778.1	2.1387	4.4478	6.5865
1.10	184.09	0.001133	0.17753	780.09	1806.3	2586.4	781.34	2000.4	2781.7	2.1792	4.3744	6.5536
1.20	187.99	0.001139	0.16333	797.29	1791.5	2588.8	798.65	1986.2	2784.8	2.2166	4.3067	6.5233
1.30	191.64	0.001144	0.15125	813.44	1777.5	2591.0	814.93	1972.7	2787.6	2.2515	4.2438	6.4953
1.40	195.07	0.001149	0.14084	828.70	1764.1	2592.8	830.30	1959.7	2790.6	2.2842	4.1850	6.4693
1.50	198.32	0.001154	0.13177	843.16	1751.3	2594.5	844.89	1947.3	2792.2	2.3150	4.1298	6.4448
2.00	212.42	0.001177	0.09963	906.44	1693.8	2600.3	908.79	1890.7	2799.5	2.4474	3.8935	6.3409

Table A.2 (continued)   Saturated Steam: Pressure Table

P (MPa)	T (°C)	Specific Volume (m³/kg)		Internal Energy (kJ/kg)			Enthalpy (kJ/kg)			Entropy (kJ/kg·K)		
		$\hat{V}^L$	$\hat{V}^V$	$\hat{U}^L$	$\Delta\hat{U}$	$\hat{U}^V$	$\hat{H}^L$	$\Delta\hat{H}$	$\hat{H}^V$	$\hat{S}^L$	$\Delta\hat{S}$	$\hat{S}^V$
2.5	223.99	0.001197	0.07998	959.11	1644.0	2603.1	962.11	1841.0	2803.1	2.5547	3.7028	6.2575
3.00	233.90	0.001217	0.06668	1004.78	1599.3	2604.1	1008.42	1795.7	2804.2	2.6457	3.5412	6.1869
3.5	242.60	0.001235	0.05707	1045.43	1558.3	2603.7	1049.75	1753.7	2803.4	2.7253	3.4000	6.1253
4	250.40	0.001252	0.04978	1082.31	1520.0	2602.3	1087.31	1714.1	2801.4	2.7964	3.2737	6.0701
5	263.99	0.001286	0.03944	1147.81	1449.3	2597.1	1154.23	1640.1	2794.3	2.9202	3.0532	5.9734
6	275.64	0.001319	0.03244	1205.44	1384.3	2589.7	1213.35	1571.0	2784.3	3.0267	2.8625	5.8892
7	285.88	0.001351	0.02737	1257.55	1323.0	2580.5	1267.00	1505.1	2772.1	3.1211	2.6922	5.8133
8	295.06	0.001384	0.02352	1305.57	1264.2	2569.8	1316.64	1441.3	2758.0	3.2068	2.5364	5.7432
9	303.40	0.001418	0.02048	1350.51	1207.3	2557.8	1363.26	1378.9	2742.1	3.2858	2.3915	5.6772
10	311.06	0.001452	0.018026	1393.04	1151.4	2544.4	1407.56	1317.1	2724.7	3.3596	2.2544	5.6141
11	318.15	0.001489	0.015987	1433.7	1096.0	2529.8	1450.1	1255.5	2705.6	3.4295	2.1233	5.5527
12	324.75	0.001527	0.014263	1473.0	1040.7	2513.7	1491.3	1193.6	2684.9	3.4962	1.9962	5.4924
13	330.93	0.001567	0.012780	1511.1	985.0	2496.1	1531.5	1130.7	2662.2	3.5606	1.8718	5.4323
14	336.75	0.001611	0.011485	1548.6	928.2	2476.8	1571.1	1066.5	2637.6	3.6232	1.7485	5.3717
15	342.24	0.001658	0.010337	1585.6	869.8	2455.5	1610.5	1000.0	2610.5	3.6848	1.6249	5.3098
16	347.44	0.001711	0.009306	1622.7	809.0	2431.7	1650.1	930.6	2580.6	3.7461	1.4994	5.2455
17	352.37	0.001770	0.008364	1660.2	744.8	2405.0	1690.3	856.9	2547.2	3.8079	1.3698	5.1777
18	357.06	0.001840	0.007489	1698.9	675.4	2374.3	1732.0	777.1	2509.1	3.8715	1.2329	5.1044
19	361.54	0.001924	0.006657	1739.9	598.1	2338.1	1776.5	688.0	2464.5	3.9388	1.0839	5.0228
20	365.81	0.002036	0.005834	1785.6	507.5	2293.0	1826.3	583.4	2409.7	4.0139	0.9130	4.9269
21	369.89	0.002207	0.004952	1842.1	388.5	2230.6	1888.4	446.2	2334.6	4.1075	0.6938	4.8013
22	373.80	0.002742	0.003568	1961.9	125.2	2087.1	2022.2	143.4	2165.6	4.3110	0.2216	4.5327
22.09	374.14	0.003155	0.003155	2029.6	0.0	2029.6	2099.3	0.0	2099.3	4.4298	0.0	4.4298

Table A.3 Superheated Vapor[†]

T	$P = 0.010\,\text{MPa}\,(45.81)$				$P = 0.050\,\text{MPa}\,(81.33)$				$P = 0.10\,\text{MPa}\,(99.63)$			
(°C)	$\hat{V}$	$\hat{U}$	$\hat{H}$	$\hat{S}$	$\hat{V}$	$\hat{U}$	$\hat{H}$	$\hat{S}$	$\hat{V}$	$\hat{U}$	$\hat{H}$	$\hat{S}$
Sat.	14.674	2437.9	2584.7	8.1502	3.240	2483.9	2645.9	7.5939	1.6940	2506.1	2675.5	7.3594
50	14.869	2443.9	2592.6	8.1749	–	–	–	–	–	–	–	–
100	17.196	2515.5	2687.5	8.4479	3.418	2511.6	2682.5	7.6947	1.6958	2506.7	2676.2	7.3614
150	19.512	2587.9	2783.0	8.6882	3.889	2585.6	2780.1	7.9401	1.9364	2582.8	2776.4	7.6134
200	21.825	2661.3	2879.5	8.9038	4.356	2659.9	2877.7	8.1580	2.172	2658.1	2875.3	7.8343
250	24.136	2736.0	2977.3	9.1002	4.820	2735.0	2976.0	8.3556	2.406	2733.7	2974.3	8.0333
300	26.445	2812.1	3076.5	9.2813	5.284	2811.3	3075.5	8.5373	2.639	2810.4	3074.3	8.2158
400	31.063	2968.9	3279.6	9.6077	6.209	2968.5	3278.9	8.8642	3.103	2967.9	3278.2	8.5435
500	35.679	3132.3	3489.1	9.8978	7.134	3132.0	3488.7	9.1546	3.565	3131.6	3488.1	8.8342
600	40.295	3302.5	3705.4	10.1608	8.057	3302.2	3705.1	9.4178	4.028	3301.9	3704.7	9.0976
700	44.911	3479.6	3928.7	10.4028	8.981	3479.4	3928.5	9.6599	4.490	3479.2	3928.2	9.3398
800	49.526	3663.8	4159.0	10.6281	9.904	3663.6	4158.9	9.8852	4.952	3663.5	4158.6	9.5652
900	54.141	3855.0	4396.4	10.8396	10.828	3854.9	4396.3	10.0967	5.414	3854.8	4396.1	9.7767
1000	58.757	4053.0	4640.6	11.0393	11.751	4052.9	4640.5	10.2964	5.875	4052.8	4640.3	9.9764
1100	63.372	4257.5	4891.2	11.2287	12.674	4257.4	4891.1	10.4859	6.337	4257.3	4891.0	10.1659
1200	67.987	4467.9	5147.8	11.4091	13.597	4467.8	5147.7	10.6662	6.799	4467.7	5147.6	10.3463
1300	72.602	4683.7	5409.7	11.5811	14.521	4683.6	5409.6	10.8382	7.260	4683.5	5409.5	10.5183

[†]Note: Number in parantheses is temperature of saturated steam in °C at the specified pressure.
$\hat{V}\,[=]\,\text{m}^3/\text{kg}$; $\hat{U}\,\&\,\hat{H}\,[=]\,\text{J}/\text{g} = \text{kJ}/\text{kg}$; $\hat{S}\,[=]\,\text{kJ}/\text{kg.K}$

Table A.3 (continued) Superheated Vapor[†]

T (°C)	P = 0.20 MPa (120.23)				P = 0.30 MPa (133.55)				P = 0.40 MPa (143.63)			
	$\hat{V}$	$\hat{U}$	$\hat{H}$	$\hat{S}$	$\hat{V}$	$\hat{U}$	$\hat{H}$	$\hat{S}$	$\hat{V}$	$\hat{U}$	$\hat{H}$	$\hat{S}$
Sat.	0.8857	2529.5	2706.7	7.1272	0.6058	2543.6	2725.3	6.9919	0.4625	2553.6	2738.6	6.8959
150	0.9596	2576.9	2768.8	7.2795	0.6339	2570.8	2761.0	7.0778	0.4708	2564.5	2752.8	6.9299
200	1.0803	2654.4	2870.5	7.5066	0.7163	2650.7	2865.6	7.3115	0.5342	2646.8	2860.5	7.1706
250	1.1988	2731.2	2971.0	7.7086	0.7964	2728.7	2967.6	7.5166	0.5951	2726.1	2964.2	7.3789
300	1.3162	2808.6	3071.8	7.8926	0.8753	2806.7	3069.3	7.7022	0.6548	2804.8	3066.8	7.5662
400	1.5493	2966.7	3276.6	8.2218	1.0315	2965.6	3275.0	8.0330	0.7726	2964.4	3273.4	7.8985
500	1.7814	3130.8	3487.1	8.5133	1.1867	3130.0	3486.0	8.3251	0.8893	3129.2	3484.9	8.1913
600	2.013	3301.4	3704.0	8.7770	1.3414	3300.8	3703.2	8.5892	1.0055	3300.2	3702.4	8.4558
700	2.244	3478.8	3927.6	9.0194	1.4957	3478.1	3927.1	8.8319	1.1215	3477.9	3926.5	8.6987
800	2.475	3663.1	4158.2	9.2449	1.6499	3662.9	4157.8	9.0576	1.2372	3662.4	4157.3	8.9244
900	2.706	3854.5	4395.8	9.4566	1.8041	3854.2	4395.4	9.2692	1.3529	3853.9	4395.1	9.1362
1000	2.937	4052.5	4640.0	9.6563	1.9581	4052.3	4639.7	9.4690	1.4685	4052.0	4639.4	9.3360
1100	3.168	4257.0	4890.7	9.8458	2.1121	4256.8	4890.4	9.6585	1.5840	4256.5	4890.2	9.5256
1200	3.399	4467.5	5147.3	10.0262	2.2661	4467.2	5147.1	9.8389	1.6996	4467.0	5146.8	9.7060
1300	3.630	4683.2	5409.3	10.1982	2.4201	4683.0	5409.0	10.0110	1.8151	4682.8	5408.8	9.8780

[†]**Note:** Number in parantheses is temperature of saturated steam in °C at the specified pressure.
$\hat{V}\,[=]\,m^3/kg;\quad \hat{U}\ \&\ \hat{H}\,[=]\,J/g = kJ/kg;\quad \hat{S}\,[=]\,kJ/kg\cdot K$

Table A.3 (continued)   **Superheated Vapor**[†]

T (°C)	$P = 0.50\,\mathrm{MPa}\,(151.86)$				$P = 0.60\,\mathrm{MPa}\,(158.85)$				$P = 0.80\,\mathrm{MPa}\,(170.43)$			
	$\hat{V}$	$\hat{U}$	$\hat{H}$	$\hat{S}$	$\hat{V}$	$\hat{U}$	$\hat{H}$	$\hat{S}$	$\hat{V}$	$\hat{U}$	$\hat{H}$	$\hat{S}$
Sat.	0.3749	2561.2	2748.7	6.8213	0.3157	2567.4	2756.8	6.7600	0.2404	2576.8	2769.1	6.6628
200	0.4249	2642.9	2855.4	7.0592	0.3520	2638.9	2850.1	6.9665	0.2608	2630.6	2839.3	6.8158
250	0.4744	2723.5	2960.7	7.2709	0.3938	2720.9	2957.2	7.1816	0.2931	2715.5	2950.0	7.0384
300	0.5226	2802.9	3064.2	7.4599	0.4344	2801.0	3061.6	7.3724	0.3241	2797.2	3056.5	7.2328
350	0.5701	2882.6	3167.7	7.6329	0.4742	2881.2	3165.7	7.5464	0.3544	2878.2	3161.7	7.4089
400	0.6173	2963.2	3271.9	7.7938	0.5137	2962.1	3270.3	7.7079	0.3843	2959.7	3267.1	7.5716
500	0.7109	3128.4	3483.9	8.0873	0.5920	3127.6	3482.8	8.0021	0.4433	3126.0	3480.6	7.8673
600	0.8041	3299.6	3701.7	8.3522	0.6697	3299.1	3700.9	8.2674	0.5018	3297.9	3699.4	8.1333
700	0.8969	3477.5	3925.9	8.5952	0.7472	3477.0	3925.3	8.5107	0.5601	3476.2	3924.2	8.3770
800	0.9896	3662.1	4156.9	8.8211	0.8245	3661.8	4156.5	8.7367	0.6181	3661.1	4155.6	8.6033
900	1.0822	3853.6	4394.7	9.0329	0.9017	3853.4	4394.4	8.9486	0.6761	3852.8	4393.7	8.8153
1000	1.1747	4051.8	4639.1	9.2328	0.9788	4051.5	4638.8	9.1485	0.7340	4051.0	4638.2	9.0153
1100	1.2672	4256.3	4889.9	9.4224	1.0559	4256.1	4889.6	9.3381	0.7919	4255.6	4889.1	9.2050
1200	1.3596	4466.8	5146.6	9.6029	1.1330	4466.5	5146.3	9.5185	0.8497	4466.1	5145.9	9.3855
1300	1.4521	4682.5	5408.6	9.7749	1.2101	4682.3	5408.3	9.6906	0.9076	4681.8	5407.9	9.5575

[†]**Note:** Number in parantheses is temperature of saturated steam in °C at the specified pressure.
$\hat{V}\,[=]\,\mathrm{m^3/kg}; \quad \hat{U}\ \&\ \hat{H}\,[=]\,\mathrm{J/g} = \mathrm{kJ/kg}; \quad \hat{S}\,[=]\,\mathrm{kJ/kg.K}$

Table A.3 (continued) Superheated Vapor†

T (°C)	P = 1.00 MPa (179.91)				P = 1.20 MPa (187.99)				P = 1.40 MPa (195.07)			
	$\hat{V}$	$\hat{U}$	$\hat{H}$	$\hat{S}$	$\hat{V}$	$\hat{U}$	$\hat{H}$	$\hat{S}$	$\hat{V}$	$\hat{U}$	$\hat{H}$	$\hat{S}$
Sat.	0.19444	2583.6	2778.1	6.5865	0.16333	2588.8	2784.8	6.5233	0.14084	2592.8	2790.0	6.4693
200	0.2060	2621.9	2827.9	6.6940	0.16930	2612.8	2815.9	6.5898	0.14302	2603.1	2803.3	6.4975
250	0.2327	2709.9	2942.6	6.9247	0.19234	2704.2	2935.0	6.8294	0.16350	2698.3	2927.2	6.7467
300	0.2579	2793.2	3051.2	7.1229	0.2138	2789.2	3045.8	7.0317	0.18228	2785.2	3040.4	6.9534
350	0.2825	2875.2	3157.7	7.3011	0.2345	2872.2	3153.6	7.2121	0.2003	2869.2	3149.5	7.1360
400	0.3066	2957.3	3263.9	7.4651	0.2548	2954.9	3260.7	7.3774	0.2178	2952.5	3257.5	7.3026
500	0.3541	3124.4	3478.5	7.7622	0.2946	3122.8	3476.3	7.6759	0.2521	3121.1	3474.1	7.6027
600	0.4011	3296.8	3697.9	8.0290	0.3339	3295.6	3696.3	7.9435	0.2860	3294.4	3694.8	7.8710
700	0.4478	3475.3	3923.1	8.2731	0.3729	3474.4	3922.0	8.1881	0.3195	3473.6	3920.8	8.1160
800	0.4943	3660.4	4154.7	8.4996	0.4118	3659.7	4153.8	8.4148	0.3528	3659.0	4153.0	8.3431
900	0.5407	3852.2	4392.9	8.7118	0.4505	3851.6	4392.2	8.6272	0.3861	3851.1	4391.5	8.5556
1000	0.5871	4050.5	4637.6	8.9119	0.4892	4050.0	4637.0	8.8274	0.4192	4049.5	4636.4	8.7559
1100	0.6335	4255.1	4888.6	9.1017	0.5278	4254.6	4888.0	9.0172	0.4524	4254.1	4887.5	8.9457
1200	0.6798	4465.6	5145.4	9.2822	0.5665	4465.1	5144.9	9.1977	0.4855	4464.7	5144.4	9.1262
1300	0.7261	4681.3	5407.4	9.4543	0.6051	4680.9	5407.0	9.3698	0.5186	4680.4	5406.5	9.2984

†Note: Number in parantheses is temperature of saturated steam in °C at the specified pressure.
$\hat{V}$ [=] m³/kg; $\hat{U}$ & $\hat{H}$ [=] J/g = kJ/kg; $\hat{S}$ [=] kJ/kg·K

Table A.3 (continued)   Superheated Vapor[†]

T (°C)	P = 1.60 MPa (201.41)				P = 1.80 MPa (207.15)				P = 2.00 MPa (212.42)			
	$\hat{V}$	$\hat{U}$	$\hat{H}$	$\hat{S}$	$\hat{V}$	$\hat{U}$	$\hat{H}$	$\hat{S}$	$\hat{V}$	$\hat{U}$	$\hat{H}$	$\hat{S}$
Sat.	0.12380	2596.0	2794.0	6.4218	0.11042	2598.4	2797.1	6.3794	0.09963	2600.3	2799.5	6.3409
225	0.13287	2644.7	2857.3	6.5518	0.11673	2636.6	2846.7	6.4808	0.10377	2628.3	2835.8	6.4147
250	0.14184	2692.3	2919.2	6.6732	0.12497	2686.0	2911.0	6.6066	0.11144	2679.6	2902.5	6.5453
300	0.15862	2781.1	3034.8	6.8844	0.14021	2776.9	3029.2	6.8226	0.12547	2772.6	3023.5	6.7664
350	0.17456	2866.1	3145.4	7.0694	0.15457	2863.0	3141.2	7.0100	0.13857	2859.8	3137.0	6.9563
400	0.19005	2950.1	3254.2	7.2374	0.16847	2947.7	3250.9	7.1794	0.15120	2945.2	3247.6	7.1271
500	0.2203	3119.5	3472.0	7.5390	0.19550	3117.9	3469.8	7.4825	0.17568	3116.2	3467.6	7.4317
600	0.2400	3293.3	3693.2	7.8080	0.2220	3292.1	3691.7	7.7523	0.19960	3290.9	3690.1	7.7024
700	0.2794	3472.7	3919.7	8.0535	0.2482	3471.8	3918.5	7.9983	0.2232	3470.9	3917.4	7.9487
800	0.3086	3658.3	4152.1	8.2808	0.2742	3657.6	4151.2	8.2258	0.2467	3657.0	4150.3	8.1765
900	0.3377	3850.5	4390.8	8.4935	0.3001	3849.9	4390.1	8.4386	0.2700	3849.3	4389.4	8.3895
1000	0.3668	4049.0	4635.8	8.6938	0.3260	4048.5	4635.2	8.6391	0.2933	4048.0	4634.6	8.5901
1100	0.3958	4253.7	4887.0	8.8837	0.3518	4253.2	4886.4	8.8290	0.3166	4252.7	4885.9	8.7800
1200	0.4248	4464.2	5143.9	9.0643	0.3776	4463.7	5143.4	9.0096	0.3398	4463.3	5142.9	8.9607
1300	0.4538	4679.9	5406.0	9.2364	0.4034	4679.5	5405.6	9.1818	0.3631	4679.0	5405.1	9.1329

Note: Number in parantheses is temperature of saturated steam in °C at the specified pressure.
$\hat{V}\,[=]\,m^3/kg;\quad \hat{U}\ \&\ \hat{H}\,[=]\,J/g = kJ/kg;\quad \hat{S}\,[=]\,kJ/kg\cdot K$

Table A.3 (continued) Superheated Vapor†

T	$P = 2.50\,\text{MPa}\ (223.99)$				$P = 3.00\,\text{MPa}\ (233.90)$				$P = 3.50\,\text{MPa}\ (242.60)$			
(°C)	$\hat{V}$	$\hat{U}$	$\hat{H}$	$\hat{S}$	$\hat{V}$	$\hat{U}$	$\hat{H}$	$\hat{S}$	$\hat{V}$	$\hat{U}$	$\hat{H}$	$\hat{S}$
Sat.	0.07998	2603.1	2803.1	6.2575	0.06668	2604.1	2804.2	6.1869	0.05707	2603.7	2803.4	6.1253
225	0.08027	2605.6	2806.3	6.2639	—	—	—	—	—	—	—	—
250	0.08700	2662.6	2880.1	6.4085	0.07058	2644.0	2855.8	6.2872	0.05872	2623.7	2829.2	6.1749
300	0.09890	2761.6	3008.8	6.6438	0.08114	2750.1	2993.5	6.5390	0.06842	2738.0	2977.5	6.4461
350	0.10976	2851.9	3126.3	6.8403	0.09053	2843.7	3115.3	6.7428	0.07678	2835.3	3104.0	6.6579
400	0.12010	2939.1	3239.3	7.0148	0.09936	2932.8	3230.9	6.9212	0.08453	2926.4	3222.3	6.8405
450	0.13014	3025.5	3350.8	7.1746	0.10787	3020.4	3344.0	7.0834	0.09196	3015.3	3337.2	7.0052
500	0.13998	3112.1	3462.1	7.3234	0.11619	3108.0	3456.5	7.2338	0.09918	3103.0	3450.9	7.1572
600	0.15930	3288.0	3686.3	7.5960	0.13243	3285.0	3682.3	7.5085	0.11324	3282.1	3678.4	7.4339
700	0.17832	3468.7	3914.5	7.8435	0.14838	3466.5	3911.7	7.7571	0.12699	3464.3	3908.8	7.6837
800	0.19716	3655.3	4148.2	8.0720	0.16414	3653.5	4145.9	7.9862	0.14056	3651.8	4143.7	7.9134
900	0.21590	3847.9	4387.6	8.2853	0.17980	3846.5	4385.9	8.1999	0.15402	3845.0	4384.1	8.1276
1000	0.2346	4046.7	4633.1	8.4861	0.19541	4045.4	4631.6	8.4009	0.16743	4044.1	4630.1	8.3288
1100	0.2532	4251.5	4884.6	8.6762	0.21098	4250.3	4883.3	8.5912	0.18080	4249.2	4881.9	8.5192
1200	0.2718	4462.1	5141.7	8.8569	0.22652	4460.9	5140.5	8.7720	0.19415	4459.8	5139.3	8.7000
1300	0.2905	4677.8	5404.0	9.0291	0.24206	4676.6	5402.8	8.9442	0.20749	4675.5	5401.7	8.8723

†**Note:** Number in parentheses is temperature of saturated steam in °C at the specified pressure.
$\hat{V}\,[=]\,\text{m}^3/\text{kg};\quad \hat{U}\ \&\ \hat{H}\,[=]\,\text{J/g} = \text{kJ/kg};\quad \hat{S}\,[=]\,\text{kJ/kg·K}$

Table A.3 (continued) Superheated Vapor[†]

T (°C)	P = 4.0 MPa (250.40)				P = 4.5 MPa (257.49)				P = 5.0 MPa (263.99)			
	$\hat{V}$	$\hat{U}$	$\hat{H}$	$\hat{S}$	$\hat{V}$	$\hat{U}$	$\hat{H}$	$\hat{S}$	$\hat{V}$	$\hat{U}$	$\hat{H}$	$\hat{S}$
Sat.	0.04978	2602.3	2801.4	6.0701	0.04406	2600.1	2798.3	6.0198	0.03944	2597.1	2794.3	5.9734
275	0.05457	2667.9	2886.2	6.2285	0.04730	2650.3	2863.2	6.1401	0.04141	2631.3	2838.3	6.0544
300	0.05884	2725.3	2960.7	6.3615	0.05135	2712.0	2943.1	6.2828	0.04532	2698.0	2924.5	6.2084
350	0.06645	2826.7	3092.5	6.5821	0.05840	2817.8	3080.6	6.5131	0.05194	2808.7	3068.4	6.4493
400	0.07341	2919.9	3213.6	6.7690	0.06475	2913.3	3204.7	6.7047	0.05781	2906.6	3195.7	6.6459
450	0.08002	3010.2	3330.3	6.9363	0.07074	3005.0	3323.3	6.8746	0.06330	2999.7	3316.2	6.8186
500	0.08643	3099.5	3445.3	7.0901	0.07651	3095.3	3439.6	7.0301	0.06857	3091.0	3433.8	6.9759
600	0.09885	3279.1	3674.4	7.3688	0.08765	3276.0	3670.5	7.3110	0.07869	3273.0	3666.5	7.2589
700	0.11095	3462.1	3905.9	7.6198	0.09847	3459.9	3903.0	7.5631	0.08849	3457.6	3900.1	7.5122
800	0.12287	3650.0	4141.5	7.8502	0.10911	3648.3	4139.3	7.7942	0.09811	3646.6	4137.1	7.7440
900	0.13469	3843.6	4382.3	8.0647	0.11965	3842.2	4380.6	8.0091	0.10762	3840.7	4378.8	7.9593
1000	0.14645	4042.9	4628.7	8.2662	0.13013	4041.6	4627.2	8.2108	0.11707	4040.4	4625.7	8.1612
1100	0.15817	4248.0	4880.6	8.4567	0.14056	4246.8	4879.3	8.4015	0.12648	4245.6	4878.0	8.3520
1200	0.16987	4458.6	5138.1	8.6376	0.15098	4457.5	5136.9	8.5825	0.13587	4456.3	5135.7	8.5331
1300	0.18156	4674.3	5400.5	8.8100	0.16139	4673.1	5399.4	8.7549	0.14526	4672.0	5398.2	8.7055

[†]**Note:** Number in parantheses is temperature of saturated steam in °C at the specified pressure.

$\hat{V}$ [=] m³/kg; $\hat{U}$ & $\hat{H}$ [=] J/g = kJ/kg; $\hat{S}$ [=] kJ/kg. K

Table A.3 (continued) Superheated Vapor[†]

T (°C)	P = 6.0 MPa (275.64)				P = 7.0 MPa (285.88)				P = 8.0 MPa (295.06)			
	$\hat{V}$	$\hat{U}$	$\hat{H}$	$\hat{S}$	$\hat{V}$	$\hat{U}$	$\hat{H}$	$\hat{S}$	$\hat{V}$	$\hat{U}$	$\hat{H}$	$\hat{S}$
Sat.	0.03244	2589.7	2784.3	5.8892	0.02737	2580.5	2772.1	5.8133	0.02352	2569.8	2758.0	5.7432
300	0.03616	2667.2	2884.2	6.0674	0.02947	2632.2	2838.4	5.9305	0.02426	2590.9	2785.0	5.7906
350	0.04223	2789.6	3043.0	6.3335	0.03524	2769.4	3016.0	6.2283	0.02995	2747.7	2987.3	6.1301
400	0.04739	2892.9	3177.2	6.5408	0.03993	2878.6	3158.1	6.4478	0.03432	2863.8	3138.3	6.3634
450	0.05214	2988.9	3301.8	6.7193	0.04416	2978.0	3287.1	6.6327	0.03817	2966.7	3272.0	6.5551
500	0.05665	3082.2	3422.2	6.8803	0.04814	3073.4	3410.3	6.7975	0.04175	3064.3	3398.3	6.7240
550	0.06101	3174.6	3540.6	7.0288	0.05195	3167.2	3530.9	6.9486	0.04516	3159.8	3521.0	6.8778
600	0.06525	3266.9	3658.4	7.1677	0.05565	3260.7	3650.3	7.0894	0.04845	3254.4	3642.0	7.0206
700	0.07352	3453.1	3894.2	7.4234	0.06283	3448.5	3888.3	7.3476	0.05481	3443.9	3882.4	7.2812
800	0.08160	3643.1	4132.7	7.6566	0.06981	3639.5	4128.2	7.5822	0.06097	3636.0	4123.8	7.5173
900	0.08958	3837.8	4375.3	7.8727	0.07669	3835.0	4371.8	7.7991	0.06702	3832.1	4368.3	7.7351
1000	0.09749	4037.8	4622.7	8.0751	0.08350	4035.3	4619.8	8.0020	0.07301	4032.8	4616.9	7.9384
1100	0.10536	4243.3	4875.4	8.2661	0.09027	4240.9	4872.8	8.1933	0.07896	4238.6	4870.3	8.1300
1200	0.11321	4454.0	5133.3	8.4474	0.09703	4451.7	5130.9	8.3747	0.08489	4449.5	5128.5	8.3115
1300	0.12106	4669.6	5396.0	8.6199	0.10377	4667.3	5393.7	8.5473	0.09080	4665.0	5391.5	8.4842

[†]**Note:** Number in parantheses is temperature of saturated steam in °C at the specified pressure.
$\hat{V}\,[=]\,\text{m}^3/\text{kg}$; $\hat{U}$ & $\hat{H}\,[=]\,\text{J}/\text{g} = \text{kJ}/\text{kg}$; $\hat{S}\,[=]\,\text{kJ}/\text{kg}\cdot\text{K}$

Table A.3 (continued)   Superheated Vapor[†]

T (°C)	P = 9.0 MPa (303.40)				P = 10.0 MPa (311.06)				P = 12.5 MPa (327.89)			
	$\hat{V}$	$\hat{U}$	$\hat{H}$	$\hat{S}$	$\hat{V}$	$\hat{U}$	$\hat{H}$	$\hat{S}$	$\hat{V}$	$\hat{U}$	$\hat{H}$	$\hat{S}$
Sat.	0.02048	2557.8	2742.1	5.6772	0.018026	2544.4	2724.7	5.6141	0.013495	2505.1	2673.8	5.4624
325	0.02327	2646.6	2856.0	5.8712	0.019861	2610.4	2809.1	5.7568	—	—	—	—
350	0.02580	2724.4	2956.6	6.0361	0.02242	2699.2	2923.4	5.9443	0.016126	2624.6	2826.2	5.7118
400	0.02993	2848.4	3117.8	6.2854	0.02641	2832.4	3096.5	6.2120	0.02000	2789.3	3039.3	6.0417
450	0.03350	2955.2	3256.6	6.4844	0.02975	2943.4	3240.9	6.4190	0.02299	2912.5	3199.8	6.2719
500	0.03677	3055.2	3386.1	6.6576	0.03279	3045.8	3373.7	6.5966	0.02560	3021.7	3341.8	6.4618
550	0.03987	3152.2	3511.0	6.8142	0.03564	3144.6	3500.9	6.7561	0.02801	3125.0	3475.2	6.6290
600	0.04285	3248.1	3633.7	6.9589	0.03837	3241.7	3625.3	6.9029	0.03029	3225.4	3604.0	6.7810
650	0.04574	3343.6	3755.3	7.0943	0.04101	3338.2	3748.2	7.0398	0.03248	3324.4	3730.4	6.9218
700	0.04857	3439.3	3876.5	7.2221	0.04358	3434.7	3870.5	7.1687	0.03460	3422.9	3855.3	7.0536
800	0.05409	3632.5	4119.3	7.4596	0.04859	3628.9	4114.8	7.4077	0.03869	3620.0	4103.6	7.2965
900	0.05950	3829.2	4364.8	7.6783	0.05349	3826.3	4361.2	7.6272	0.04267	3819.1	4352.5	7.5182
1000	0.06485	4030.3	4614.0	7.8821	0.05832	4027.8	4611.0	7.8315	0.04658	4021.6	4603.8	7.7237
1100	0.07016	4236.3	4867.7	8.0740	0.06312	4234.0	4865.1	8.0237	0.05045	4228.2	4858.8	7.9165
1200	0.07544	4447.2	5126.2	8.2556	0.06789	4444.9	5123.8	8.2055	0.05430	4439.3	5118.0	8.0987
1300	0.08072	4662.7	5389.2	8.4284	0.07265	4460.5	5387.0	8.3783	0.05813	4654.8	5381.4	8.2717

[†]**Note:** Number in parantheses is temperature of saturated steam in °C at the specified pressure.
$\hat{V}\,[=]\,m^3/kg;\;\;\hat{U}\,\&\,\hat{H}\,[=]\,J/g = kJ/kg;\;\;\hat{S}\,[=]\,kJ/kg.K$

Table A.3 (continued) Superheated Vapor†

T (°C)	P = 15.0 MPa (342.24)				P = 17.5 MPa (354.75)				P = 20.0 MPa (365.81)			
	$\hat{V}$	$\hat{U}$	$\hat{H}$	$\hat{S}$	$\hat{V}$	$\hat{U}$	$\hat{H}$	$\hat{S}$	$\hat{V}$	$\hat{U}$	$\hat{H}$	$\hat{S}$
Sat.	0.010337	2455.5	2610.5	5.3098	0.007920	2390.2	2528.8	5.1419	0.005834	2293.0	2409.7	4.9269
350	0.011470	2520.4	2692.4	5.4421	—	—	—	—	—	—	—	—
400	0.015649	2740.7	2975.5	5.8811	0.012447	2685.0	2902.9	5.7213	0.009942	2619.3	2818.1	5.5540
450	0.018445	2879.5	3156.2	6.1404	0.015174	2844.2	3109.7	6.0184	0.012695	2806.2	3060.1	5.9017
500	0.02080	2996.6	3308.6	6.3443	0.017358	2970.3	3274.1	6.2383	0.014768	2942.9	3238.2	6.1401
550	0.02293	3104.7	3448.6	6.5199	0.019288	3083.9	3421.4	6.4230	0.016555	3062.4	3393.5	6.3348
600	0.02491	3208.6	3582.3	6.6776	0.02106	3191.5	3560.1	6.5866	0.018178	3174.0	3537.6	6.5048
650	0.02680	3310.3	3712.3	6.8224	0.02274	3296.0	3693.9	6.7357	0.019693	3281.4	3675.3	6.6582
700	0.02861	3410.9	3840.1	6.9572	0.02434	3398.7	3824.6	6.8736	0.02113	3386.4	3809.0	6.7993
800	0.03210	3610.9	4092.4	7.2040	0.02738	3601.8	4081.1	7.1244	0.02385	3592.7	4069.7	7.0544
900	0.03546	3811.9	4343.8	7.4279	0.03031	3804.7	4335.1	7.3507	0.02645	3797.5	4326.4	7.2830
1000	0.03875	4015.4	4596.6	7.6348	0.03316	4009.3	4589.5	7.5589	0.02897	4003.1	4582.5	7.4925
1100	0.04200	4222.6	4852.6	7.8283	0.03597	4216.9	4846.4	7.7531	0.03145	4211.3	4840.2	7.6874
1200	0.04523	4433.8	5112.3	8.0108	0.03876	4428.3	5106.6	7.9360	0.03391	4422.8	5101.0	7.8707
1300	0.04845	4649.1	5376.0	8.1840	0.04154	4643.5	5370.5	8.1093	0.03636	4638.0	5365.1	8.0442

†**Note:** Number in parantheses is temperature of saturated steam in °C at the specified pressure.
$\hat{V}\,[=]\,\mathrm{m^3/kg}$; $\hat{U}$ & $\hat{H}\,[=]\,\mathrm{J/g}=\mathrm{kJ/kg}$; $\hat{S}\,[=]\,\mathrm{kJ/kg\cdot K}$

Table A.3 (continued)   Superheated Vapor[†]

T	$P = 25.0\,\text{MPa}$				$P = 30.0\,\text{MPa}$				$P = 35.0\,\text{MPa}$			
(°C)	$\hat{V}$	$\hat{U}$	$\hat{H}$	$\hat{S}$	$\hat{V}$	$\hat{U}$	$\hat{H}$	$\hat{S}$	$\hat{V}$	$\hat{U}$	$\hat{H}$	$\hat{S}$
375	0.001973	1798.7	1848.0	4.0320	0.001789	1737.8	1791.5	3.9305	0.001700	1702.9	1762.4	3.8722
400	0.006004	2430.1	2580.2	5.1418	0.002790	2067.4	2151.1	4.4728	0.002100	1914.1	1987.6	4.2126
425	0.007881	2609.2	2806.3	5.4723	0.005303	2455.1	2614.2	5.1504	0.003428	2253.4	2373.4	4.7747
450	0.009162	2720.7	2949.7	5.6744	0.006735	2619.3	2821.4	5.4424	0.004961	2498.7	2672.4	5.1962
500	0.011123	2884.3	3162.4	5.9592	0.008678	2820.7	3081.1	5.7905	0.006927	2751.9	2994.4	5.6282
550	0.012724	3017.5	3335.6	6.1765	0.010168	2970.3	3275.4	6.0342	0.008345	2921.0	3213.0	5.9026
600	0.014137	3137.9	3491.4	6.3602	0.011446	3100.5	3443.9	6.2331	0.009527	3062.0	3395.5	6.1179
650	0.015433	3251.6	3637.4	6.5229	0.012596	3221.0	3598.9	6.4058	0.010575	3189.8	3559.9	6.3010
700	0.016646	3361.3	3777.5	6.6707	0.013661	3335.8	3745.6	6.5606	0.011533	3309.8	3713.5	6.4631
800	0.018912	3574.3	4047.1	6.9345	0.015623	3555.5	4024.2	6.8332	0.013278	3536.7	4001.5	6.7450
900	0.021045	3783.0	4309.1	7.1680	0.017448	3768.5	4291.9	7.0718	0.014883	3754.0	4274.9	6.9886
1000	0.02310	3990.9	4568.5	7.3802	0.019196	3978.8	4554.7	7.2867	0.016410	3966.7	4541.1	7.2064
1100	0.02512	4200.2	4828.2	7.5765	0.020903	4189.2	4816.3	7.4845	0.017895	4178.3	4804.6	7.4057
1200	0.02711	4412.0	5089.9	7.7605	0.022589	4401.3	5079.0	7.6692	0.019360	4390.7	5068.3	7.5910
1300	0.02910	4626.9	5354.4	7.9342	0.024266	4616.0	5344.0	7.8432	0.020815	4605.1	5333.6	7.7653

[†]**Note:** Number in parentheses is temperature of saturated steam in °C at the specified pressure.
$\hat{V}\,[=]\,\text{m}^3/\text{kg};\quad \hat{U}\,\&\,\hat{H}\,[=]\,\text{J}/\text{g} = \text{kJ}/\text{kg};\quad \hat{S}\,[=]\,\text{kJ}/\text{kg.K}$

Table A.3 (continued) Superheated Vapor†

T (°C)	P = 40.0 MPa				P = 50.0 MPa				P = 60.0 MPa			
	$\hat{V}$	$\hat{U}$	$\hat{H}$	$\hat{S}$	$\hat{V}$	$\hat{U}$	$\hat{H}$	$\hat{S}$	$\hat{V}$	$\hat{U}$	$\hat{H}$	$\hat{S}$
375	0.001641	1677.1	1742.8	3.8290	0.001559	1638.6	1716.6	3.7639	0.001503	1609.4	1699.5	3.7141
400	0.001908	1854.6	1930.9	4.1135	0.001731	1788.1	1874.6	4.0031	0.001634	1745.4	1843.4	3.9318
425	0.002532	2096.9	2198.1	4.5029	0.002007	1959.7	2060.0	4.2734	0.001817	1892.7	2001.7	4.1626
450	0.003693	2365.1	2512.8	4.9459	0.002486	2159.6	2284.0	4.5884	0.002085	2053.9	2179.0	4.4121
500	0.005622	2678.4	2903.3	5.4700	0.003892	2525.5	2720.1	5.1726	0.002956	2390.6	2567.9	4.9321
550	0.006984	2869.7	3149.1	5.7785	0.005118	2763.6	3019.5	5.5485	0.003956	2658.8	2896.2	5.3441
600	0.008094	3022.6	3346.4	6.0114	0.006112	2942.0	3247.6	5.8178	0.004834	2861.1	3151.2	5.6452
650	0.009063	3158.0	3520.6	6.2054	0.006966	3093.5	3441.8	6.0342	0.005595	3028.8	3364.5	5.8829
700	0.009941	3283.6	3681.2	6.3750	0.007727	3230.5	3616.8	6.2189	0.006272	3177.2	3553.5	6.0824
800	0.011523	3517.8	3978.7	6.6662	0.009076	3479.8	3933.6	6.5290	0.007459	3441.5	3889.1	6.4109
900	0.012962	3739.4	4257.9	6.9150	0.010283	3710.3	4224.4	6.7882	0.008508	3681.0	4191.5	6.6805
1000	0.014324	3954.6	4527.6	7.1356	0.011411	3930.5	4501.1	7.0146	0.009480	3906.4	4475.2	6.9127
1100	0.015642	4167.4	4793.1	7.3364	0.012496	4145.7	4770.5	7.2184	0.010409	4124.1	4748.6	7.1195
1200	0.016940	4380.1	5057.7	7.5224	0.013561	4359.1	5037.2	7.4058	0.011317	4338.2	5017.0	7.3083
1300	0.018229	4594.3	5323.5	7.6969	0.014616	4572.8	5303.6	7.5808	0.012215	4551.4	5284.3	7.4837

†**Note:** $\hat{V}$ [=] m³/kg; $\hat{U}$ & $\hat{H}$ [=] J/g = kJ/kg; $\hat{S}$ [=] kJ/kg·K

Table A.4 Compressed Liquid†

T	P = 5 MPa (263.99)				P = 10 MPa (311.06)				P = 15 MPa (324.24)			
(°C)	$\hat{V}$	$\hat{U}$	$\hat{H}$	$\hat{S}$	$\hat{V}$	$\hat{U}$	$\hat{H}$	$\hat{S}$	$\hat{V}$	$\hat{U}$	$\hat{H}$	$\hat{S}$
Sat.	0.0012859	1147.8	1154.2	2.9202	0.0014524	1393.0	1407.6	3.3596	0.0016581	1585.6	1610.5	3.6848
0	0.0009977	0.04	5.04	0.0001	0.0009952	0.09	10.04	0.0002	0.0009928	0.15	15.05	0.0004
20	0.0009995	83.65	88.65	0.2956	0.0009972	83.36	93.33	0.2945	0.0009950	83.06	97.99	0.2934
40	0.0010056	166.95	171.97	0.5705	0.0010034	166.35	176.38	0.5686	0.0010013	165.76	180.78	0.5666
60	0.0010149	250.23	255.30	0.8285	0.0010127	249.36	259.49	0.8258	0.0010105	248.51	263.67	0.8232
80	0.0010268	333.72	338.85	1.0720	0.0010245	332.59	342.83	1.0688	0.0010222	331.48	346.81	1.0656
100	0.0010410	417.52	422.72	1.3030	0.0010385	416.12	426.50	1.2992	0.0010361	414.74	430.28	1.2955
120	0.0010576	501.80	507.09	1.5233	0.0010549	500.08	510.64	1.5189	0.0010522	498.40	514.19	1.5145
140	0.0010768	586.76	592.15	1.7343	0.0010737	584.68	595.42	1.7292	0.0010707	582.66	598.72	1.7242
160	0.0010988	672.62	678.12	1.9375	0.0010953	670.13	681.08	1.9317	0.0010918	667.71	684.09	1.9260
180	0.0011240	759.63	765.25	2.1341	0.0011199	756.65	767.84	2.1275	0.0011159	753.76	770.50	2.1210
200	0.0011530	848.1	853.9	2.3255	0.0011480	844.5	856.0	2.3178	0.0011433	841.0	858.2	2.3104
220	0.0011866	938.4	944.4	2.5128	0.0011805	934.1	945.9	2.5039	0.0011748	929.9	947.5	2.4953
240	0.0012264	1031.4	1037.5	2.6979	0.0012187	1026.0	1038.1	2.6872	0.0012114	1020.8	1039.0	2.6771
260	0.0012749	1127.9	1134.3	2.8830	0.0012645	1121.1	1133.7	2.8699	0.0012550	1114.6	1133.4	2.8576
280					0.0013216	1220.9	1234.1	3.0548	0.0013084	1212.5	1232.1	3.0393
300					0.0013972	1328.4	1342.3	3.2469	0.0013770	1316.6	1337.3	3.2260
320									0.0014724	1431.1	1453.2	3.4247
340									0.0016311	1567.5	1591.9	3.6546

†Note: Number in parantheses is temperature of saturated liquid in °C at the specified pressure.

$\hat{V}\,[=]\,m^3/kg;\quad \hat{U}\ \&\ \hat{H}\,[=]\,J/g = kJ/kg;\quad \hat{S}\,[=]\,kJ/kg.\,K$

Table A.4 (continued)   Compressed Liquid†

T (°C)	$\hat{V}$	$\hat{U}$	$\hat{H}$	$\hat{S}$	$\hat{V}$	$\hat{U}$	$\hat{H}$	$\hat{S}$	$\hat{V}$	$\hat{U}$	$\hat{H}$	$\hat{S}$
	P = 20 MPa (365.81)				P = 30 MPa				P = 50 MPa			
Sat.	0.002036	1785.6	1826.3	4.0139								
0	0.0009904	0.19	20.01	0.0004	0.0009856	0.25	29.82	0.0001	0.0009766	0.20	49.03	0.0014
20	0.0009928	82.77	102.62	0.2923	0.0009886	82.17	111.84	0.2899	0.0009804	81.00	130.02	0.2848
40	0.0009992	165.17	185.16	0.5646	0.0009951	164.04	193.89	0.5607	0.0009872	161.86	211.21	0.5527
60	0.0010084	247.68	267.85	0.8206	0.0010042	246.06	276.19	0.8154	0.0009962	242.98	292.79	0.8502
80	0.0010199	330.40	350.80	1.0624	0.0010156	328.30	358.77	1.0561	0.0010073	324.34	374.70	1.0440
100	0.0010337	413.39	434.06	1.2917	0.0010290	410.78	441.66	1.2844	0.0010201	405.88	456.89	1.2703
120	0.0010496	496.76	517.76	1.5102	0.0010445	493.59	524.93	1.5018	0.0010348	487.65	539.39	1.4857
140	0.0010678	580.69	602.04	1.7193	0.0010621	576.88	608.75	1.7098	0.0010515	569.77	622.35	1.6915
160	0.0010885	665.35	687.12	1.9204	0.0010821	660.82	693.28	1.9096	0.0010703	652.41	705.92	1.8891
180	0.0011120	750.95	773.20	2.1147	0.0011047	745.59	778.73	2.1024	0.0010912	735.69	790.25	2.0794
200	0.0011388	837.7	860.5	2.3031	0.0011302	831.4	865.3	2.2893	0.0011146	819.7	875.5	2.2634
220	0.0011693	925.9	949.3	2.4870	0.0011590	918.3	953.1	2.4711	0.0011408	904.7	961.7	2.4419
240	0.0012046	1016.0	1040.0	2.6674	0.0011920	1006.9	1042.6	2.6490	0.0011072	990.7	1049.2	2.6158
260	0.0012462	1108.6	1133.5	2.8459	0.0012303	1097.4	1134.3	2.8243	0.0012034	1078.1	1138.2	2.7860
280	0.0012965	1204.7	1230.6	3.0248	0.0012755	1190.7	1229.0	2.9986	0.0012415	1167.2	1229.3	2.9537
300	0.0013596	1306.1	1333.3	3.2071	0.0013304	1287.9	1327.8	3.1741	0.0012860	1258.7	1323.0	3.1200
320	0.0014437	1415.7	1444.6	3.3979	0.0013997	1390.7	1432.7	3.3539	0.0013388	1353.3	1420.2	3.2868
340	0.0015684	1539.7	1571.0	3.6075	0.0014920	1501.7	1546.5	3.5426	0.0014032	1452.0	1522.1	3.4557
360	0.0018226	1702.8	1739.3	3.8772	0.0016265	1626.6	1675.4	3.7494	0.0014838	1556.0	1630.2	3.6291
380					0.0018691	1781.4	1837.5	4.0012	0.0015884	1667.2	1746.6	3.8101

†**Note:** Number in parantheses is temperature of saturated liquid in °C at the specified pressure.
$\hat{V}[=]\,\mathrm{m^3/kg};\;\;\hat{U}\,\&\,\hat{H}[=]\,\mathrm{J/g}=\mathrm{kJ/kg};\;\;\hat{S}[=]\,\mathrm{kJ/kg\cdot K}$

Appendix B

Thermodynamic Properties of Refrigerant-134a (1,1,1,2-tetrafluoroethane)

Table B.1 - Saturated R-134a

Table B.2 - Superheated R-134a

Source: G.J. Van Wylen, R.E. Sonntag, and C. Borgnakke, *Fundamentals of Classical Thermodynamics*, 4th Ed., John Wiley & Sons, Inc., New York, 1994. Used with permission.

Table B.1   Saturated R-134a

T (°C)	P (MPa)	Specific Volume (m³/kg)			Enthalpy (kJ/kg)			Entropy (kJ/kg·K)		
		$\hat{V}^L$	$\Delta\hat{V}$	$\hat{V}^V$	$\hat{H}^L$	$\Delta\hat{H}$	$\hat{H}^V$	$\hat{S}^L$	$\Delta\hat{S}$	$\hat{S}^V$
−33	0.0737	0.000718	0.25574	0.25646	157.417	220.491	377.908	0.8346	0.9181	1.7528
−30	0.0851	0.000722	0.22330	0.22402	161.118	218.683	379.802	0.8499	0.8994	1.7493
−26.25	0.1013	0.000728	0.18947	0.19020	165.802	216.360	382.162	0.8690	0.8763	1.7453
−25	0.1073	0.000730	0.17956	0.18029	167.381	215.569	382.950	0.8754	0.8687	1.7441
−20	0.1337	0.000738	0.14575	0.14649	173.744	212.340	386.083	0.9007	0.8388	1.7395
−15	0.1650	0.000746	0.11932	0.12007	180.193	209.004	389.197	0.9258	0.8096	1.7354
−10	0.2017	0.000755	0.098454	0.099209	186.721	205.564	392.285	0.9507	0.7812	1.7319
−5	0.2445	0.000764	0.081812	0.082576	193.324	202.016	395.340	0.9755	0.7534	1.7288
0	0.2940	0.000773	0.068420	0.069193	200.000	198.356	398.356	1.0000	0.7262	1.7262
5	0.3509	0.000783	0.057551	0.058334	206.751	194.572	401.323	1.0243	0.6995	1.7239
10	0.4158	0.000794	0.048658	0.049451	213.580	190.652	404.233	1.0485	0.6733	1.7218
15	0.4895	0.000805	0.041326	0.042131	220.492	186.582	407.075	1.0725	0.6475	1.7200
20	0.5728	0.000817	0.035238	0.036055	227.493	182.345	409.838	1.0963	0.6220	1.7183
25	0.6663	0.000829	0.030148	0.030977	234.590	177.920	412.509	1.1201	0.5967	1.7168
30	0.7710	0.000843	0.025865	0.026707	241.790	173.285	415.075	1.1437	0.5716	1.7153

Table B.1 (continued) Saturated R-134a

T (°C)	P (MPa)	Specific Volume (m³/kg)			Enthalpy (kJ/kg)			Entropy (kJ/kg·K)		
		$\hat{V}^L$	$\Delta\hat{V}$	$\hat{V}^V$	$\hat{H}^L$	$\Delta\hat{H}$	$\hat{H}^V$	$\hat{S}^L$	$\Delta\hat{S}$	$\hat{S}^V$
35	0.8876	0.000857	0.022237	0.023094	249.103	168.415	417.518	1.1673	0.5465	1.7139
40	1.0171	0.000873	0.019147	0.020020	256.539	163.282	419.821	1.1909	0.5214	1.7123
45	1.1602	0.000890	0.016499	0.017389	264.110	157.852	421.962	1.2145	0.4962	1.7106
50	1.3180	0.000908	0.014217	0.015124	271.830	152.085	423.915	1.2381	0.4706	1.7088
55	1.4915	0.000928	0.012237	0.013166	279.718	145.933	425.650	1.2619	0.4447	1.7066
60	1.6818	0.000951	0.010511	0.011462	287.794	139.336	427.130	1.2857	0.4182	1.7040
65	1.8898	0.000976	0.008995	0.009970	296.088	132.216	428.305	1.3099	0.3910	1.7009
70	2.1169	0.001005	0.007653	0.008657	304.642	124.468	429.110	1.3343	0.3627	1.6970
75	2.3644	0.001038	0.006453	0.007491	313.513	115.939	429.451	1.3592	0.3330	1.6923
80	2.6337	0.001078	0.005368	0.006446	322.794	106.395	429.189	1.3849	0.3013	1.6862
85	2.9265	0.001128	0.004367	0.005495	332.644	95.440	428.084	1.4117	0.2665	1.6782
90	3.2448	0.001195	0.003412	0.004606	343.380	82.295	425.676	1.4404	0.2266	1.6670
95	3.5914	0.001297	0.002432	0.003729	355.834	64.984	420.818	1.4733	0.1765	1.6498
101.15	4.0640	0.001969	0	0.001969	390.977	0	390.977	1.5658	0	1.5658

Table B.2 Superheated R-134a[†]

T (°C)	P = 0.10 MPa (−26.45)			P = 0.15 MPa (−17.23)			P = 0.20 MPa (−10.08)		
	$\hat{V}$	$\hat{H}$	$\hat{S}$	$\hat{V}$	$\hat{H}$	$\hat{S}$	$\hat{V}$	$\hat{H}$	$\hat{S}$
−25	0.19400	383.212	1.75058	—	—	—	—	—	—
−20	0.19860	387.215	1.76655	—	—	—	—	—	—
−10	0.20765	395.270	1.79775	0.13603	393.839	1.76058	0.10013	392.338	1.73276
0	0.21652	403.413	1.82813	0.14222	402.187	1.79171	0.10501	400.911	1.76474
10	0.22527	411.668	1.85780	0.14828	410.602	1.82197	0.10974	409.500	1.79562
20	0.23393	420.048	1.88689	0.15424	419.111	1.85150	0.11436	418.145	1.82563
30	0.24250	428.564	1.91545	0.16011	427.730	1.88041	0.11889	426.875	1.85491
40	0.25102	437.223	1.94355	0.16592	436.473	1.90879	0.12335	435.708	1.88357
50	0.25948	446.029	1.97123	0.17168	445.350	1.93669	0.12776	444.658	1.91171
60	0.26791	454.986	1.99853	0.17740	454.366	1.96416	0.13213	453.735	1.93937
70	0.27631	464.096	2.02547	0.18308	463.525	1.99125	0.13646	462.946	1.96661
80	0.28468	473.359	2.05208	0.18874	472.831	2.01798	0.14076	472.296	1.99346
90	0.29303	482.777	2.07837	0.19437	482.285	2.04438	0.14504	481.788	2.01997
100	0.30136	492.349	2.10437	0.19999	491.888	2.07046	0.14930	491.424	2.04614

[†]**Note:** Number in parantheses is temperature of saturated R-134a in °C at the specified pressure.
$\hat{V}\,[=]\,m^3/kg;$ $\hat{H}\,[=]\,J/g = kJ/kg;$ $\hat{S}\,[=]\,kJ/kg \cdot K$

Table B.2 (continued) Superheated R-134a†

T (°C)	P = 0.25 MPa (−4.33)			P = 0.30 MPa (0.63)			P = 0.40 MPa (8.88)		
	$\hat{V}$	$\hat{H}$	$\hat{S}$	$\hat{V}$	$\hat{H}$	$\hat{S}$	$\hat{V}$	$\hat{H}$	$\hat{S}$
0	0.082637	399.579	1.74284	—	—	—	—	—	—
10	0.086584	408.357	1.77440	0.071110	407.171	1.75637	0.051681	404.651	1.72611
20	0.090408	417.151	1.80492	0.074415	416.124	1.78744	0.054362	413.965	1.75844
30	0.094139	425.997	1.83460	0.077620	425.096	1.81754	0.056926	423.216	1.78947
40	0.097798	434.925	1.86357	0.080748	434.124	1.84684	0.059402	432.465	1.81949
50	0.101401	443.953	1.89195	0.083816	443.234	1.87547	0.061812	441.751	1.84868
60	0.104958	453.094	1.91980	0.086838	452.442	1.90354	0.064169	451.104	1.87718
70	0.108480	462.359	1.94720	0.089821	461.763	1.93110	0.066484	460.545	1.90510
80	0.111972	471.754	1.97419	0.092774	471.206	1.95823	0.068767	470.088	1.93252
90	0.115440	481.285	2.00080	0.095702	480.777	1.98495	0.071022	479.745	1.95948
100	0.118888	490.955	2.02707	0.098609	490.482	2.01131	0.073254	489.523	1.98604
110	0.122318	500.766	2.05302	0.101498	500.324	2.03734	0.075468	499.428	2.01223
120	0.125734	510.720	2.07866	0.104371	510.304	2.06305	0.077665	509.464	2.03809

†**Note:** Number in parantheses is temperature of saturated R-134a in °C at the specified pressure.
$\hat{V}$ [=] m³/kg; $\hat{H}$ [=] J/g = kJ/kg; $\hat{S}$ [=] kJ/kg.K

Table B.2 (continued) Superheated R-134a†

T (°C)	P = 0.50 MPa (15.70)			P = 0.60 MPa (21.51)			P = 0.70 MPa (26.65)		
	$\hat{V}$	$\hat{H}$	$\hat{S}$	$\hat{V}$	$\hat{H}$	$\hat{S}$	$\hat{V}$	$\hat{H}$	$\hat{S}$
20	0.042256	411.645	1.73420	—	—	—	—	—	—
30	0.044457	421.221	1.76632	0.036094	419.093	1.74610	0.030069	416.809	1.72770
40	0.046557	430.720	1.79715	0.037958	428.881	1.77786	0.031781	426.933	1.76056
50	0.048581	440.205	1.82696	0.039735	438.589	1.80838	0.033392	436.895	1.79187
60	0.050547	449.718	1.85596	0.041447	448.279	1.83791	0.034929	446.782	1.82201
70	0.052467	459.290	1.88426	0.043108	457.994	1.86664	0.036410	456.655	1.85121
80	0.054351	468.942	1.91199	0.044730	467.764	1.89471	0.037848	466.554	1.87964
90	0.056205	478.690	1.93921	0.046319	477.611	1.92220	0.039251	476.507	1.90743
100	0.058035	488.546	1.96598	0.047883	487.550	1.94920	0.040627	486.535	1.93467
110	0.059845	498.518	1.99235	0.049426	497.594	1.97576	0.041980	496.654	1.96143
120	0.061639	508.613	2.01836	0.050951	507.750	2.00193	0.043314	506.875	1.98777
130	0.063418	518.835	2.04403	0.052461	518.026	2.02774	0.044633	517.207	2.01372
140	0.065184	529.187	2.06940	0.053958	528.425	2.05322	0.045938	527.656	2.03932

†**Note:** Number in parentheses is temperature of saturated R-134a in °C at the specified pressure.
$\hat{V}\,[=]\,m^3/kg;\quad \hat{H}\,[=]\,J/g = kJ/kg;\quad \hat{S}\,[=]\,kJ/kg.K$

Table B.2 (continued) Superheated R-134a[†]

T	$P = 0.80\,\text{MPa}\,(31.28)$			$P = 0.90\,\text{MPa}\,(35.50)$			$P = 1.00\,\text{MPa}\,(39.36)$		
(°C)	$\hat{V}$	$\hat{H}$	$\hat{S}$	$\hat{V}$	$\hat{H}$	$\hat{S}$	$\hat{V}$	$\hat{H}$	$\hat{S}$
40	0.027113	424.860	1.74457	0.023446	422.642	1.72943	0.020473	420.249	1.71479
50	0.028611	435.114	1.77680	0.024868	433.235	1.76273	0.021849	431.243	1.74936
60	0.030024	445.223	1.80761	0.026192	443.595	1.79431	0.023110	441.890	1.78181
70	0.031375	455.270	1.83732	0.027447	453.835	1.82459	0.024293	452.345	1.81273
80	0.032678	465.308	1.86616	0.028649	464.025	1.85387	0.025417	462.703	1.84248
90	0.033944	475.375	1.89427	0.029810	474.216	1.88232	0.026497	473.027	1.87131
100	0.035180	485.499	1.92177	0.030940	484.441	1.91010	0.027543	483.361	1.89938
110	0.036392	495.698	1.94874	0.032043	494.726	1.93730	0.028561	493.736	1.92682
120	0.037584	505.988	1.97525	0.033126	505.088	1.96399	0.029556	504.175	1.95371
130	0.038760	516.379	2.00135	0.034190	515.542	1.99025	0.030533	514.694	1.98013
140	0.039921	526.880	2.02708	0.035241	526.096	2.01611	0.031495	525.305	2.00613
150	0.041071	537.496	2.05247	0.036278	536.760	2.04161	0.032444	536.017	2.03175

[†]**Note:** Number in parantheses is temperature of saturated R-134a in °C at the specified pressure.
$\hat{V}\,[=]\,\text{m}^3/\text{kg};\quad \hat{H}\,[=]\,\text{J/g} = \text{kJ/kg};\quad \hat{S}\,[=]\,\text{kJ/kg.K}$

Table B.2 (continued)   Superheated R-134a[†]

T	P = 1.20 MPa (46.27)			P = 1.40 MPa (52.36)			P = 1.60 MPa (57.85)		
(°C)	$\hat{V}$	$\hat{H}$	$\hat{S}$	$\hat{V}$	$\hat{H}$	$\hat{S}$	$\hat{V}$	$\hat{H}$	$\hat{S}$
50	0.017243	426.845	1.72373	—	—	—	—	—	—
60	0.018439	438.210	1.75837	0.015032	434.079	1.73597	0.012392	429.322	1.71349
70	0.019530	449.179	1.79081	0.016083	445.720	1.77040	0.013449	441.888	1.75066
80	0.020548	459.925	1.82168	0.017040	456.944	1.80265	0.014378	453.722	1.78466
90	0.021512	470.551	1.85135	0.017931	467.931	1.83333	0.015225	465.145	1.81656
100	0.022436	481.128	1.88009	0.018775	478.790	1.86282	0.016015	476.333	1.84695
110	0.023329	491.702	1.90805	0.019583	489.589	1.89139	0.016763	487.390	1.87619
120	0.024197	502.307	1.93537	0.020362	500.379	1.91918	0.017479	498.387	1.90452
130	0.025044	512.965	1.96214	0.021118	511.192	1.94634	0.018169	509.371	1.93211
140	0.025874	523.697	1.98844	0.021856	522.054	1.97296	0.018840	520.376	1.95908
150	0.026691	534.514	2.01431	0.022579	532.984	1.99910	0.019493	531.427	1.98551
160	0.027495	545.426	2.03980	0.023289	543.994	2.02481	0.020133	542.542	2.01147
170	0.028289	556.443	2.06494	0.023988	555.097	2.05015	0.020761	553.735	2.03702

[†]**Note:** Number in parantheses is temperature of saturated R-134a in °C at the specified pressure.
$\hat{V}\,[=]\,m^3/kg$; $\hat{H}\,[=]\,J/g = kJ/kg$; $\hat{S}\,[=]\,kJ/kg.K$

Table B.2 (continued) Superheated R-134a†

T (°C)	$P = 1.80\,\mathrm{MPa}\ (62.84)$			$P = 2.00\,\mathrm{MPa}\ (67.43)$			$P = 2.50\,\mathrm{MPa}\ (77.53)$		
	$\hat{V}$	$\hat{H}$	$\hat{S}$	$\hat{V}$	$\hat{H}$	$\hat{S}$	$\hat{V}$	$\hat{H}$	$\hat{S}$
70	0.011341	437.562	1.73085	0.009581	432.531	1.71011	–	–	–
80	0.012273	450.202	1.76717	0.010550	446.304	1.74968	0.007221	433.797	1.70180
90	0.013099	462.164	1.80057	0.011374	458.951	1.78500	0.008157	449.499	1.74567
100	0.013854	473.741	1.83202	0.012111	470.996	1.81772	0.008907	463.279	1.78311
110	0.014560	485.095	1.86205	0.012789	482.693	1.84866	0.009558	476.129	1.81709
120	0.015230	496.325	1.89098	0.013424	494.187	1.87827	0.010148	488.457	1.84886
130	0.015871	507.498	1.91905	0.014028	505.569	1.90686	0.010694	500.474	1.87904
140	0.016490	518.659	1.94639	0.014608	516.900	1.93463	0.011208	512.307	1.90804
150	0.017091	529.841	1.97314	0.015168	528.224	1.96171	0.011698	524.037	1.93609
160	0.017677	541.068	1.99936	0.015712	539.571	1.98821	0.012169	535.722	1.96338
170	0.018251	552.357	2.02513	0.016242	550.963	2.01421	0.012624	547.399	1.99004
180	0.018814	563.724	2.05049	0.016762	562.418	2.03977	0.013066	559.098	2.01614
190	0.019369	575.177	2.07549	0.017272	573.950	2.06494	0.013498	570.841	2.04177

†**Note:** Number in parantheses is temperature of saturated R-134a in °C at the specified pressure.

$\hat{V}\,[=]\,\mathrm{m^3/kg}; \quad \hat{H}\,[=]\,\mathrm{J/g} = \mathrm{kJ/kg}; \quad \hat{S}\,[=]\,\mathrm{kJ/kg.K}$

Table B.2 (continued)   Superheated R-134a†

T	$P = 3.00\,\text{MPa}\,(86.17)$			$P = 3.50\,\text{MPa}\,(93.69)$			$P = 4.00\,\text{MPa}\,(100.32)$		
(°C)	$\hat{V}$	$\hat{H}$	$\hat{S}$	$\hat{V}$	$\hat{H}$	$\hat{S}$	$\hat{V}$	$\hat{H}$	$\hat{S}$
90	0.005755	436.193	1.69950	–	–	–	–	–	–
100	0.006653	453.731	1.74717	0.004839	440.433	1.70386	–	–	–
110	0.007339	468.500	1.78623	0.005667	459.211	1.75355	0.004277	446.844	1.71480
120	0.007924	482.043	1.82113	0.006289	474.697	1.79346	0.005005	465.987	1.76415
130	0.008446	494.915	1.85347	0.006813	488.771	1.82881	0.005559	481.865	1.80404
140	0.008926	507.388	1.88403	0.007279	502.079	1.86142	0.006027	496.295	1.83940
150	0.009375	519.618	1.91328	0.007706	514.928	1.89216	0.006444	509.925	1.87200
160	0.009801	531.704	1.94151	0.008103	527.496	1.92151	0.006825	523.072	1.90271
170	0.010208	543.713	1.96892	0.008480	539.890	1.94980	0.007181	535.917	1.93203
180	0.010601	555.690	1.99565	0.008839	552.185	1.97724	0.007517	548.573	1.96028
190	0.010982	567.670	2.02180	0.009185	564.430	2.00397	0.007837	561.117	1.98766
200	0.011353	579.678	2.04745	0.009519	576.665	2.03010	0.008145	573.601	2.01432

†**Note:** Number in parantheses is temperature of saturated R-134a in °C at the specified pressure.
$\hat{V}\,[=]\,\text{m}^3/\text{kg};\quad \hat{H}\,[=]\,\text{J}/\text{g} = \text{kJ}/\text{kg};\quad \hat{S}\,[=]\,\text{kJ}/\text{kg}\cdot\text{K}$

Appendix C

Constants and Conversion Factors

Physical Constants

Gas constant (R)
$$\begin{cases} = 82.05\,\text{cm}^3.\,\text{atm}/\,\text{mol}.\,\text{K} \\ = 0.08205\,\text{m}^3.\,\text{atm}/\,\text{kmol}.\,\text{K} \\ = 1.987\,\text{cal}/\,\text{mol}.\,\text{K} \\ = 8.314\,\text{J}/\,\text{mol}.\,\text{K} \\ = 8.314 \times 10^{-6}\,\text{MPa}.\,\text{m}^3/\,\text{mol}.\,\text{K} \\ = 8.314 \times 10^{-3}\,\text{kPa}.\,\text{m}^3/\,\text{mol}.\,\text{K} \\ = 8.314 \times 10^{-5}\,\text{bar}.\,\text{m}^3/\,\text{mol}.\,\text{K} \\ = 8.314 \times 10^{-2}\,\text{bar}.\text{L}/\,\text{mol}.\,\text{K} \\ = 8.314 \times 10^{-2}\,\text{bar}.\,\text{m}^3/\,\text{kmol}.\,\text{K} \\ = 83.14\,\text{bar}.\,\text{cm}^3/\,\text{mol}.\,\text{K} \end{cases}$$

Acceleration of gravity (g) $\begin{cases} = 9.8067\,\text{m}/\,\text{s}^2 \\ = 32.1740\,\text{ft}/\,\text{s}^2 \end{cases}$

Avogadro's number $\qquad 6.0221415 \times 10^{23}$ entities (atoms or molecules)/ mol

Conversion Factors

Density	$1\,kg/m^3 = 10^{-3}\,g/cm^3 = 10^{-3}\,kg/L$ $1\,kg/m^3 = 0.06243\,lb/ft^3$
Energy, Heat, Work	$1\,J = 1\,W.s = 1\,N.m = 10^{-3}\,kJ = 10^{-5}\,bar.m^3$ $= 10\,bar.cm^3$ $1\,cal = 4.184\,J$ $1\,kJ = 2.7778 \times 10^{-4}\,kW.h = 0.94783\,Btu$
Heat capacity	$1\,kJ/kg.K = 0.239\,cal/g.K$ $1\,kJ/kg.K = 0.239\,Btu/lb.°R$
Force	$1\,N = 1\,kg.m/s^2 = 10^5\,g.cm/s^2$ (dyne) $1\,N = 0.2248\,lbf = 7.23275\,lb.ft/s^2$ (poundals)
Length	$1\,m = 100\,cm = 10^6\,\mu m = 10^9\,nm$ $1\,m = 39.370\,in = 3.2808\,ft$
Mass	$1\,kg = 1000\,g$ $1\,kg = 2.2046\,lb$
Power	$1\,W = 1\,J/s = 10^{-3}\,kW$ $1\,kW = 3412.2\,Btu/h = 1.341\,hp$
Pressure	$1\,Pa = 1\,N/m^2$ $1\,kPa = 10^3\,Pa = 10^{-3}\,MPa$ $1\,bar = 10^5\,Pa = 100\,kPa = 0.98692\,atm$ $1\,atm = 1.01325\,bar = 101.325\,kPa = 760\,mmHg$ $1\,atm = 14.696\,lbf/in^2$
Temperature	$1\,K = 1.8\,°R$ $T(°F) = 1.8\,T(°C) + 32$
Volume	$1\,m^3 = 1000\,L$ $1\,m^3 = 6.1022 \times 10^4\,in^3 = 35.313\,ft^3 = 264.17\,gal$

Notation

Symbols

A	area
a	acceleration
B	availability function
$\mathcal{C}$	number of components, phase rule
C_P, C_V	heat capacity at constant pressure and at constant volume
C_P^*, C_V^*	ideal gas heat capacity
COP	coefficient of performance
D	diameter
E	total energy
E_K	kinetic energy
E_P	potential energy
E_v	friction loss
F	force
F_c	conservative force
F_{nc}	nonconservative force
$\mathcal{F}$	degrees of freedom, phase rule
f	Fanning friction factor
g	acceleration of gravity
g_c	gravitational conversion factor
H	enthalpy
k	spring constant
k	Boltzmann's constant
L	length
M	molecular weight
m	mass
$\dot{m}$	mass flow rate

$\mathcal{N}$	Avogadro's number
n	number of moles
$\dot{n}$	molar flow rate
P	pressure
$\mathcal{P}$	number of phases, phase rule
P_c	critical pressure
P_{ex}	external pressure
P^{vap}	vapor (saturation) pressure
Q	heat
$\dot{Q}$	heat transfer rate
$\dot{\mathcal{Q}}$	volumetric flow rate
R	gas constant
r_c	compression ratio
r_e	cutoff ratio
r_p	pressure ratio
S	entropy
S_{gen}	entropy generation
T	temperature
T_b	boiling temperature
T_c	critical temperature
T_m	melting (or freezing) temperature
T_s	sublimation temperature
T^{sat}	saturation temperature
t	time
U	internal energy
V	volume
v	velocity
$\mathbf{W}$	weight
W	work
$\dot{W}$	rate of work (power)
W_c	conservative work
W_c	compressor work
W_{nc}	nonconservative work
W_p	pump work
W_s	shaft work
W_t	turbine work
$\dot{W}_s$	shaft power
x	quality
x, y, z	rectangular coordinates

Greek symbols

γ	ratio of heat capacities, $\widetilde{C}_P^*/\widetilde{C}_V^*$
η	efficiency (thermal or isentropic)
η_c	isentropic efficiency of a compressor
η_t	isentropic efficiency of a turbine
η^*	thermodynamic efficiency
ρ	density
μ	Joule-Thomson coefficient

Operators

δ	denotes a path dependent variation
$\Delta\varphi$	$\varphi_{final} - \varphi_{initial}$ (closed system); $\varphi_{out} - \varphi_{in}$ (open system)
d	total differential
∂	partial differential
$\int$	integral
$\ln$	natural logarithm
$\log$	common logarithm
$\prod$	cumulative product
$\sum$	cumulative summation

Overlines

$\widetilde{\varphi}$	per mole
$\widehat{\varphi}$	per unit mass

Subscripts

atm	atmosphere
in	incoming stream
$irrev$	irreversible process
out	exiting stream
ref	reference
rev	reversible process
$surr$	surroundings
sys	system

Superscripts

fus	property change on fusion (or melting)
IG	ideal gas property
L	liquid phase
S	solid phase
sat	saturated condition at phase equilibrium
sub	property change on sublimation
V	vapor phase
vap	property change on vaporization

Index

Printed in the United States
By Bookmasters